Earth's Fury

Earth's Fury

The Science of Natural Disasters

ALEXANDER GATES
Department of Earth & Environmental Science
Rutgers University

WILEY Blackwell

This edition first published 2022
© 2022 John Wiley & Sons Ltd

All rights reserved. No part of this publication may be reproduced, stored in a retrieval system, or transmitted, in any form or by any means, electronic, mechanical, photocopying, recording or otherwise, except as permitted by law. Advice on how to obtain permission to reuse material from this title is available at http://www.wiley.com/go/permissions.

The right of Alexander Gates to be identified as the author of this work has been asserted in accordance with law.

Registered Offices
John Wiley & Sons, Inc., 111 River Street, Hoboken, NJ 07030, USA
John Wiley & Sons Ltd, The Atrium, Southern Gate, Chichester, West Sussex, PO19 8SQ, UK

Editorial Office
9600 Garsington Road, Oxford, OX4 2DQ, UK

For details of our global editorial offices, customer services, and more information about Wiley products visit us at **www.wiley.com**.

Wiley also publishes its books in a variety of electronic formats and by print-on-demand. Some content that appears in standard print versions of this book may not be available in other formats.

Limit of Liability/Disclaimer of Warranty
While the publisher and authors have used their best efforts in preparing this work, they make no representations or warranties with respect to the accuracy or completeness of the contents of this work and specifically disclaim all warranties, including without limitation any implied warranties of merchantability or fitness for a particular purpose. No warranty may be created or extended by sales representatives, written sales materials or promotional statements for this work. The fact that an organization, website, or product is referred to in this work as a citation and/or potential source of further information does not mean that the publisher and authors endorse the information or services the organization, website, or product may provide or recommendations it may make. This work is sold with the understanding that the publisher is not engaged in rendering professional services. The advice and strategies contained herein may not be suitable for your situation. You should consult with a specialist where appropriate. Further, readers should be aware that websites listed in this work may have changed or disappeared between when this work was written and when it is read. Neither the publisher nor authors shall be liable for any loss of profit or any other commercial damages, including but not limited to special, incidental, consequential, or other damages.

Library of Congress Cataloging-in-Publication Data

Names: Gates, Alexander E., 1957- author.
Title: Earth's fury : the science of natural disasters / Alexander Gates.
Description: Hoboken, NJ : Wiley-Blackwell, 2022. | Includes
  bibliographical references and index.
Identifiers: LCCN 2021970026 (print) | LCCN 2021970027 (ebook) | ISBN
  9781119546597 (paperback) | ISBN 9781119546801 (adobe pdf) | ISBN
  9781119546832 (epub)
Subjects: LCSH: Natural disasters.
Classification: LCC GB5014 .G38 2022 (print) | LCC GB5014 (ebook) | DDC
  363.34–dc23/eng20220521
LC record available at https://lccn.loc.gov/2021970026
LC ebook record available at https://lccn.loc.gov/2021970027

Cover Design: Wiley
Cover Image: © KalypsoWorldPhotography/Alamy Stock Photo

Set in 9.5/12.5pt SourceSansPro by Straive, Pondicherry, India
Printed and bound by CPI Group (UK) Ltd, Croydon, CR0 4YY

C9781119546597_270622

Contents

PREFACE vii
ACKNOWLEDGMENTS ix
ABOUT THE COMPANION WEBSITE xi

1 Introduction to Natural Disasters 1
2 Moving Continents 15
3 How Does Rock Melt? 31
4 Types of Volcanoes 47
5 Volcanic Hazards 61
6 Causes of Earthquakes 101
7 Earthquakes 101 111
8 Earthquake Hazards 125
9 Killer Tsunamis 155
10 Predicting Earthquakes and Reducing Hazards 185
11 Avalanches and Landslides 201
12 Weather and Storms 221
13 Ocean Circulation and Coastal Systems 249
14 Hurricanes, Cyclones, and Typhoons 265
15 Tornadoes and Supercells: Terrors of the Plains 295
16 Devastating Floods and Their Aftermath 321
17 Droughts and Desertification 339
18 Impacts: Collisions from Space 361
19 Climate Change Dynamics 377

INDEX 391

Preface

Most courses on natural disasters are attended primarily by non-science majors. These students are not used to the style of science textbooks and have difficulty with them. Much of the problem is that chapters must be read in one sitting or else students lose their train of thought and are forced to start over or give up. Wading through the scientific concepts and information at college level is also foreign to them. After years of struggling with several textbooks on natural disasters, several colleagues and I attempted to use an encyclopedia on earthquakes and volcanoes instead of a textbook. As a result, student evaluations on the usefulness of the textbook improved dramatically. Students even commented on the textbook in their written assessments of the course, which is rare. They found that reading about science in short, contained packets was much more convenient and easier to digest.

The problem is that the information in an encyclopedia is organized alphabetically rather than to develop an understanding of a topic. Therefore, the book can only be used to support the class notes which provide the entire structure rather than as a stand-alone resource. Encyclopedias are also not as useful in reviewing for exams because of this lack of topical organization. They tend also not to contain high-quality graphics and many non-science students are visual learners.

This book attempts to combine the best features of a scientific textbook and an encyclopedia into a single natural disasters book that is more appropriate to non-science majors. It retains the organization of a textbook and adopts the highly illustrative graphics of some of the newer and more effective textbooks. However, most of the scientific content is delivered through a plethora of case studies. These are short, self-contained, and well-illustrated stories of specific natural disasters that are highly engaging for both science and non-science majors. The stories incorporate the scientific concepts and vocabulary into the event so that students appreciate and remember it as part of the story. By relating the event to the impact on society and human lives, this method of presentation also places the science into the context of what is personally important to the student. The case studies are short and self-contained so that chapters can be read in pieces without losing continuity. However, the chapters are organized to deliver the material in the logical manner of the topical development. Therefore, the book can be used as a stand-alone resource for a course and be effective for exam preparation.

The other important feature is the visual support of the case studies. A number of artistic, highly detailed illustrations of disaster-producing features serve as lead-ins to the case studies. The features include volcanoes, earthquakes, tsunamis, supercells and hurricanes. These illustrations show the spatial and temporal relations of the natural hazards of the event. Each case study then describes a specific, premier example of the natural disaster. The case study can then be related back to the illustration to place it in context. For example, stratovolcanoes can produce deadly lahars, *nuée ardentes*, pyroclastic flows, and lateral blasts, among others. Case studies on interesting disasters involving each of these hazards are presented in two-page spreads richly illustrated with diagrams and photographs of the event and resulting damage.

The book is organized to introduce the scientific basis of the disasters either early in the chapter or in a separate leading chapter including a list of easily referenced definitions to enhance understanding. The background can be interspersed with case studies, but all are followed by a group of case studies to give the reader a full flavor of the character of the processes and devastation. This organization coupled with the exciting and highly illustrated stories is intended to enhance interest and understanding.

There are many types of natural disasters, and courses tend to emphasize certain types. This book contains as many types of natural disasters as could be included in a single book. It also focuses on recent disasters so students may be able to relate to them more easily. It is not meant that the natural disaster courses should cover all of the topics, but that professors can choose the disasters that best fit their course.

Acknowledgments

The contributions of the following people is acknowledged and appreciated. Dr. David Valentino tested the encyclopedia idea, acted as a consultant and reviewed some of the chapters, providing helpful feedback. Dr. Ismael Calderon reviewed most of the chapters and provided extensive feedback. Cary Lu reviewed several chapters for ease of understanding.

About the Companion Website

This book is accompanied by a companion website:

www.wiley.com/go/gates/earthsfury

This website includes:

- Powerpoints of all figures and tables from the book for downloading
- Web links from the book for downloading

CHAPTER 1

Introduction to Natural Disasters

CHAPTER OUTLINE

1.1 What Is a Natural Disaster? 3

1.2 Why Do People Live in Dangerous Areas? 5

1.3 What Can We Do to Stay Safe? 8

1.4 Societal Response to Natural Disasters 9

1.5 How This Book Is Organized 14

Earth's Fury: The Science of Natural Disasters, First Edition. Alexander Gates.
© 2022 John Wiley & Sons Ltd. Published 2022 by John Wiley & Sons Ltd.
Companion website: www.wiley.com/go/gates/earthsfury

Words You Should Know:

Earthquake – An event in which the breakage of rock along a fault within the earth causes the release of energy in the form of seismic waves.

Disaster reduction engineering – Construction practices that are designed to reduce the impact of a natural disaster.

Hurricane – A highly energetic, individual tropical storm that develops over warm ocean waters.

Natural disaster – A natural event that causes loss of life or property, either directly or indirectly, as a result of the aftermath.

Population growth – The increase in human population over time.

Tsunami – Meaning "harbor wave" in Japanese, a tsunami is an energetic ocean wave generated by a physical event (earthquake, volcano, landslide) on the earth's surface.

Volcano – The geographic feature resulting from the release of lava and/or solid ejecta from a vent in the earth.

1.1 What Is a Natural Disaster?

Humans seek to protect their lives at all costs. Major efforts and enormous amounts of money are devoted to medications and medical procedures that can increase human life by even a short amount of time. These efforts have increased the average life span worldwide and protected people from many natural threats. This increased protection has led to a sense of security among many of not having to worry about the threats, especially in the developed world. Increased feelings of security have also given humans the courage to inhabit areas of the planet that might otherwise be too subject to threats of nature to be safe. But natural disasters are not controlled by human will and continue to be threats to safety and security. On the other hand, some people do not have the finances or resources to live in areas other than those that are subject to natural disasters and face risks without expectations of security.

Very few people want to be involved in a natural disaster, but most people are mesmerized by them and marvel at their destructive force. In an attempt to find reasons for these catastrophic events, some cultures choose to explain them through folklore and religious beliefs, whereas scientists seek to understand the processes behind them and ultimately to forecast them and reduce their destruction.

Natural disasters are defined as any catastrophic loss of life and/or property caused by a natural event or situation. This definition could include biologic issues such as disease, injurious viral or bacterial colonization, invasion of dangerous plants, and infestations of insects and other vermin. However, this book solely focuses on disasters involving the physical properties of the shallow earth, the atmosphere, and those from extraterrestrial sources. These disasters include earthquakes, volcanoes, tsunamis, avalanches, landslides, tropical storms, tornadoes, floods, and extraterrestrial impacts (Figure 1.1). The difference between these physical threats and the biological threats is that humans have virtually no control over the physical threats, whereas they may have some control on the biological threats. All people can do to prepare for these physical threats is to choose safe locations, engineer structures to withstand the impact, and/or evacuate areas before the threat produces a disaster.

In terms of the planet earth and its history, the processes that produce natural disasters are not extraordinary events. They occur on a daily basis and have done so since the beginning of time. Many large events go unnoticed because they do not impact humans. Unlike the way many ancient and primitive cultures regarded these events as being imposed by supernatural forces as a punishment, they are part of the natural sequence of events of planet earth. They are governed by the physical and chemical processes that define the planet and can be explained in those terms. That is one part of natural disasters. The other part is the impact on humans and response of society to these natural events.

Single natural disasters have changed the course of human history in profound ways. The dual components of natural disasters make them complex issues in modern society. In any evaluation of natural disasters, both aspects must be considered. This means that an understanding of the controlling science is the first part of appreciating a natural disaster. The human aspect is more difficult to generalize because each situation is different. The best way to appreciate the full meaning of a natural disaster is through case studies in the context of the controlling science.

CHAPTER 1 Introduction to Natural Disasters

FIGURE 1.1 Collage of natural disasters which come in a number of forms that have varying abilities to wreak havoc on human communities.
Source: Images from top, left to right courtesy of US National Park Service, Langevin Jacques/Getty Images, Getty Images, NOAA, David Mabe/Almay Stock Photo, Phys.org.

CASE STUDY 1.1 464 BCE Sparta, Greece Earthquake

Earthquake that Changed History

A major earthquake took place in 464 BCE that forever changed the course of Greek history. The estimated magnitude of this earthquake is 7.2, and the destruction from ground shaking is estimated at X on the Modified Mercalli Scale of earthquake intensity as interpreted from historical records of the damage it caused. A magnitude-7.2 earthquake is considered to be strong. The strongest earthquake ever measured was of magnitude 9.6. The shaking intensity of X is on a scale of XII and is designated "extreme." The probable result of the earthquake was a 20-km-long north-south normal fault scarp or small cliff that lies a few miles east of Sparta (Figure 1.2). The earthquake is estimated to have produced a death toll of 20 000 people in the Greek city-state of Sparta. Prior to this, the Spartans established themselves as the most formidable army in the region by defeating the invading Persian (now Iran) force which was previously considered the most fearsome. However, the earthquake so weakened the Spartans that they allowed troops from the nearby and allied city-state of Athens to help with the rescue and relief operations. The Athenians learned that the Spartans had partially enslaved the Helots, another group of Greeks, by forcing them into an agrarian existence to supply the Spartans with food. The Athenians protested based on a regional agreement that all Greeks should be free and, as a result, were immediately expelled from Sparta. The removal of the Athenians, however, did not end the Spartans' troubles. The Messenians, yet another group of Greeks, came to the aid of the Helots in a revolt against the weakened Spartans which lasted for years. Resentment from these issues eventually led to the Peloponnesian War, a major and defining event in Greek history.

FIGURE 1.2 Map of ancient Greece showing the cities of Sparta and Athens as well as the location of the Messenians.

1.2 Why Do People Live in Dangerous Areas?

Certainly, by this point in modern society, the geographic areas dominated by specific natural disasters are well known. Some areas are even informally named for the most common disasters such as "Tornado Alley" in the Midwestern United States (Figure 1.3). In addition, systems to warn people of impending disasters have been established worldwide so that residents can make preparations and escape harm. Some of these preparations include engineering solutions to reduce the impact of the disaster. This is in sharp contrast with the vast majority of human history where people were commonly completely unprepared for natural disasters and surprised by their appearance. Only modern science has changed the situation. However, in terms of preventing natural disasters from occurring, humans have made no progress.

There are two main reasons why people are now inhabiting areas that are subject to natural disasters. The first is human population growth. The human population worldwide has quickly ballooned from about 1 billion in 1800 to 7.7 billion today. In order to accommodate this large number, urban areas have sprawled into less safe locations. Settlements now encroach the damage zones of volcanoes with recent activity, inviting disaster. For example, despite the 79 AD (CE) Mount Vesuvius eruption that completely destroyed the cities of Pompeii and Herculaneum, the city of Naples has now reached nearly 3.1 million people and encroached the danger area of the volcano to a much greater degree than in ancient Rome (Figure 1.4). Populations of many urban areas have increased to such densities that events that may have not been so catastrophic in the past are now disasters. New Orleans, Louisiana has been struck repeatedly by major hurricanes throughout its history, yet Hurricane Katrina in 2004 was more devastating than previous storms. The population and development sprawl occurred dramatically between the last two major storms. A major earthquake struck Port-au-Prince, Haiti in 2010 that is estimated to have killed between 100 000 and 316 000 people, making it one of the deadliest of all natural disasters. It caused such chaos that the actual death toll may never be known. The reason for so much devastation was the population of Port-au-Prince boomed to more than 10 million people despite previous earthquakes that destroyed the city in 1770 and 1842 and a tsunami that devastated the harbor area in 1946.

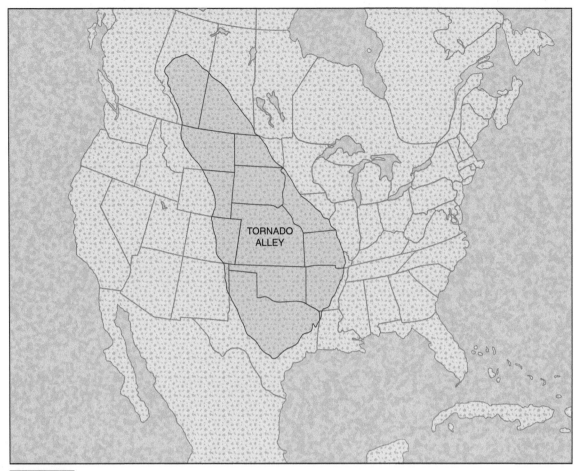

FIGURE 1.3 The approximate location of Tornado Alley in North America.

FIGURE 1.4 Night satellite image of the Mount Vesuvius area in Naples in Italy including the ancient towns of Pompeii and Herculaneum. The lighted lines, areas, and spots show human habitation. The entire human population of the area is under threat by an eruption of the Mount Vesuvius active volcano. *Source*: Photograph courtesy of NASA.

The other reason that people are tending to live in areas that are more prone to natural disasters is psychological. Humans have attempted to control nature with ever-increasing efforts and have done so in very significant ways. If a natural disaster has not occurred in a prone area for several decades, residents quickly become complacent and assume safety. They overdevelop areas that may not be safe and occupy them with an attitude of impunity and entitlement. Miami, Florida has grown dramatically in recent years with extensive development right on the beaches (Figure 1.5). In 1926, less than 100 years ago, the city was destroyed by a major hurricane. It is estimated that if a similar storm strikes today, it will cause damages of more than $235 billion, making it, by far, the costliest natural disaster in the history of the United States. In 1928, the much more powerful Okeechobee hurricane just missed the Miami area. The potential for damage and loss of life in Miami from a major hurricane is very high, yet fear does not appear to be a limiting factor.

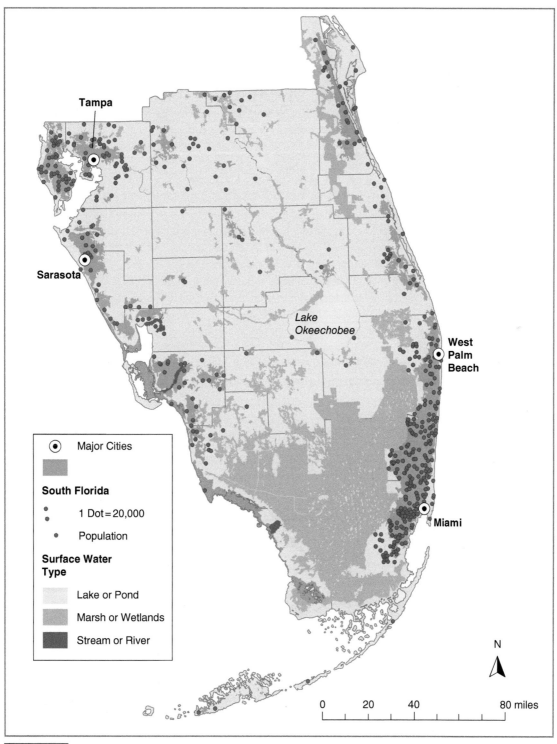

FIGURE 1.5 Population distribution map of South Florida showing the high population density in the Miami area.
Source: Adapted from Hickford (2015), http://alanhickford.blogspot.com/2015/03/module-10-dot-density-mapping.html.

CASE STUDY 1.2 Death Tolls

Horrifying death tolls have been reported for many natural disasters. If the numbers are low, typically, the death toll can be considered relatively accurate. Large events tend to be less accurate because multiple sources of casualties must be tallied, which leaves multiple sources for error. In ancient events, survival was such a concern that there wasn't time to count all of the bodies, so the death toll could be in error by as much as ten-fold or more. Death tolls for recent events tend to be more accurate. However, even with recent events, some disasters are subject to more counting errors. In inland earthquakes where rescue and relief efforts are done properly, the numbers are more accurate. In floods and coastal events such as hurricanes, typhoons, and tsunamis, there can be large errors even today. In case of an earthquake, the collapsed houses remain in place as do the bodies, so counting is easier. With many floods and flooding events, the bodies get displaced along with the houses and even whole neighborhoods. Even the town hall may be washed away, removing the residency and birth records. In these cases, numbers tend to be very inaccurate. Volcanic eruptions can be equally inaccurate with the severe destruction caused by the explosions and lava flows. Translations from other languages may also lead to errors which can then disseminate throughout the literature. For example, one of the worst earthquake disasters on record occurred in China in the 1920s. Most literature report a death toll of 200 000. A recently released more reliable source states that 200 000 livestock perished in this event, but only 35 000 people died. Death tolls may be purposefully exaggerated by officials for political reasons. The Tangshan earthquake of China has an official death toll of 249 000 people. However, recent studies show that it was likely closer to 655 000. The communist government at the time did not want to look weakened to its enemies, so it reported a deflated death toll. On the other hand, if there is international relief aid available, some governments inflate the death tolls to get more aid.

1.3 What Can We Do to Stay Safe?

There are many ways for humans to be safe from natural disasters. The most reasonable of these solutions is not to inhabit locations that are subject to frequent natural disasters. That would eliminate many inhabited areas and still not guarantee full safety because events can still occur outside of expected areas. The next best solutions are to plan for them by engineering solutions or warning and evacuation systems or some combination of these.

The simplest of these is an early warning system. These traditionally have consisted of multiple far-reaching sirens, television and radio announcements, and cell phone alerts. In some areas, there are marked escape routes for evacuation, and in others, emergency drills are conducted, such as in middle and high schools. The more challenging aspect of warning is the detection of the impending dangerous events far enough in advance for residents to take action. Volcanoes are so destructive that those in populated areas are well monitored using advanced technology, and warnings are typically issued well ahead of the eruptions. Hurricane movement is typically slow enough that they can be accurately tracked by national weather services and, in most cases, residents can be safely evacuated ahead of the danger. Tsunamis can also be tracked and the systems provide ample warning in many cases (Figure 1.6).

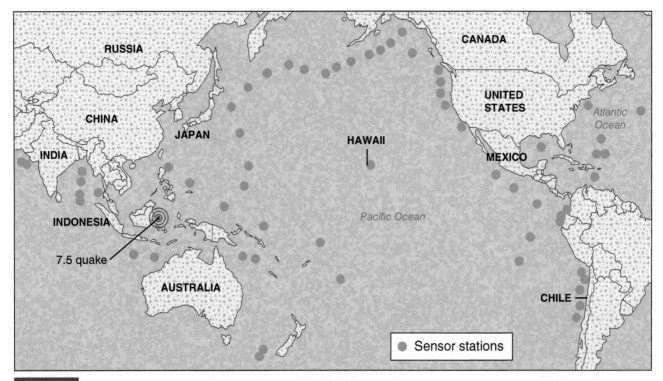

FIGURE 1.6 Location of tsunami sensors in the Pacific Ocean and surrounding areas.

FIGURE 1.7 Sharp contrast of earthquake engineered Transamerica building (left structure) with a surviving older building (right structure). *Source*: Courtesy US Library of Congress.

The sensing systems, however, are more complex and difficult to effectively install. Systems were installed in the Pacific Ocean after a major earthquake in 1964 and in the Indian Ocean after a major earthquake in 2004, but few other ocean basins contain them. Tornadoes, on the other hand, strike so quickly that they may not be as easy to identify or predict by national or local weather services. In most cases, there is little forewarning and, consequently, people must take refuge in a home tornado shelter or in a reinforced, safer part of a house.

The other main way people can protect themselves from natural disasters is through engineering. This involves different solutions that depend on the specific threat. In general, construction practices have improved with time in advanced countries as the result of updated building codes. California updates its building codes after each large earthquake, which is a great example of a proactive policy against one type of natural disaster. The Transamerica building in San Francisco has a recognizable shape that was constructed to be earthquake resistant (Figure 1.7). Japan also has excellent earthquake engineering including the installation of walls and gates in critical coastal locations to protect against tsunamis. If a tsunami is sensed in time, the gates slide shut preventing the wave from penetrating inland. The Netherlands has developed one of the most advanced technologies to protect against flooding to which they are so prone. Even large skyscrapers along the East Coast of the United States are typically designed to flex in response to strong winds.

These may be big-scale engineering examples, but there are many smaller and simpler methods as well. The ancient complex of Machu Picchu in the mountains of Peru actually included earthquake reduction techniques in its construction. In Sicily, towns and residents built rock walls along the slopes of the Mount Etna volcano to hold back lava flows from reaching inhabited areas. In many disaster-prone areas, public buildings are constructed with extra reinforcement so that they can act as emergency shelters during a natural disaster. Residents are informed and alerted to be evacuated to them in case of an emergency. In the Midwestern United States, many people build tornado shelters or cellars in their homes. Even local building codes may include some provisions for disaster reduction in home construction.

1.4 Societal Response to Natural Disasters

In ancient civilizations, natural disasters could mean the downfall of cities and empires. If two groups were battling or even competitive, a natural disaster weakened one of the groups, making it susceptible to being overthrown. Large earthquakes, tsunamis, storms, and volcanic eruptions have been responsible for several of the great and enigmatic historical catastrophes. This is especially true for the Eastern Mediterranean region where tectonic activity tends to be episodic. Large earthquakes, for example, tend to occur over a 50–100-year interval followed by a period of quiet that may last hundreds of years. The active periods are termed *seismic crises*. Because ancient cities were commonly protected by walls, a large earthquake could breach them making the city vulnerable. Such attacks might be from: (i) external enemies during active conflicts such as Joshua at Jericho and the Arab attack on Jerusalem in 31 BCE, (ii) neighbors in long-term conflict like Mycenae's fall in 1200 BCE and Saul's battle around 1020 BCE, and (iii) uprisings of enslaved people like Sparta and the Helots in 464 BCE, Hattusas around 1200 BCE, and Teotihuacan around 700 AD. If a small earthquake occurred during a modest conflict, damage would take a few tens of years to repair with minor if any shift of power. However, if a large earthquake occurred during a major military conflict, it would take hundreds of years to rebuild the damage and the city would be plunged into a dark age with all efforts centered on pure survival. It is during these incidents that the outcomes of conflicts could be decided by earthquakes and radical shifts of power possible.

For that reason, afflicted regions downplayed the impact of the damage and casualties from natural disasters. This practice existed throughout most of history and is not completely gone today. As late as 1976, communist China was struck by two major earthquakes of the Tangshan event in a few hours and in a highly populated area. The official number of casualties listed by the government was 252 000 and that remains the official total today. This was the height of the Cold War and showing weakness could have been deadly. Later studies, after the Cold War subsided, estimated the death toll from the Tangshan event at 655 000, making it potentially the worst natural disaster ever.

CASE STUDY 1.3 1949 Khait Earthquake, USSR

Downplaying Casualties Not to Show Weakness

On 10 July 1949 at 9:45 a.m., a powerful earthquake struck the Gharm Oblast region of Tajikistan (USSR at the time). The magnitude was 7.4–7.5, and the maximum shaking intensity was IX (out of XII) on the Modified Mercalli Scale. It had two foreshock earthquakes on 8 July that were 12 minutes apart and had magnitudes of 5.1 and 5.6. The main shock was powerful enough to be destructive on its own, but the real damage was caused by the massive landslides it generated (Figure 1.8).

Eastern Tajikistan contains the Pamir Mountains, known as the "Roof of the World," and the high Pamir Plateau. This rugged terrain generates numerous rockfalls, landslides, floods, and avalanches even without earthquakes. The reason the terrain is rugged is that it is in a very complex plate tectonic zone which also produces the strong earthquakes. The Khait earthquake occurred at the southern margin of the Tian Shan belt which extends westward from China. The belt is being generated by the ongoing collision between the Indian and Eurasian continents forming the Himalayan orogeny to the south. This area is marked by a combination of dextral strike-slip faults which move the land laterally and thrust faults which move it vertically. There are a series of north-south to northeast-oriented thrust faults that overlap the Tian Shan faults. The Khait earthquake appears to be the result of movement on the Vakhsh fault.

This part of the Pamir is known for seismic activity and mass movement. In 1911, an earthquake generated a landslide that blocked and flooded a river valley creating earthquake Lake Sarez, 36 mi (60 km) long and more than 1625 ft (500 m) deep. This activity led American seismologists to take an interest in the area. Through an unprecedented cooperative study with Soviet seismologists, they established a seismic network across the region and a research center in the town of Gharm on the Surkhob River.

July 1949 had heavy rains that soaked the soils, loess deposits, and sediments that mantled the slopes. Loess is fine dust blown by wind from deserts and deposited on some slopes. Earthquake activity then increased causing minor landslides and contributing to the instability on many of the slopes. When the Khait earthquake struck on 10 July, the area was primed for massive landslides, especially on the loess slopes. Hundreds of small loess slides coalesced into a huge flow with a volume of 320.4 million cubic yards (245 million m³) that traveled 12.4 mi (20 km) in the Yasman Valley on a shallow slope of 2°. In the adjacent Yarhich and Obidara-Khauz Valleys, the Khait landslide began as a rockslide that entrained water-saturated loess deposits (Figure 1.8A). It traveled 4.6 mi (7.41 km) and descended 4662 ft (1421 m) at an average velocity of 97.5 ft (30 m) per second. In some sections, it achieved a velocity of 325 ft (100 m) per second. The estimates of the volume of this slide range from 327 million m³ (250 million m³) to as little as 98.1 million cubic yards (75 million m³). While moving, these landslides and avalanches made a deafening noise and were accompanied by strong winds and huge dust clouds.

All of the buildings in an area 37–41 mi (60–65 km) long by 4–5 mi (6–8 km) wide were destroyed by the earthquake and landslides. The town of Khait and around 20 neighboring towns from the two valleys were buried in soil and debris to a depth of 100 ft (30 m) in some places (Figure 1.8A). Overall, about 150 towns and villages were impacted. With the complete burial of so many towns, it was difficult to determine the death toll of the disaster. The minimum count is about 7200 people, but other estimates are 12 000–28 000 deaths.

In addition to the burial, an another factor that complicated the death toll was the secrecy that surrounded the disaster. The Khait earthquake occurred at the beginning of the Cold War during the rule of Joseph Stalin. The USSR viewed disasters as a sign of weakness that could encourage enemies to attack. The government attempted to cover up and disavow the Khait disaster, not admitting to any fatalities

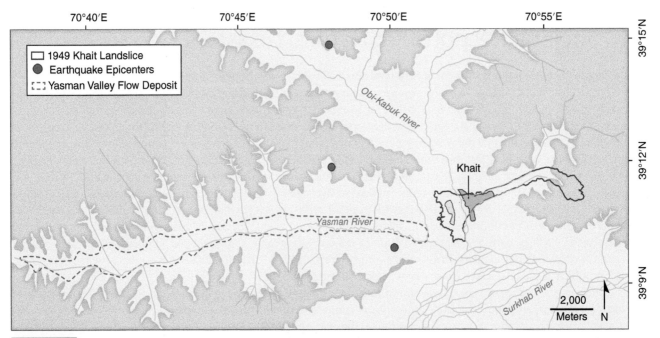

FIGURE 1.8 Map showing the locations of the 1949 earthquake epicenters, landslides and former town of Khait in the Gharm Oblast region of Tajikistan. Adapted from Evans et al. (2009), DOI:10.1016/j.enggeo.2009.08.007.

at all for many years. The true extent of the disaster was not revealed until the past decade.

Today, between the towns of Gharm and Jirgital, there is a monument of white marble that commemorates the tragedy. A statue of a woman with her head down in grief overlooks the Yarhich River Valley where the largest mass burial remains unexcavated. This is where the thriving towns of Khait, Badimlog, Turatal, Deplan, Karakol, Mazar, Aitalob, and many others once stood. It is now a grim reminder of the power of mass movements and the danger inherent to life in the Pamir.

FIGURE 1.8A Photo of a major 1949 Khait landslide. *Source*: Courtesy of the USGS.

Recent world politics has changed the response to natural disasters among many nations. In recent history, many of the larger and richer countries began to provide aid to countries that had been impacted by natural disasters. Even private groups developed ways to generate significant aid. In 1970, ex-Beatle George Harrison organized the first benefit concert to help the people of Bangladesh after they had been devastated by the Bhola typhoon. It generated an enormous amount of aid and many other benefit concerts followed. This new attitude led countries to more accurately report casualties or even to exaggerate them in hopes of receiving more aid. In 2010, a major earthquake struck Port-au-Prince, Haiti. It did extensive damage and caused an enormous number of casualties. Aid poured into the poor island from around the world. The official government estimate of casualties remains at 316 000, making it one of the worst natural disasters ever. However, independent studies conducted after the event estimate the casualties at 100 000–160 000, which is devastating but far less than the government claim.

CASE STUDY 1.4 2010 Port-au-Prince, Haiti Earthquake

Poor Construction and Possible Exaggerated Death Tolls

California updates their building codes after every strong earthquake, and it is reflected in the outcomes. In 1989, the Loma Prieta earthquake struck California at a magnitude of 6.9 located a mere 44 mi (70 km) from San Francisco and at a depth of 12 mi (19 km) producing a strong shaking intensity of IX (Modified Mercalli Scale) in the city. Yet, only 63 people lost their lives, 42 of whom perished because of a single highway collapse. This result stands in stark contrast to the 2010 Port-au-Prince earthquake in Haiti and illustrates the striking impact that construction practices play in preventing the loss of life. The faults that generated the two earthquakes are similar, both being marked by lateral movement. However, California, and especially San Francisco, is among the most affluent areas in the world and can afford to prepare for earthquakes, whereas Haiti is among the poorest and cannot.

After a decade of some of the worst natural disasters ever, the second decade of the twenty-first century began with more devastation. On Tuesday, 12 January 2010 at 04:53 p.m., a major earthquake struck near Port-au-Prince, Haiti. The magnitude was 7.0 with a shallow focus of 8.1 mi (13 km), a mere 15 mi (25 km) west-southwest of the city. Even though Hispaniola is seismically active having experienced earthquakes that destroyed Port-au-Prince in 1751 and 1842 and an earthquake of magnitude 6.8 in the Dominican Republic as recently as 1985, the city was ill-prepared for the event.

The source of the earthquake was the Enriquillo–Plantain Garden fault system which has produced several large historical earthquakes in 1860, 1770, and 1751. The east-west fault system is one of

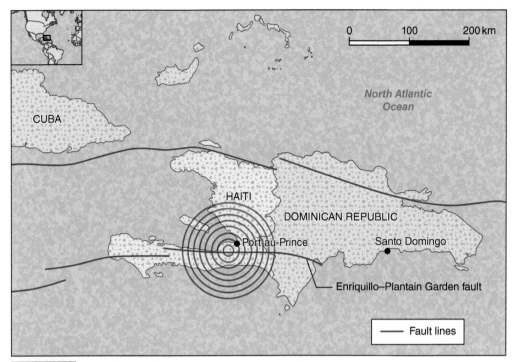

FIGURE 1.9 Map of the location of the 2010 Port-au-Prince, Haiti earthquake in Hispaniola, and the generating Enriquillo-Plaintain Garden fault.

two that cross the island (Figure 1.9). These systems are part of the boundary region that separates the Caribbean tectonic plate from the North America tectonic plate. This transform plate boundary connects the Caribbean arc and the Central American arc and experiences left-lateral strike-slip or lateral motion and uplift. The movement on the boundary is about 0.8 in. (2.0 cm) per year. This boundary has produced numerous earthquakes including the 1692 Port Royal and 1907 Kingston earthquakes in Jamaica and the 1972 Motagua earthquake in Guatemala that killed 23 000 people.

Surface waves from the main shock produced ground shaking with a Modified Mercalli Intensity of IX (of a maximum of XII) in Port-au-Prince and its suburbs (Figure 1.9A). Shaking was felt throughout the region in Cuba at intensity III, in Jamaica at intensity II, in Venezuela at intensity II, in Puerto Rico at intensity II–III, and in the bordering Dominican Republic at intensity III. It also produced a local tsunami that killed at least four people.

The reason this earthquake was so devastating is that it was the worst in more than 200 years and Haiti is not prepared for earthquakes.

FIGURE 1.9A Collapsed buildings and crushed cars in Port-au-Prince, Haiti after the 2010 earthquake. *Source*: Courtesy of the USGS.

FIGURE 1.9B Aerial view of the destruction in downtown Port au Prince, Haiti as the result of the 2010 earthquake. *Source*: Courtesy of the USGS.

The country is the poorest in the Western Hemisphere, with four of five people living in poverty and more than 50% in abject poverty. Haiti has substandard building codes, and few builders follow them anyway. Most buildings can barely survive under normal conditions, much less the shaking from a strong earthquake. It was these collapsing buildings that caused the most loss of life. The United Nations estimates were 105 000 homes destroyed and 208 000 damaged, mostly in Port-au-Prince (Figure 1.9B).

Estimates are that upwards of 3 000 000 people were directly affected by the earthquake. The Haitian government estimates that 316 000 people were killed, 300 000 injured, and 1 000 000 made homeless. If the 316 000 deaths are correct, officially, it would be the worst earthquake disaster in modern times and the second worst ever. The death toll, however, was estimated by other international organizations to be much lower, between 92 000 and 160 000. The Inter-American Development Bank estimated that the total cost of the disaster was between $7.8 billion and $8.5 billion, based on a death toll of 200 000–250 000.

The international response to the disaster was immediate. On 13 January 2010, Secretary-General Ban Ki-moon of the United Nations released a statement of the humanitarian disaster in Haiti. President Barack Obama immediately pledged $1.15 billion toward the rescue and relief efforts. After that, aid poured in from around the world. Venezuela pledged $1.32 billion, and the International Monetary Fund canceled Haiti's $268 million debt. Other countries made similarly large pledges and sent troops, rescue personnel, and supplies. Former president Bill Clinton and the Haitian prime minister headed up a 26-person team that oversaw the $5.3 billion raised in the first two years. Many countries and benefit fund raising efforts continued to generate relief funding for many years.

Although the world leaders pledged generous aid, delivering on the promises was a logistical nightmare. In the days following the earthquake, aid arrived in a trickle, electric power was unavailable, and telecommunications were spotty. Ships could not bring their cargos of supplies and were forced to remain offshore because of Haiti's damaged port (Figure 1.9C). The airport was functional but with severe limitations on flights both in and out of the country. All main roads and even most secondary roads were blocked by debris and throngs of survivors.

The Haitian government developed an aggressive plan to begin the reconstruction phase of the recovery process. They used the plan at a March 2010 donor conference to solicit almost $10 billion in pledges. The rebuilding began in a very disorganized manner and was slow and inefficient with accusations of misuse of funds. The only bright spot in this disaster is that the government intends to inspect all new construction to be sure that it meets building codes. Permits are now required for those who want to repair damaged houses deemed as salvageable by the Haitian Ministry of Public Works, Transport and Communications. Building inspectors have even suspended reconstruction projects that they deemed questionable. There is at least the appearance of action to avoid such a terrible tragedy in the future.

Another consideration in the study of natural disasters is that ancient and modern events are not always comparable. There are differences in hazards such as fires which could be responsible for the majority of the damage in older events. Fires did the majority of the damage in the 1906 San Francisco earthquake and the 1923 Great Kantō earthquake in Japan. This was because fire was used for heating and cooking in the past, but typically is not anymore. The other problem is that there can be significant errors in reporting and recording of the details of events. As described, some details are misreported on purpose. However, in other cases, the misreporting can be by mistake.

FIGURE 1.9C Collapse of cliff face in Haiti as the result of the 2010 earthquake. *Source*: Courtesy of the USGS.

CASE STUDY 1.5 The 1737 Calcutta Earthquake

The Earthquake that Wasn't: Reliability of Ancient Accounts

The 1737 Calcutta, India earthquake is included in a number of lists of the world's greatest natural disasters. It was estimated to have had a death toll of 300 000, placing it among the top three or four of all time. Recent analyses of the event, however, not only cast doubt on the number of deaths but also suspect whether there was an earthquake at all. The problem is that on 11 October 1737, the day of the earthquake, a major cyclone struck the Calcutta area. It was such an intense storm that it sunk eight of the nine British ships and virtually all of the boats operated by the natives that were moored in the Ganges River. Many deaths were attributed to this storm.

However, the British East India Company did not compile a report of the disaster until survivors reached the headquarters in London six months later. Their report listed 3000 fatalities with no mention of an earthquake at all. However, the following year, stories in *The Gentleman's Magazine* and *The London Magazine* claimed 300 000 deaths as the result of an earthquake. This caused the confusion about the event. Historical records indicate that the urban population of Calcutta was less than 20 000 in 1737. There were other settlements outside of Calcutta, but it is doubtful that there were 300 000 people in the entire area much less than many killed. Although there are small to moderate earthquakes in the Calcutta area, they are not the large events that strike the Himalayan region to the north. For this reason, it is unlikely that such a huge earthquake could strike Calcutta. It is most likely that the 1737 Calcutta earthquake is the earthquake that wasn't.

1.5 How This Book Is Organized

There is science controlling natural disasters that must be understood to fully appreciate these catastrophic events. Several of the chapters are the pure science of the controlling processes in order to set the stage for understanding the natural disasters. However, most of the processes of the disaster itself are conveyed through case studies of individual exemplary events. These case studies include much of the science of the event itself that provide the specific background. As case studies, however, the human aspect is included.

CHAPTER 2

Moving Continents

CHAPTER OUTLINE

2.1 Control of Plate Margins on Earthquakes and Volcanoes 17

2.2 Architecture of the Earth 17

2.3 Divergent Margins 20

2.4 Convergent Margins 22

2.5 Transform Margins 23

2.6 Escape Tectonics 27

2.7 Plate Assembly and Disassembly 30

Earth's Fury: The Science of Natural Disasters, First Edition. Alexander Gates.
© 2022 John Wiley & Sons Ltd. Published 2022 by John Wiley & Sons Ltd.
Companion website: www.wiley.com/go/gates/earthsfury

Words You Should Know:

Asthenosphere – The mechanical layer beneath the lithosphere upon which the plates float. The asthenosphere is soft and flows.
Continental collision – The collision between two tectonic plates that contain only continental crust.
Continental rift – A divergent margin that develops on continental crust.
Convergent margin – The boundary between two tectonic plates where they are driven into each other.
Core – The core is the innermost chemical layer of the earth and is composed of iron/nickel. It contains two mechanical layers, the liquid outer core and the solid inner core.
Crust – The outermost chemical layer of the solid earth. There are two types of crust, oceanic and continental.
Divergent margin – The boundary between two tectonic plates where they are driven apart from each other.
Lithosphere – The mechanical outermost shell of the solid earth including crust and the rigid uppermost mantle. Tectonic plates are composed of lithosphere and are relatively stiff.
Mantle – The chemical layer of the earth beneath the crust. It is composed of several mechanical layers and is very dense.
Mid-ocean ridge – The submarine mountain range that marks a divergent margin.
Orogeny – A mountain-building event.
Subduction zone – A convergent margin in which a lithospheric plate containing ocean crust is driven beneath the facing plate and into the mantle.
Transform margin – The boundary between two tectonic plates in which the motion between them is lateral.

2.1 Control of Plate Margins on Earthquakes and Volcanoes

The natural disasters that originate from processes in the crust of the earth are primarily controlled by plate tectonics. For most of the history of geology as a science, the location of events that cause natural disasters were known, but there was no unified reason for these locations. The theory of plate tectonics was developed in the 1950s and 1960s, and it revolutionized the science, sometimes called "the glue that holds geology together." It not only explains the locations of most earth-generated natural disasters but also most of the distribution of rocks and sediments on earth. The planet went from being a fixed, static rock to a dynamic system with continuous and predictable activity.

To understand the theory of plate tectonics requires thinking about the earth in a completely new way. Rather than the earth being rock solid underfoot, it must be viewed as flexible and mobile. In this way, not only can natural disasters be placed into proper context but plate tectonics also explains the locations of mountains, oceans, and continents, the bathymetry of the ocean floor, and the locations of most island chains and mineral and petroleum deposits. The types, features, and locations of volcanoes, earthquakes, and most tsunamis are also controlled by plate tectonics. All of the science and descriptions of these events will be set within this paradigm.

2.2 Architecture of the Earth

Even though the earth seems like a solid sphere, it actually contains distinct layers that are both mechanical and chemical in nature. Some of these two types of layering coincide, but some do not. The envelope of gases that are held to the earth by gravity is called the atmosphere. Although there are many layers to the atmosphere, for the purposes of natural disasters, only the troposphere, or the layer closest to the surface, is of interest. It is both chemical and mechanical in nature, but it is not involved in plate tectonics. Below the atmosphere but also above the solid planet is the hydrosphere which is also both chemical and mechanical and constitutes all surface water bodies but primarily the oceans. This layer also produces many natural disasters, but is not involved in plate tectonics either. Only the solid earth is relevant to plate movements.

The solid earth contains chemically distinct layers of the crust, mantle, and core (Figure 2.1). The core is divided into

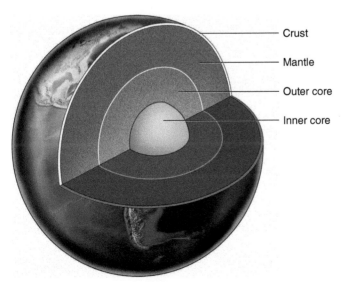

FIGURE 2.1 Diagram showing the chemical layers of the earth.

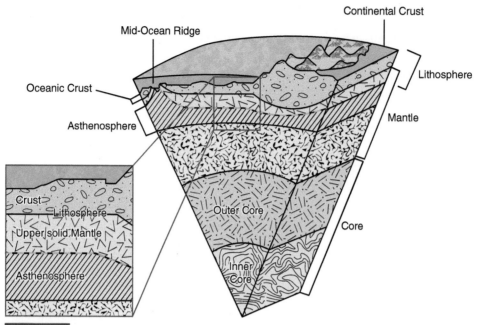

FIGURE 2.2 Diagram showing the lithosphere and asthenosphere (mechanical layers) in the earth relative to the layers.

two mechanical parts, a solid inner core and a liquid outer core. The core is composed of an iron/nickel mixture that is very dense and produces the earth's magnetic field as a result of the earth's rotation. The core is surrounded by the mantle which is a thick layer of dense iron-magnesium silicate rocks and is divided into an upper and a lower mantle based on mineral density. The outermost layer is the crust. There are two types of crust, continental and oceanic. Ocean crust is younger, thinner, and denser than continental crust. Gravity draws the ocean crust deeper into the planet because of its density, so with its thinness, the elevation is much lower than continental crust. Because gravity also draws water to the lowest point, ocean water covers the ocean crust. Continental crust is much thicker and lighter than ocean crust, so it floats up and forms the landmasses. Continental crust can contain sedimentary, igneous, and metamorphic rocks, and it can be very old in comparison with ocean crust.

The mechanical outer layers of the earth are the main controllers of the elevation of the ocean versus continental crust. The outer mechanical layer of the earth is the lithosphere which contains ocean and continental crust and a layer of rigid mantle attached beneath it (Figure 2.2). The lithosphere is broken into fragments or plates of various sizes and compositions around the planet. The boundary between the crust and the rigid mantle is called the Mohorovičić discontinuity or "Moho." At a certain depth beneath the Moho, the mantle becomes gum-like, which defines the next lower layer called the asthenosphere. The depth of the asthenosphere primarily varies according to the thickness of the lithosphere and the topography of its surface. The rigid lithospheric plates float on the softer asthenosphere beneath them. This is why the density of the plates dictates how high or low they sit on the earth. This balance among density, thickness, and surface elevation is called isostasy and it operates continuously (Figure 2.3). A hard-boiled egg with the shell cracked makes a visual illustration of this relationship. The hard-shell fragments represent the lithospheric plates which can be pressed down into the rubbery egg white representing the asthenosphere.

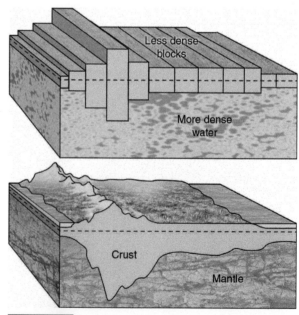

FIGURE 2.3 Diagram showing the concept of isostasy. Top diagram shows isostasy of wooden blocks floating in water. Bottom diagram shows mountains in the lithosphere with deep roots beneath them into the asthenosphere.

CASE STUDY 2.1 Lake Champlain

Depressing the Crust with Ice

During the last ice age from about 22 000 to 12 000 years ago, mile-thick continental ice sheets flowed southward from the poles to now temperate latitudes. In the Northeastern United States, continental ice sheets extended all the way to New York City. The sediment transported southward by the glacial movement now makes up Long Island. In New York City, the ice sheet was about 2000 ft thick (610 m), but just a few miles to the north, it was one mile thick (1.6 km). Because a mile thick of ice is so heavy, it pressed the underlying lithosphere downward into the asthenosphere and the land surface was significantly depressed (Figure 2.4). Once the glacier retreated (by global warming and resulting melting), the earth surface remained depressed for thousands of years before the surface could "bob" back up. Movement of continents and the mantle in response to the weight changes is a slow process.

The surface of the earth around the current large Lake Champlain on the New York State–Vermont border was so depressed that it was below sea level. This depression caused the area to open to the Atlantic Ocean through its northern end through the St. Lawrence Seaway (Figure 2.5). Whales and other marine life swam through this passage and inhabited this pre-Lake-Champlain body of water. Slowly, the lithosphere floated back upward by the process called isostatic rebound. Consequently, the land surface slowly increased its elevation. Eventually, the passage from Champlain into the St. Lawrence Seaway was cut off by the rising land surface and the body of water became Lake Champlain in its current form (Figure 2.5A). With the input of rain and surface runoff, the waters of the lake became fresh. It is now an odd situation that relatively recent whale bones can be found around a freshwater lake.

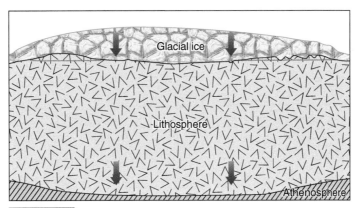

FIGURE 2.4 Diagram showing the depression of the lithosphere into the asthenosphere as the result of the weight of overlying ice.

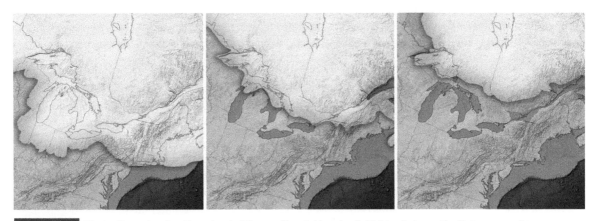

FIGURE 2.5 Illustrations showing the retreat of the continental ice sheet. Right cell shows the St. Lawrence Seaway connected to Lake Champlain.

The analogy with the hard-boiled egg becomes less appropriate when considering the positioning of the lithospheric plates. The positions of the broken shell fragments on a hard-boiled egg are fixed, but the seven major and several smaller lithospheric plates on the earth can move. They move because the mantle flows underneath them, driven by thermal currents, and the plates are carried along like rafts (Figure 2.6). The mantle flows in big convection cells where hotter areas rise up (upwell) because the hot rocks are less dense. When the hot material reaches the top of the asthenosphere, it spreads out away from the hot area and cools. Cooling causes it to become denser and it is driven back down into the deeper mantle by gravity acting on this increased density and interaction with other cells. There, it can then be reheated and go through the convection cell in a repetitive process.

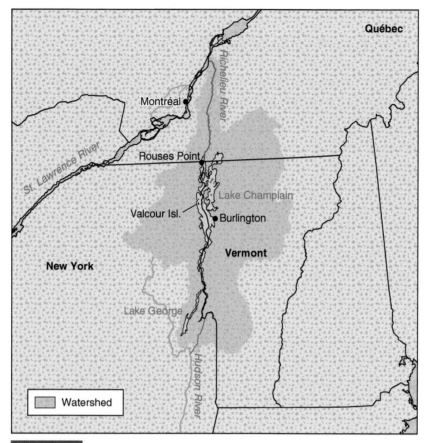

FIGURE 2.5A Current map of Lake Champlain after glacial rebound was complete.

As a consequence of these numerous flowing cells, the lithospheric plates do not all move in the same direction but instead grind against each other at their margins. The sources of the events that cause most solid earth natural disasters are the sites where plates grind against each other. There are three possible interactions among plates: convergence, divergence, and transform. Convergence results from mantle flow forces pushing plates into each other in a collision. Divergence results from forces pushing plates away from each other and transform results from forces causing plates to slide past each other in lateral motion. All of these zones of interaction are called margins or boundaries, and each produces a distinct group of potential natural hazards.

2.3 Divergent Margins

Lithospheric plates move away from each other at divergent margins (Figures 2.6 and 2.7). Beneath them, hotter mantle upwells along a linear zone and splits, flowing directly away in both directions. This forms adjacent convection cells with opposite directions of rotation. Solid earth natural disasters are common above the point of upwelling, but they are only devastating under certain conditions. In many divergent margins, few or no natural disasters occur.

The best way to describe divergent boundaries is to follow the development from continental rift to mid-ocean ridge, the end members of the process. Mantle upwelling may begin, as a line of mantle plumes under continental crust. The mantle plumes are individual points of rising hot mantle material. As the plumes coalesce into a linear zone of upwelling, the flow in the mantle starts to pull the overlying lithospheric plate apart forming a continental rift. The bottom of the plate stretches and gets thinned like stretching gum, but the crust above gets fractured and breaks into a series of faults. This breakage can produce significant earthquakes. The upwelling of hot mantle in the plumes and along the zones generates magma in the upper mantle, which rises through the crust above producing low flat volcanoes and fissure eruptions that emit lava from cracks within the continental rift. The heat from this magma melts the crust producing steep-sided, explosive volcanoes as well. The thinning of the crust causes the surface to subside into a depression characterized by series of basins that fill with lakes and sediments. The Basin and Range Province of the Southwestern United States is an excellent example of a continental rift as is the East African Rift System.

As divergence continues, the depression continues to subside and, eventually, the continental plate breaks in two with the depression in between (Figure 2.7A). Once the depression becomes sufficiently deep, it floods and a long, narrow sea forms

Divergent Margins   21

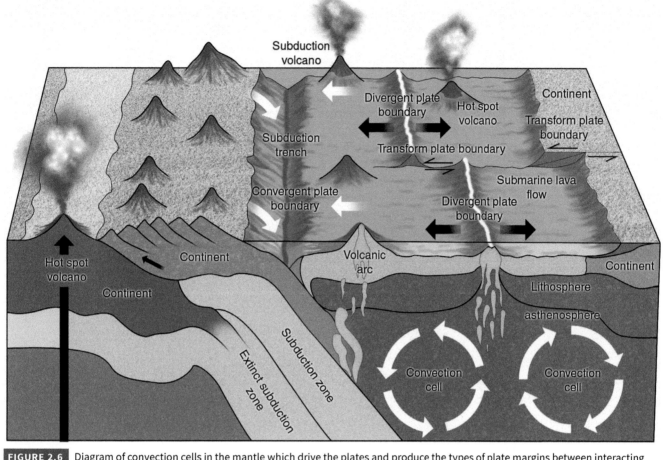

FIGURE 2.6 Diagram of convection cells in the mantle which drive the plates and produce the types of plate margins between interacting plates as shown.

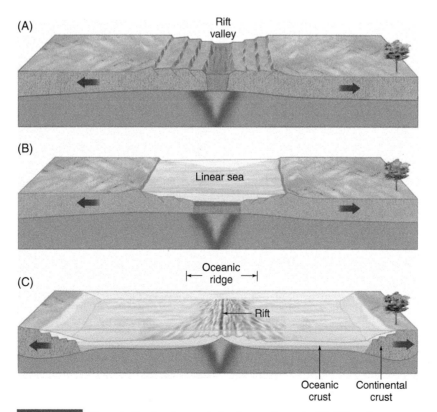

FIGURE 2.7 Stages in the development of a divergent margin; (A) rifting of continental crust, (B) narrow ocean basin, and (C) full ocean basin.

(Figure 2.7B). The narrow sea is typically open to the ocean at least at one end, but circulation within it can be poor because of the narrow opening. Igneous activity becomes even more intense and begins to form dense oceanic crust within this depression. The volcanic activity under the narrow sea spews gases and fluids into the poorly circulating water. It becomes hypersaline, depositing salts and metal compounds on the seafloor. Earthquakes still affect the coasts of the narrow sea, but the volcanoes are mostly submarine, so do not commonly affect the land around it. Small tsunamis can also be generated from the earthquakes and volcanic eruptions. The Red Sea is an example of this stage in the development of a divergent boundary.

The final stage in a divergent boundary is the formation of a full ocean basin (Figure 2.7C). As the two halves of the original continent (now two continents) are driven apart, the narrow ocean basin widens into a full ocean basin. Once the basin becomes established, the divergent margin develops into a mid-ocean ridge. This feature is an underwater mountain range that is elevated because it is volcanic, hot, and, as a result, less dense. The rocks are new ocean crust formed by the igneous activity. Basically, as the plates separate, the cracks are filled with the magma that feeds the volcanoes. This builds the ocean crust which is constantly forming at the mid-ocean ridge. Once formed, it too cracks apart and even newer ocean crust fills the crack. This means that ocean crust is youngest at the mid-ocean ridge and is progressively older in both directions away from it. All of the earthquake and volcanic activity is right at the ridge in the middle of the ocean. The land around the ocean basin has no activity. As a result, the shorelines are termed passive margins. The Atlantic Ocean is an example of a full ocean basin with the volcanic mid-ocean ridge at the middle as illustrated by Iceland where the mid-ocean ridge comes out of the water and forms land. It also contains passive margins both in eastern North America and in Western Africa and Europe.

2.4 Convergent Margins

If new ocean crust is being produced at the mid-ocean ridge, other crust must be destroyed somewhere else or the planet would grow larger. Convergent margins are where ocean crust is destroyed (Figures 2.6 and 2.8). Adjacent convection cells in the mantle form currents that drive plates away from each other at divergent margins. By contrast, the plates are driven together at other end of the convection cells where they abut the next cell. The flowing material sinks deeper into the mantle at these zones of convergence. The plates above them are driven into each other in a linear or arcuate convergent boundary or margin. There are three types of convergent margins depending upon the types of crust colliding with each other: ocean–ocean, ocean–continent, and continent–continent.

As two lithospheric plates are pushed together, one goes under and the other rides above. The control on which plate rides above is density. Ocean crust has a density of about $3.4\,g/cm^3$, whereas continental crust density is about $2.65\,g/cm^3$. Where the two plates push together, the ocean crust is driven beneath the continental crust in a subduction zone (Figure 2.8A). These form trenches on the seafloor, which are the deepest places on the planet. As the oceanic plate subducts, it carries rocks and sediments to great depth where it is highly pressurized but not heated as much as it should be at that depth. Water escapes from these rocks and melts the mantle above it, producing magma. The magma rises through the crust and produces extensive plutonism and volcanism directly above it. The feature is called a magmatic arc. The plutonism, volcanism, and deformation produce a towering mountain range. The bending of the oceanic plate

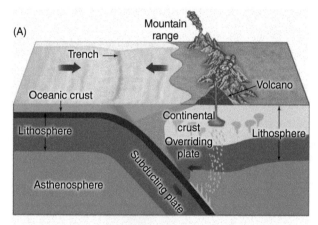

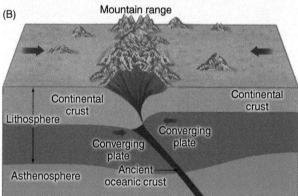

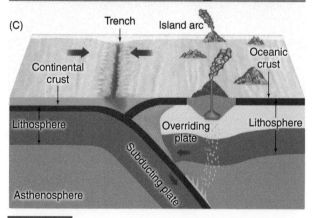

FIGURE 2.8 Diagram of convection cells in the mantle which drive the plates and produce the types of plate margins between interacting plates as shown.

as it subducts and the deformation of the overriding continental plate produce very powerful earthquakes and tsunamis. The subducted oceanic plate is consumed in the mantle and is recycled to its source. The best example of an ocean–continent collision is the Andean orogeny along the entire western margin of South America. In this case, the Nazca plate of the Pacific Ocean subducts beneath the South American plate.

Many lithospheric plates contain both continental crust and oceanic crust. For example, the North American plate contains the North American continental crust and the attached Atlantic Ocean crust eastward to the mid-Atlantic ridge. As subduction of ocean crust under a continental plate continues, at some point, it runs out and the trailing attached continent collides with the continent on the overriding plate. For example, Australia is the trailing continent in an ocean-continent collision and will eventually collide with Indonesia. This is how a continent–continent collision is produced (Figure 2.8B). Although it tries to follow the ocean crust, continental crust is too light to be driven into the dense mantle, making the collision similar to a car accident. The crust on the subducting plate drives under the overriding plate, forming a large fault between the two. Other similar faults form, as the crust is crumpled producing a very high and broad mountain range and plateau on the overriding continent. The volcanism stops soon after the continents make contact, but very strong earthquakes abound. The best example of a continent–continent collision is the Himalayan orogeny. The northern margin of India is on the subducting plate and colliding with the overriding southern margin of Asia where the the Himalayan Mountains and Tibetan Plateau are produced.

The final type of convergent margin is an ocean–ocean collision (Figure 2.8C). In this case, ocean crust is subducted beneath other ocean crust and the geometry is similar to an ocean–continent collision. A subduction zone forms with an arcuate chain of volcanic islands, called an island arc or volcanic arc which is analogous to a magmatic arc. Regional plate tectonics typically controls which plate goes over and which goes under. Similar to the ocean–continent margin, these are zones of intense volcanic eruptions and earthquakes that commonly produce tsunamis. A good example of an island arc is the Aleutian Islands of Alaska.

2.5 Transform Margins

The primary margin types are divergent and convergent. However, because the earth is spherical and the convection cells in the mantle are not perfectly aligned or symmetrical, these margins cannot be completely parallel. Any misalignment or difference in plate velocity results in the development of a transform margin to compensate. Transform margins are large faults with primarily lateral motion in contrast to divergent margins where faults drop the surface down and convergent margins where faults push the surface up. Transform margins must connect two other types of margins. Therefore, the possible types of transform margins are named by the margins they connect: divergent–divergent, divergent–convergent, and convergent–convergent (Figure 2.9A–C). These are zones of strong earthquake activity but not volcanism.

The most common transform margin, by far, is the divergent–divergent type. Almost all are on ocean crust (Figure 2.9A). Mid-ocean ridges are clearly visible on bathymetric maps of the ocean basins of the world as long submarine mountain ranges. An example is the Mid-Atlantic Ridge which lies north to south along the center of the Atlantic Ocean. On close inspection, there are hundreds of crossing lines perpendicular to the ridge. All of these lines are divergent–divergent transform margins. They compensate for every bend in the ridge and lateral change in spreading velocity and the shape of the margins.

Convergent–convergent transform margins commonly form offsets on island arcs, so are also primarily marine as well (Figure 2.9B,C). An exception is an excellent example of a convergent–convergent transform margin on land that crosses the entire South Island of New Zealand. This margin is expressed by the South Alpine Fault which produces regular major earthquakes and the breathtaking topography. Convergent–divergent margins are similarly rare (Figure 2.9A). They are most commonly associated with smaller tectonic plates especially where young ocean crust is subducted. They are also primarily on ocean crust. An exception is in California where the San Andreas Fault is the surface expression of the margin (Figure 2.9D). The divergent margin is the Gulf of California in Mexico and the convergent margin is the Cascadia subduction zone. It is responsible for the volcanoes in the Cascade Mountains from Northern California to southern Canada. The San Andreas is well-known for regular earthquake activity.

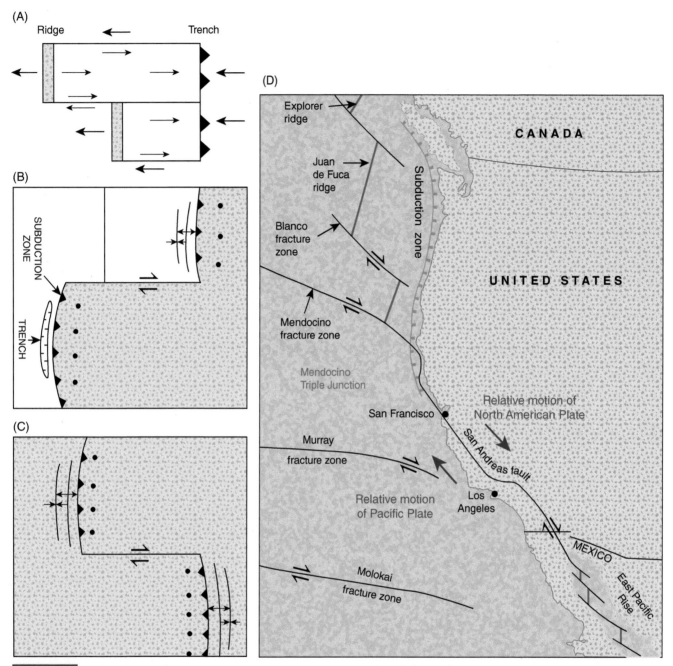

FIGURE 2.9 Diagrams of transform boundaries; (A) divergent–divergent transform (left side), (B) convergent–convergent transform, (C) convergent–convergent transform, (D) map of San Andreas transform and the plate boundary connections of all transform faults on the West Coast of the United States.

CASE STUDY 2.2 The 1906 Great San Francisco Earthquake

A Transform Margin Earthquake

The San Andreas Fault is more than just a fault; it is a divergent–convergent transform margin that separates the North American Plate to the east from the Pacific Plate to the west (Figure 2.9D). Because it is a plate margin, it will continue to be active for longer than humans will be on the earth. It is also one of the most seismically active faults in North America and it is, by far, the fault of greatest concern. This is because it underlies several of the major cities of California including Los Angeles and San Francisco. There was an old high school video from the 1960s entitled "San Francisco: City that waits to die." The most infamous example of this danger was the 1906 San Francisco earthquake.

On the morning of 18 April 1906, at 5:12 a.m., a major earthquake of magnitude 7.9 erupted from the San Andreas Fault and struck the city of San Francisco. There was a strong foreshock about 25 seconds before the main shock which had a duration of about 45 seconds. The strong surface waves were emitted from a shallow focus, 5 mi (8 km) deep, beneath the epicenter offshore near the Golden Gate, producing a maximum shaking intensity of XI on the Modified Mercalli Scale which is ranked as extreme. The rupture on the fault that caused the earthquake propagated both north and south of the focus and extended about 300 mi (477 km). Maximum offset on the fault was about 28 ft (8.5 m) in right lateral, strike-slip movement (Figure 2.10).

The earthquake literally demolished the city with more than 80% of the buildings destroyed (Figure 2.10A). Although the surface waves demolished many buildings, the main destruction was in knocking over heating and cooking fires and rupturing gas lines. This caused the entire city to go up in flames. As much as 90% of the total destruction of San Francisco was the result of the fires rather than the earthquake itself (Figure 2.10B). More than 30 fires destroyed some 25 000

FIGURE 2.10 Right lateral offset of a fence across the San Andreas Fault as a result of the 1906 San Francisco earthquake. *Source*: Courtesy of the US Geological Survey.

FIGURE 2.10A Photograph of downtown San Francisco in flames as a result of the 1906 San Francisco earthquake. *Source:* Courtesy of the US Geological Survey.

buildings on 490 city blocks across the city over the next three days. The military destroyed city blocks using dynamite in an attempt to contain the rampaging flames. The destruction of so many buildings left 300 000 people homeless out of a population of 410 000 people (Figure 2.10C). The best estimate of a death toll is about 3000 which makes it the second worst natural disaster in US history. However, the estimate has varied significantly over the years. The amount of property destruction is estimated at an equivalent of $11.4 billion in 2019 dollars.

Perhaps the most prominent part of the story was the rebuilding of San Francisco. The enthusiasm and eagerness of the city to restore itself rose to international headlines and was covered and even followed by numerous newspapers. Aid poured in from across the United States and local businessmen took a personal interest in bringing the residents together in this effort. In the end, San Francisco was restored to its pre-earthquake splendor in short order which amazed the world. It was not the first earthquake to impact the city, nor will it be the last, but to date, it is the most famous.

FIGURE 2.10B Photograph of destruction of downtown San Francisco as a result of the 1906 San Francisco earthquake. *Source*: Courtesy of the US Geological Survey.

FIGURE 2.10C Destroyed downtown area in San Francisco as a result of the 1906 earthquake. *Source*: Courtesy of the US Geological Survey.

2.6 Escape Tectonics

Transform margins are not the only source of faults with lateral (strike-slip) motion. Another source is from escape or extrusion tectonics which is formed by continent–continent collisions. To understand escape tectonics, place a piece of clay in your hand and smash it with your fist. The force of your fist causes the clay to squirt out of the sides as it is impacted. Substitute a continent on the subducting plate for your fist and the continent on the plate with the magmatic arc for the clay and it is a plate tectonic situation.

The best current example of escape tectonics is in the Himalayan orogeny (Figure 2.11A). India is on the subducting plate and is impacting Southern Asia. It is the *rigid indenter* in escape tectonic terms (Figure 2.11B). No land can move to the west because there is too much continental mass. To the east, however, is the Pacific Ocean with several subduction zone margins. That is considered a *free face* because land can move relatively freely in that direction. When India began colliding with Asia, the land along Asia's southern edge to the east of the collision was squeezed and extruded southeastward 600 mi (1000 km) and became Indochina. Once this land was out of the way of the collision, the next fragment of land was squeezed out of the way and extruded several hundred miles to form the South China Sea and jagged coastline. This is the reason for the extensive earthquake activity in China. Escape tectonics is also responsible for the extrusion of Turkey into the Mediterranean Sea and the extensive earthquake activity in that area.

Although the large faults with lateral offset that are produced by escape tectonics are not technically plate margins, they are not easily categorized. They have the same movement sense as transform margins and they typically connect two other plate boundaries. Therefore, they can satisfy the definition for transform margins, even though they are not technically margins.

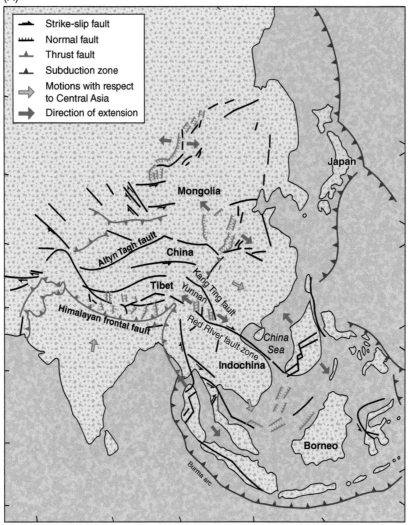
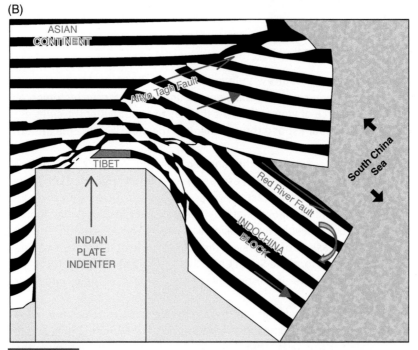

FIGURE 2.11 Diagrams showing escape or extrusion tectonics. (A) Fault and tectonic map of Eastern Asia showing the relative movement of large crustal blocks (open arrows) and local blocks (filled arrows). (B) Analog model of Eastern Asia using a brick (India) and layered clay (black and white striped) to explain the tectonic geometry.
Source: Rasoul Sorkhabi (2013) ; Paul Tapponier et al. (1982).

CASE STUDY 2.3 1970 Tonghai Earthquake, China

Escape Tectonic Earthquake

The impacts of the 1970 Tonghai earthquake were unknown in the West until 1988 because of Cold War secrecy in China. The Tonghai earthquake struck the Yunnan Province, Southern China at about 1 p.m. on 4 January 1970 (Figure 2.12). It had a magnitude of 7.5–7.7 and the shaking lasted at least 50 seconds. It was felt strongly over an area of 5456 mi² (8781 km²) causing people to flee from their homes as far away as Hanoi, Vietnam, 300 mi (483 km) away. The epicenter was in Wuxing near Tonghai, about 75 mi (121 km) southwest of the city of Kunming. The focus was at 12 mi (20 km) depth.

It was first determined that the Tonghai earthquake occurred on the Honghe Fault of the Red River Fault System (Figure 2.11). The Red River Fault is a major fault in China and had a history of intense activity including about 600 mi (1000 km) of offset over 25 million years. It is the type example of a strike-slip fault accommodating eastward extrusion of a block of crust from escape tectonics. The fault is generated by the Himalayan orogeny to the southwest. The extruded block moved eastward along the Red River Fault out of the direct impingement by India. Once the land was out of the way, the Red River Fault ceased earthquake activity, but the rock around it formed a series of fragmented crustal blocks separated by faults. Stress from the continuing Himalayan orogeny periodically causes these blocks to move, resulting in earthquakes on the faults.

The Tonghai earthquake actually resulted from movement on the Qujiang Fault, which is a possible branch of the Red River Fault. The earthquake ruptured a 56-mi (90 km) length of the fault. The average left-lateral strike-slip displacement was 6.9 ft (2.1 m) with a maximum lateral movement of 8.9 ft (2.7 m). It also included a dip-slip uplift of about 1.5 ft (0.5 m). The earthquake produced a continuous surface rupture of 30 mi (48 km) length. Previously, the faults of this area were considered to be inactive, but new GPS data indicate 0.28 in. (7 mm) per year of left-lateral strike-slip movement across the Qujiang Fault Zone. The Tonghai earthquake was the first major event in 300 years in the area, but there have been several strong earthquakes on other segments of the fault including a magnitude 8 event in 1833 and a magnitude 7.8 event in 1733.

The Tonghai earthquake was devastating. Virtually all buildings in the epicentral area were damaged or destroyed because, after 300 years of the assumption of inactivity, they were not designed or built to withstand the strong shaking. The death toll for the event is 15 621 people with 26 783 injured. Rough damage estimates range from US$5 million to US$25 million.

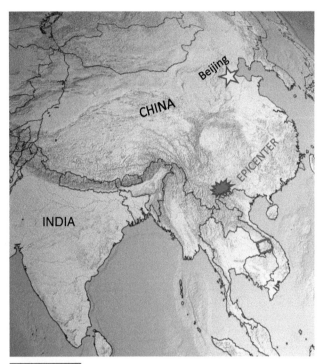

FIGURE 2.12 Map of Eastern Asia showing the location of the 1970 Tonghai earthquake.

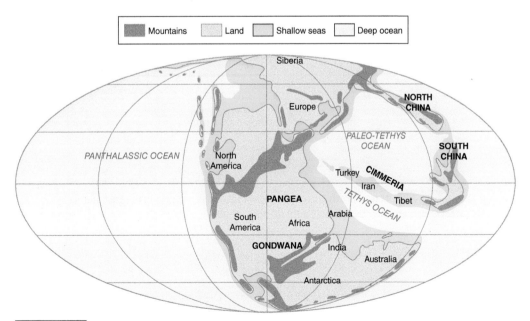

FIGURE 2.13 Diagram of plate configuration for the Triassic Period showing the supercontinent Pangea.
Source: Adapted from Alan Logan, Triassic Period. Encyclopædia Britannica, Inc.

The political reaction to the Tonghai earthquake was unusual. It was not even publicized in China and the amount of aid offered to the survivors by the government was termed "pathetically small" by a Beijing newspaper. The local leaders refused to take donations and returned all that were sent. The reason for this response is that the earthquake occurred at the peak of the Cultural Revolution in China. Chairman Mao Tse-tung pushed ideals of self-determination, self-reliance, and hard work to overcome adversity. The only donations residents accepted were Chairman Mao's red book, badges, and numerous letters of condolence. Premier Zhou Enlai did not even visit the area because he was busy, but sent a written apology instead.

2.7 Plate Assembly and Disassembly

Tectonic plates constantly move around the earth, collide with each other and combine, and then they break apart again. At times, all or almost all continents join together to form a single supercontinent before breaking apart again. The last supercontinent was Pangea which broke apart 200 to 65 million years ago (Figure 2.13). Continents tend to break apart in approximately the same place as they suture together during collision. The reason for this is not clear, but the pressing of the root of high mountains into the mantle beneath is suspected as a contributing factor. If continents tend to break apart along the same zones as they collided, the next collision tends to be in the same place as the previous one. This repetition of developing margins means that there are zones that record multiple periods of tectonic activity. These zones are called mobile belts. The Appalachian orogeny of eastern North America is such a belt recording at least four periods of collision. Its current position of facing onto the Atlantic Ocean means that its next collision will be in the same place (Figure 2.14).

This tendency also means that much of the continent will never be involved in containing a plate boundary. These are all areas interior to the mobile belts. They are called stable interiors or cratons. These may contain older mobile belts, but they have been quiet for a significant period of time. The interior of the United States and Canada has been stable for at least one billion years and it is bounded to the east and west by repeatedly active mobile belts. This is the type example for the theory.

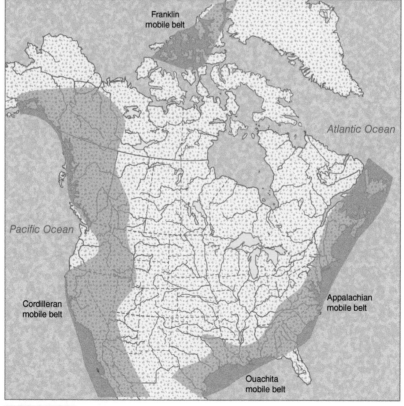

FIGURE 2.14 Map of North America showing the mobile belts along the East and West Coasts and the stable craton interior in the middle.

References

Sorkhabi, R. (2013). Folds and folding. Part I. *GeoExpro*, 10 (3), 92–96; Part II. *Geo Expro*, 10 (4), 60–64.

Tapponnier, P., Peltzer. G., Le Dain, A. Y., Armijo, R., Cobbold, P. (1982). Propagating extrusion tectonics in Asia: New insights from simple experiments with plasticine. *Geology* 10 (12): 611–616.

CHAPTER 3

How Does Rock Melt?

CHAPTER OUTLINE

3.1 If the Earth Is Solid, Why Do Rocks Melt? 33

3.2 Types of Magma 33

3.3 Melting at Divergent Margins 35

3.4 Melting at Convergent Margins 37

3.5 Melting at Mantle Plumes 38

3.6 Contamination of Magma 41

3.7 Magma to Lava 45

3.8 Physical and Chemical Constraints on Volcanic Hazards 45

Earth's Fury: The Science of Natural Disasters, First Edition. Alexander Gates.
© 2022 John Wiley & Sons Ltd. Published 2022 by John Wiley & Sons Ltd.
Companion website: www.wiley.com/go/gates/earthsfury

Words You Should Know

Andesite – A volcanic rock with intermediate chemical composition and which is common from volcanoes in subduction zones (convergent margin).

Assimilation – As magma intrudes through the crust, it can melt the rock it passes through and/or remove pieces of it (xenoliths) and mix it in, all of which change the composition of the original magma.

Basalt – Volcanic rock with mafic composition and by far the most abundant volcanic rock on the earth. Basalt is most common from volcanoes in divergent margins.

Dacite – An intermediate volcanic rock similar to andesite but which is enriched in quartz and feldspar.

Decompression melting – Melting of solid rock to magma purely by a drop in pressure.

Devolatilization – The loss of volatile components (water, gases) from magma or lava.

Felsic – A compositional term for igneous rocks that are light in color and rich in feldspar ("fel") and silica ("si") or quartz, such as granite and rhyolite.

Fractional crystallization or fractionation – As magma cools, it crystallizes the higher melting temperature Fe–Mg minerals (mafic) first. If these minerals are removed from the liquid, the composition of it progressively changes to that resembling the lower temperature minerals (more felsic).

Intermediate – A compositional term for igneous rocks that indicates chemistry between mafic and felsic.

Lava – Molten rock on the surface of the earth.

Mafic – A compositional term for igneous rocks meaning rich in magnesium (ma) and iron (fic).

Magma – Molten rock that is underground and still containing its volatile components.

Partial melting – When rock is heated up enough to melt some of the minerals with lower melting temperatures, but the rest of the minerals remain solid.

Pluton – The body of a crystallized magma chamber of any shape or size.

Rhyolite – A light-colored volcanic rock with felsic composition.

Volatile – A substance that readily evaporates to a gas.

Wet melting – Melting of a rock due to the presence of water which depresses the melting temperature.

3.1 If the Earth Is Solid, Why Do Rocks Melt?

Even though some movies portray the earth as full of liquid rock that is ready to escape, besides the liquid core, the interior of the earth is almost entirely solid rock. In some parts of the crust and mantle, the rocks have a gummy consistency, but they are all considered solid. Only a few volumetrically small pockets of melted rock or magma exist at some plate boundaries or associated with mantle plumes. Considering that the liquid core plays no role in the generation of lava or magma, the situation seems to be a paradox. Where does the huge amount of lava that comes out of volcanoes come from?

The answer is that magma is locally produced as needed from the interplay of composition, pressure, and temperature of rocks under a specific set of conditions. The guiding force in most of these situations is plate tectonics. That is why most volcanoes and plutons occur at plate boundaries or at developing plate boundaries. The only real exception to this correlation is mantle plumes which occur in the interior of the plates at various locations both on land and in the ocean. Under the right set of conditions and processes, the solid rocks in the earth melt and produce vast quantities of magma. That is how magma is produced on a local basis; it does not sit in large reservoirs within the earth waiting to be released by a volcano.

This chapter describes the processes of how most common types of magma are produced and modified from solid rock based on the specific sets of conditions. The magmas can emerge from volcanoes as lava if they reach the surface or crystallize underground as plutons. This chapter serves as preparation for constraining the locations and frequency of the various types of volcanoes and volcanic hazards.

3.2 Types of Magma

Magma is melted or liquid rock that is present within the earth. It can be a pure liquid or a crystal mush containing a mixture of liquid rock and crystals and even rock fragments. Once it reaches the surface, it undergoes chemical and physical changes and becomes lava. Most of the chemical changes involve devolatilization or loss of fluids and gases, but otherwise, the bulk chemistry of the magma and the lava is basically the same. The physical changes involve faster cooling, integration of surface materials, and gas release features. For that reason, plutonic and volcanic rocks are grouped together based upon their general compositions. There are three main compositional categories and several subordinate ones. The three main categories are felsic, intermediate, and mafic (Figure 3.1). For the purposes of this book, these three are sufficient to understand the main types of volcanoes on the earth. There are, however, several others and many have important variations.

If magma cools underground, it is a slow process that allows mineral crystals to grow large and interlocking. Rocks formed in this process are plutonic because they comprise plutons which are underground bodies of igneous rock. They are also referred to as intrusive because they intrude into the preexisting bedrock known as country rock. Plutons might be former magma chambers or fractures (fissures) that fed volcanoes or they may not have been associated with volcanoes at all.

Plutonic rocks have large mineral grains that can all be identified with the naked eye. If the rocks are volcanic, they cooled from lava or pyroclastic material on the surface of the earth. In these rocks, some of the minerals might be identifiable with the naked eye if they grew in plutons prior to eruption, but most require an optical microscope for identification if they are minerals at all. Some components of volcanic rock can be glass, known as obsidian and pumice (like hardened foam), neither of which has minerals. Glass is supercooled liquid with no chemical bonding structure as with crystalline solids.

Comp.	Felsic	Intermediate	Mafic
Volcanic	Rhyolite	Andesite	Basalt
Plutonic	Granite	Diorite	Gabbro
Example	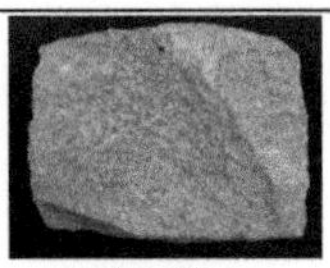 Rhyolite	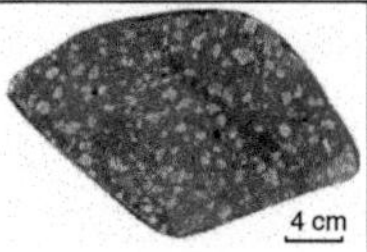 Andesite	 Basalt
Temp.	750 °C	1,000°C	1,400°C
Viscosity	Very Sticky	Sticky	Fluid
Ejecta	Very High	Mixed	Low
Shape			
Water Content	High	Very High	Low
Contamination	Very High	High	Low

FIGURE 3.1 Igneous rock types, corresponding volcanoes and their characteristics. *Source*: Images courtesy of the US Geological Survey.

3.2.1 Felsic

The word felsic is a derivative of *feldspar* and *silicic* because these are the primary compositional components of rocks in this group (Figure 3.1). Felsic rocks are primarily composed of the minerals quartz and feldspar. Of these, they always contain potassium feldspar, and, in most instances, they also contain sodic plagioclase feldspar, both of which are light in color under these conditions. Felsic rocks are, therefore, almost exclusively light in color. If cooled underground in plutons, felsic rocks are called granite. If they come to the surface as lava and consequently have fine mineral grains and possibly glass, they are called rhyolites.

3.2.2 Intermediate

Intermediate rock compositions span the range between felsic and mafic compositions (Figure 3.1). These rocks are generally medium gray in color and tone, but they are variable from lighter to darker gray. They must contain plagioclase feldspar which is medium gray at these compositions. The coarse-grained, plutonic varieties of intermediate rocks range from diorite on the mafic side to granodiorite on the felsic side. The fine-grained, volcanic intermediate rocks range from andesite on the mafic side to dacite on the felsic side.

3.2.3 Mafic

The word *mafic* is a composite term that means rich in magnesium (Ma) and iron (Fe) (Figure 3.1). These rocks are primarily composed of calcic plagioclase (very dark gray) with black to greenish pyroxene and possible olivine. They are therefore very dark gray to black in color and quite dense. The names for mafic rocks are gabbros if they are plutonic and coarse-grained and basalt if they are volcanic and fine-grained. They are the most common igneous rocks in the crust of the earth.

3.2.4 Other Igneous Rocks

There are many other types of igneous rocks in addition to those listed, but they are less voluminous and less important in volcanic activity. The most important of these is the group called ultramafic rocks which are typically included in the general classification with those listed in the preceding text. They never occur in volcanoes, only plutons. The most common ultramafic rock is peridotite, but there are many other varieties. Ultramafic rocks are important because the mantle of the earth is composed of them.

3.3 Melting at Divergent Margins

There are numerous volcanoes at divergent margins and at intracontinental areas that are in the early stages of divergence, known as continental rifts. Melting in these areas is driven by the same mantle circulation that moves the tectonic plates. Directly beneath a divergent boundary or a continental rift, the rock in the asthenosphere of the mantle upwells in a long, narrow zone. As this flowing mantle nears the lithosphere, it diverges driving the plates on either side away from each other. This divergence thins and cracks the lithosphere (mantle and crust) directly above the zone of upwelling. The surface subsides and creates a basin by rifting on land. The thinning and cracking of the lithosphere at the rift zone or in the ocean at a mid-ocean ridge allow magma to penetrate the crust and feed lava to volcanoes at the surface directly above. These volcanoes can be very active and produce large volcanic provinces.

The upwelling mantle brings very hot material from point A deep in the mantle closer to the surface at point B (Figure 3.2). The thermal conductivity of most materials, and certainly rocks, is such that they heat and cool relatively slowly, but react to pressure changes immediately. Therefore, the mantle rocks remain at nearly the same hot temperature as they move upward, but the pressure reduces dramatically. The temperature versus pressure graph for rocks in the mantle that shows the conditions where they are liquid versus where they are solid is called a phase diagram because it shows the same rock in different phases (solid versus liquid) (Figure 3.3). The boundary between the fields where the rock is completely solid and where there is some liquid (melting) for an ultramafic rock with no fluids is at an angle on the graph. This means that rocks can melt with additional heat as expected, but pressure changes can also cause melting.

This situation may not be intuitive for many people. Normally, only increasing temperature (heating) is thought of as being able to melt or boil a substance like water (ice, liquid, steam). The reason that it seems like only temperature matters is because it is more difficult to change pressure. Although pressure typically does not play as significant a role as temperature, it can still contribute to phase changes.

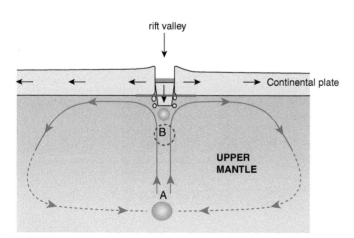

FIGURE 3.2 Diagram showing the circulation in the asthenosphere and movement of overlying plates in a divergent margin. The circulation drives mantle material from a deep point A to a shallow point B resulting in a sharp decrease in pressure.

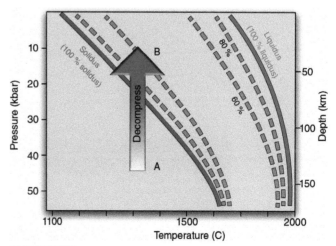

FIGURE 3.3 Pressure–temperature graph of a phase diagram for the melting of dry mantle rocks. Points A and B correspond to the locations on Figure 3.2. As the pressure decreases (decompresses), the rock crosses the solidus (blue line), and melts.

Decreasing only the pressure (or depth) as shown by the arrow on Figure 3.2 from point A to point B can make the mantle rock cross the line separating the solid and liquid fields (solidus) and make it partially melt. The point A to point B decompression path on the graph in Figure 3.3 reflects the corresponding upwelling path from point A to point B on Figure 3.2 in the earth. The partial melting creates newly formed liquid, or magma, that migrates and collects to produce a body of magma. This melting zone occurs right beneath the surface.

The original mantle rock is ultramafic, but it partially melts through this process to form mafic magma. This process is called decompression melting because it is driven completely by decreasing pressure with no change in temperature. The newly formed mafic liquid is less dense than the surrounding mantle rock, and buoyancy drives it quickly upward through cracks in the overlying lithosphere and to the surface where it produces basalt volcanoes or crystallizes below the surface to produce gabbro.

CASE STUDY 3.1 Giant's Causeway, Ireland

Columnar Joints

Massive basalt flows on the surface are called flood basalts because they flood over the landscape like water. Basalt lava is so fluid that the flows stay thin and spread out to cover vast areas. As they spread over the countryside, they are so hot that they fuse to the land surface below them. Once they begin to cool and harden, a problem develops because of thermal shrinkage; cooling tends to shrink solids. The flow cannot shrink because it is firmly adhered to the ground below it. As a result, the flow forms regular vertical cracks called columnar joints to absorb the size change and thermal stress.

One of the best examples of columnar joints is on the coast of Ireland and it has been dubbed the Giant's Causeway (Figure 3.4A). About 250 million years ago, all of the continents came together to form a single supercontinent called Pangea. Having all of the landmasses in one place is not a gravitationally stable situation and Pangea soon broke apart into the continents we see today. The splitting of Europe and Africa from the Americas to form the Atlantic Ocean did not happen all at once, but instead it formed in segments. The main part of the ocean formed in a continental rift beginning about 220 million years ago, but the North Sea rifting did not occur until about 60 million years ago. As it did, a series of mantle plumes that formed such igneous complexes as the Isles of Mull and Skye coalesced to form the rift. Flood basalts formed as the rift developed and spread over the soon-to-be coastal regions of Ireland. The cooling and shrinkage of the flows produced columnar joints that look as if they are from a medieval tale of magic and sorcery. They are a popular tourist attraction.

Columnar joints are direct evidence that a basalt flow occurred on the surface of the earth regardless of where it is currently observed. Most surface flows exhibit them and there are many examples across the United States from Devils Postpile in California (Figure 3.4B) to the

FIGURE 3.4A Photo of the Giant's Causeway, Ireland. *Source*: Hugh Rooney/Eye Ubiquitous/Universal Images Group via Getty Images.

FIGURE 3.4B Photo of the Devils Postpile, California. *Source*: US National Park Service.

Watchung basalts in New Jersey where the tops have been likened to the plates on the shell of a turtle. Even shallow intrusive rocks can have columnar joints. The palisades that face New York City are formed by a shallow tabular pluton called a sill with columnar joints. A palisade is a fence with vertical elements. Those vertical elements in the sill are formed by columnar joints.

In a continental rift zone, the very hot mafic magma passes through cooler continental crust. Unlike the mantle, which is composed of dense, high-temperature minerals, continental rocks are mostly composed of light, low-temperature minerals like quartz and feldspar. Mafic rocks crystallize at very high temperatures and if they are completely liquid, they must be at temperatures of about 1400 °C or more. On the other hand, the quartz and feldspar in the rocks through which they are passing melt at much lower temperatures of about 650–750 °C. The hot infiltrating mafic magma heats broad areas of continental crust and partially melts it to form a liquid. This liquid or magma is rich in quartz and feldspar which means it has a felsic composition. The felsic magma is also squeezed to the surface through buoyancy and high pressures where it spawns rhyolite volcanoes that can be active at the same time as the basalt volcanoes, though they are separate and appear quite different. This is an odd situation having two distinctly different types of volcanism happening at the same time with virtually no mixing between the two.

3.4 Melting at Convergent Margins

In most convergent margins, ocean crust, which is largely composed of mafic rock, is subducted beneath continental crust or other ocean crust forming a magmatic arc or an island arc, respectively. Subduction involves a rapid increase in pressure, as the rocks of the ocean crust at point A in Figure 3.5 are pushed into the mantle to point B with a much slower increase in temperature. The reason for this, as in decompression melting, is the thermal conductivity properties of rocks which cause heating to be relatively slow in comparison to pressurization.

Clearly, with this increasing pressure, decompression melting, like that at divergent margins, is impossible. On the other hand, the ocean crust has spent its entire existence at the bottom of the ocean, so unlike the dry mantle rocks in a divergent boundary, it is quite wet. Water in rocks lowers their melting temperatures and, as a result, a different phase diagram applies. The "wet" phase diagram shows a lower temperature solidus (melting curve) at an opposite slope compared to the dry melting solidus. The path in pressure and temperature of a subducting rock on this phase diagram (Figure 3.6) from the point A to point B shown in Figure 3.5 indicates initial strongly increasing pressure with little temperature increase followed by an even and regular increase in both temperature and pressure.

As the burial path for the rock continues, it crosses the solidus or the line between solid rock and partially melted rock and a liquid is formed. This liquid is light and buoyant relative to the dense and highly pressurized mantle rocks around it. The liquid

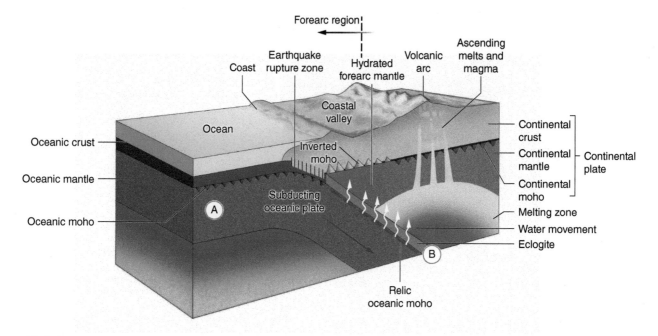

FIGURE 3.5 Diagram of a subduction zone showing the movement of wet rock at point A at the surface to Point B at depth where the magma is produced.

is squeezed through the surrounding rock and collects to form a body of magma. That body is still buoyant and migrates into the crust where it can crystallize as a pluton and/or feed volcanoes at the surface.

When the process was first determined, it was assumed that it was the wet ocean crust that was melting to form the magma. More recently, however, it was proposed that it is actually the mantle above the wet ocean crust that melts because of the water being added to it through its dewatering. This has become the accepted mechanism making the path shown in Figure 3.6 a bit misleading. The pressure–temperature path is that of the subducting ocean crust, but when it crosses the solidus, it is mantle that is melting, not the ocean crust.

The first melt in igneous activity related to a subduction zone is generally mafic, but it is marginally so and very close to andesite/diorite. By far, the predominant igneous activity in subduction zones is intermediate regardless of whether they are ocean–ocean or ocean–continent collisions. The volcanoes produce intermediate lava, andesite and dacite, but also a significant amount of ash and pumice known as ejecta. The water that is incorporated into the magma during generation is released when ascends the pipe within the volcano and converts to lava. This loss of all of the water and other volatiles during devolatilization makes these volcanoes very dangerous because they tend to explode.

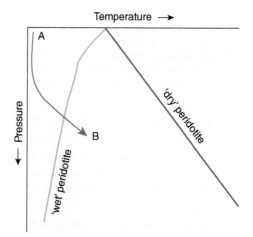

FIGURE 3.6 Pressure–temperature graph of a phase diagram for the melting of wet mantle rocks. Points A and B correspond to the locations on Figure 3.5. As the pressure and temperature increase, the rock crosses the wet solidus and melts.

3.5 Melting at Mantle Plumes

Mantle plumes are the third major environment for igneous activity. Although covering only a small area of the earth and producing volumetrically small amounts of igneous rocks relative to the other two types, they can be locations of intense and continuous activity. Mantle plumes are similar to divergent margins in that they involve the upwelling of mantle material. The difference is they are a single point of upwelling that does not move location in contrast to a long zone in a divergent margin. They also do not apply enough force to the bottom of the plates to move them in any way. They are more passive generators of igneous activity over which the tectonic plates move.

Mantle plumes involve partial melting of the mantle and, like divergent margins, produce magma with mafic composition. Magma is also formed by decompression melting, though the temperature in the mantle is usually significantly elevated in them, so it is less of a factor. It is for this reason that they are also called hot spots. The depth at which the partial melting takes place controls the minor variations in the composition of the magma. Usually, small amounts of magma are produced slowly and deeply, but large amounts

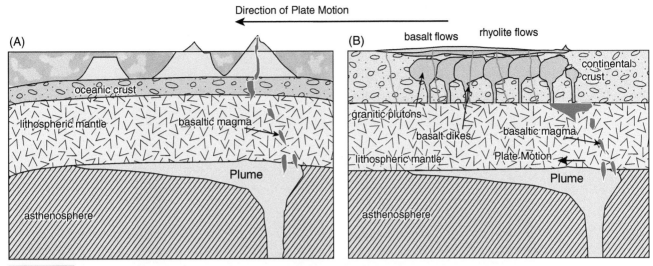

FIGURE 3.7 Diagrams of mantle plume igneous activity. (A) Mantle plume under ocean crust producing mafic magma and basalt volcanoes. (B) Mantle plume under continental crust producing both mafic magma from the mantle and felsic magma from partial melting of the continental crust.

are produced shallowly and at a high rate. As the mafic magma ascends, it can leave behind some of the crystals that have already formed through a process called crystal settling. These crystals are composed of the high-temperature minerals, primarily olivine. In this way, the magma can change composition to some degree. Considering that high-temperature minerals are common to mafic rocks, removing them makes the magma become more felsic because it contains a higher component of low-temperature minerals which are characteristically felsic. This process of removing crystals from magma through settling is called fractional crystallization or fractionation, and it can occur in virtually any magma, but is more common in mafic rocks. Hawaii is a good example of a mantle plume on ocean crust, and it is all basalt (Figure 3.7A). Fractionation is common in these rocks, which leads to a variation in compositions.

CASE STUDY 3.2 Volcanoes in Hawaii

Pillow Basalts

Just like the columnar joints on land, basalt erupted into water forms distinctive features called pillow basalts or pillow lavas. If melted wax is dripped into a glass of water, each drop will form a sphere of wax as it settles to the bottom. The spherical shape is the result of surface tension. Lava behaves the same way. As lava spills into water or is erupted in a submarine volcano, it forms rounded to spherical blobs that settle to the seafloor or lake floor. The lava, however, remains hot and relatively soft even after settling. Unlike the wax spheres which keep their shape, the lava spheres flatten once they settle. As a result, they tend to resemble flat round couch pillows by the time they harden. This is where they derive their name (Figure 3.8).

FIGURE 3.8 Pillow lavas on the ocean floor in Hawaii. *Source*: Photograph courtesy of NOAA.

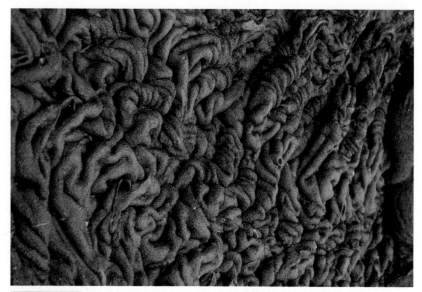

FIGURE 3.8A Photograph of aa lava in Hawaii. *Source*: NOAA.

Pillow lavas can be produced by any volcano in water, and they are probably most common in divergent boundaries and best observed in oceanic mantle plume volcanoes like Hawaii. There are even underwater videos of pillow lavas forming in Hawaii. However, not all of the Hawaiian flows form pillow lavas. First, the flow must be voluminous enough to reach the ocean. Second, it must be the right kind of flow. The Hawaiians named two different types of lava: aa and pahoehoe. Aa typically forms at a distant point on a flow. It is a thick, slow moving, blocky flow that avalanches its way forward downhill (Figure 3.8B). Aa never forms pillow lavas.

Pahoehoe is a thin, fast moving, fluid lava flow. In pahoehoe, the surface of the flow tends to harden and form a skin that is twisted and carried along by the flow forming a ropey textured surface (Figure 3.8A). Farther along the flow, this hardened surface becomes thicker and no longer deforms. Instead, it forms a roof over the fast-flowing lava beneath. The liquid lava winds up flowing through a tunnel of hardened lava called a lava tube which allows it to maintain its temperature and fluidity over long distances. This makes it easier for the lava to reach the ocean where it spills into the water and forms pillow lava. When it contacts the ocean, the intense heat can cause phreatic explosions where the water flash boils, exploding and generating a lot of steam. The pillows react violently as they interact with the water and settle to the seafloor.

Pillow basalts are ubiquitous on the ocean floor. Pieces of ocean crust that wind up on land through tectonic processes are called ophiolites and contain excellent examples of pillow basalts. One of the best examples is in the Sultanate of Oman. The Oman ophiolite contains classic examples of ancient pillow lavas that are now part of a national park.

FIGURE 3.8B Photograph of pahoehoe lava in Hawaii. *Source*: NOAA.

If a mantle plume occurs under a continent, it typically produces a large basalt province, but it can also cause partial melting of the crust above it (Figure 3.7B). This process works similar to an intracontinental rift where infiltrating mafic magma is the source of the heat causing large-scale melting in the crust. In this case, rhyolite volcanoes can form above the mantle plume. Some of these volcanoes can be quite significant. This is the case with Yellowstone which is situated above a mantle plume and has had very large rhyolite eruptions in the distant past. It has been termed a "supervolcano" for its ability to cause devastating eruptions. The same mantle plume formed the Columbia River Basalt Province at an earlier time.

3.6 Contamination of Magma

One of the main reasons for variations in the composition of magma is crustal contamination. When magma is first formed, it is hot and buoyant because it is less dense than the partially melted rock surrounding it. The high pressure from the depth in the earth squeezes the magma through cracks and weak spots in the rock above it and it ascends to shallower levels in the crust as a pluton. As it ascends, it interacts with the wall rocks in the magma conduit, sometimes breaking fragments off and entraining them (Figure 3.10). The magma is so hot that it can partially or completely melt these fragments or xenoliths and the resulting liquid mixes into it. This process is known as assimilation because foreign rocks are assimilated into the magma and alter its composition.

If the composition of the xenolith is the same as the magma, there won't be much change in the composition of the magma. In most oceanic basalt provinces, that is generally the case. Deviations from purely mafic volcanic and plutonic rocks are rare and volumetrically insignificant. Hawaii, for example, has all mafic rocks, whereas Iceland has a few non-mafic volcanoes as a result of contamination. They are typically easy to identify because they form steep cones and produce a lot of ash.

CASE STUDY 3.3 2010 Eyjafjallajökull Eruption

Volcanoes and Air Traffic

Eyjafjallajökull volcano is one of the few largely intermediate composition volcanoes in Iceland which is overwhelmingly mafic. It is not the product of subduction like most intermediate volcanoes, but rather the product of crustal contamination of mafic magma. In January 2010, Eyjafjallajökull experienced inflation of the volcano surface by 1.6 in. (4 cm) and a swarm of earthquakes that reached 3000 tremors per day with magnitudes up to 3.1 by March 2010. The eruption began on 20 March with the appearance of a glowing cloud and lava fountains (Figure 3.9). As this first phase of the eruption began to wane, a new eruption began from another vent of the volcano. The eruption was

FIGURE 3.9 Photograph showing the eruption column of the Eyjafjallajökull volcano with lightning. *Source*: © Sigurður Hrafn Stefnisson.

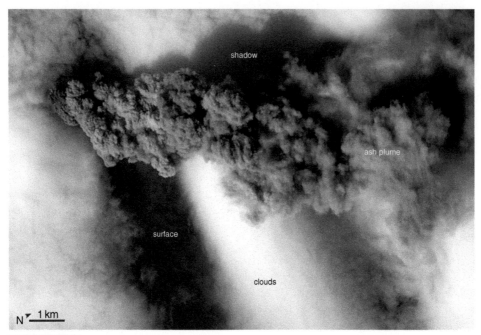

FIGURE 3.9A Satellite image showing the plume emerging from the Eyjafjallajökull volcano. *Source*: Courtesy of NASA.

heralded by a sudden increase in the water level of the streams and rivers in the area as the glacial ice at the peak of the volcano began to melt. Flooding occurred and the water level in one local river rose 33 in. (84 cm). On 14 April, the Icelandic Department of Civil Protection and Emergency Management consequently ordered the evacuation of the Eyjafjallajökull area, which involved the relocation of about 800 people.

This explosive eruption produced andesite rather than the basalt of the earlier phase. It also produced an eruption column up

FIGURE 3.9B Satellite image showing the plume from the Eyjafjallajökull volcano being blown across the Atlantic Ocean. *Source*: Courtesy of NASA.

FIGURE 3.9C Photograph showing the eruption of the Eyjafjallajökull volcano. *Source*: Sigurdur Stefnisson/USGS.

to 36 000 ft (11 km) high (Figure 3.9A). It is estimated that more than 330 million cubic yards (250 m³) of ash were produced by this eruption phase. On the morning of 14 April, a plume of ash and steam extended 40 mi (70 km) due east of the volcano and by the afternoon it was 60 mi (100 km) long. A strong jet stream quickly extended the plume toward the east where it began to impact air travel in Europe (Figure 3.9B).

By 16 April, the eruption cloud drifted across Europe and forced the closure of most airports resulting in the cancelation of 17 000 flights. Millions of passengers were stranded worldwide because connecting flights were also impacted. It developed into the worst peacetime air travel disruption in history and the first time ever that British airspace had been closed. By 18 April, over five million travelers were stranded. Some people waited a week or more to find available seats. The cancelation of flights cost Britain $1.56 billion (£1 billion) by then, and they were losing $358 million (£230 million) every day. By 19 April, the number of canceled flights exceeded 63 000.

A decrease in activity on 19 April allowed airports in Scotland and Northern England to begin limited service on 20 April, but the break appeared short-lived as a new heavy ash cloud emitted the same day (Figure 3.9C). The next day, however, the volcano melted through the overlying glacier and the amount of ash being emitted was greatly reduced. This allowed more air travel and the backlog of stranded travelers finally began to decrease. By 22 April, the eruption had subsided to about 10% of the maximum output. The resulting plume reached an altitude of more than 1.9 mi (3 km), which was still a concern but not nearly that of the previous week. During the following week, the less-intense eruption continued producing ash plumes up to 17 000 ft (5.3 km), but the winds had shifted and there was minimal impact on Europe. It closed airports in Iceland and forced the rerouting of trans-Atlantic flights, but the large-scale closing of airports had ended.

During the first week of May, the eruption reenergized producing ash clouds 30 000 ft (9 km) high and extending more than 1200 mi (2000 km) across Europe. It disrupted air travel throughout Ireland and Britain shutting down numerous airports. On 9 May, the plume disrupted air travel in Spain and Portugal, canceling some 5000 flights. Fears grew that it would eliminate air travel in Europe again. Fortunately, the winds shifted, and the eruption intensity decreased ending the fears. The eruption continued until 21 May 2010 with variable amounts of activity and periodic airport closures in the United Kingdom and even several other European countries. The eruption was finally declared over in October 2010.

Mafic and intermediate magmas that penetrate continental crust to shallow levels can be contaminated enough to change their compositions. Continental crust is rich in quartz and feldspar both of which have low melting temperatures. As the magma assimilates quartz and feldspar, it slowly changes to more felsic compositions (Figure 3.10). The thicker the continental crust, the more interaction of magma with crustal rocks and the more partial melting of quartz and feldspar, and as a result, the magma becomes more felsic.

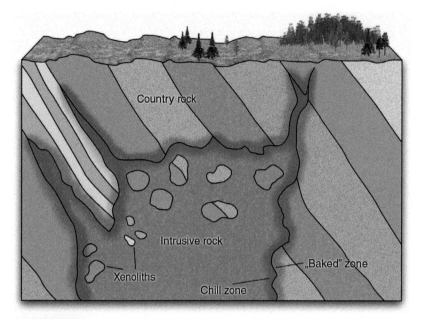

FIGURE 3.10 Intrusion of magma into continental crust pulling in xenoliths from the country rock and assimilating it into the magma. The magma is contaminated and changes composition becoming more felsic.

CASE STUDY 3.4 Andesite and the Andes

The Andes Mountains of South America are the ultimate example of an ocean–continent collision. Well-developed ocean–continent collisions worldwide are termed "Andean margins" for this reason (Figure 3.11). The Andes are formed by the subduction of the Nazca ocean plate under the South American plate. Organization of plate motions toward creating the Andes may have begun as early as 200 million years ago, but organized volcanism probably did not begin until about 65 million years ago. The rapid rise of the mountains to their current majestic height as the second highest mountain range in the world did not begin until about 25–30 million years ago.

The volcanic rock *andesite*, which is the most common in convergent margins, is named for the Andes Mountains. Early volcanic rocks before the rise of the Andes were andesite, but they are volumetrically small. During the past 25 million years, the crust beneath the Andes has become so thick that magma now must pass through 70 km of crustal rock to be erupted from a volcano. This increased interaction with crustal rocks has resulted in increased assimilation. The contamination changed the general composition of the lava from andesite to dacite, the more felsic variety of intermediate volcanic rock. By far, most of the volcanic rocks in the Andes are dacites and not andesites, creating a bit of a paradox.

The contamination is so extreme in parts of the Andes that volcanic rock compositions commonly border on rhyolite (rhyodacite). Further, they have actually been found containing primary minerals that are common to metamorphic rocks, not igneous rocks. Some volcanic rocks contain andalusite, a purely metamorphic mineral. This is the only documentation in the world of such an occurrence and it reflects the extreme amount of contamination.

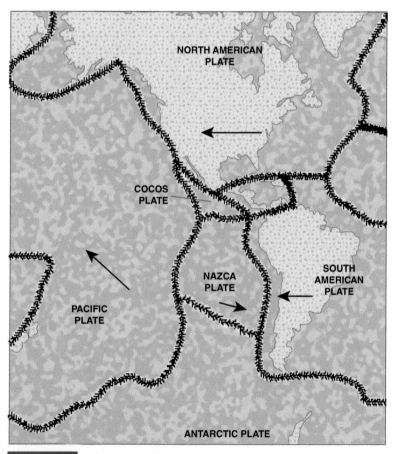

FIGURE 3.11 Tectonic map of Nazca plate and South America.

3.7 Magma to Lava

When magma ascends the conduit in a volcano and is extruded onto the surface, some major changes take place. The first is that it moves from a subsurface location to a surface location. The subsurface is under great pressure and high temperatures that are relatively constant. At the surface, the temperature is much lower, and the liquid begins to cool rapidly. Underground, mineral crystals grow slowly and unimpeded. At the surface, cooling is so quick that much of the liquid just hardens to volcanic glass as obsidian and pumice.

The release of pressure as the magma moves to the top of the volcano allows the release of the volatile components which they had been holding in solution through their entire subsurface history. Basically, the magma holds these components like CO_2 is held in soda or seltzer. As long as the pressure is maintained, the CO_2 is held in the liquid. Once the pressure is released by removing the cap, the CO_2 can come out of solution and foam, explosively in some cases. This analogy explains the processes in an erupting volcano with a high volatile content. The volatile components, primarily water, are released from the magma upon eruption and expand dramatically and instantaneously. Depending upon the conditions, 1 l of water released from magma can expand over 1600 times its volume as vapor. This expansion is explosive and explains the explosive eruptions of water-charged magmatic systems.

The lava that flows down the volcano is therefore quite different than its magmatic counterpart. The minerals are different in both composition and texture. Glass and pumice can be included in the rock and the lava undergoes very rapid and radical changes unlike the more stable magma. The main difference, however, is the lack of volatile components in lava, which changes the chemical composition significantly from the magma. The solid bulk chemistry and even minor element chemistry of the magma are basically the same as the volcanic counterpart, so they can be related, but the other differences are extreme.

3.8 Physical and Chemical Constraints on Volcanic Hazards

Both the chemical and physical states of magma and lava play controlling roles on the types and severity of the hazards that are produced by the erupting volcano (Figure 3.1). Considering the control that bulk composition has on just the shape of a volcano (mafic is flat, felsic is very steep), it is easy to understand the importance for volcanic hazards. The steepness alone controls how volcanic rocks will tumble down the slopes in pyroclastic flows. The bulk composition also determines the viscosity of the lava. Mafic lavas are very fluid and can flow long distances, whereas more felsic lavas tend to be sticky and resist flow. From this, it would seem that mafic lava is more dangerous than the more felsic varieties, but that is not true. As long as the flow does not impact humans or human settlements, mafic flows are typically not big-scale threats. Sticky felsic and intermediate lavas, on the other hand, can clog up the volcanic vent. Considering the power of a volcano, preventing it from erupting can only be temporary. Eventually, the pressure overcomes the clog with violent and disastrous results. Explosive eruptions can cause widespread devastation for great distances around the volcano. Such eruptions have changed the course of history on a global basis.

The other factors with a major impact on the severity of volcanic hazards are the amount and type of volatile components in the magma. Certainly, the most dangerous volatile component is water. The rapid expansion as liquid water turns to steam in a fraction of a second is as devastating as the explosion of any bomb created by humans. Fortunately, the vast majority of these eruptions shoot straight up in the air. These eruption columns can achieve heights of 10 mi (16 km) or more while carrying tens of tons of solid ejecta. Considering that the force of gravity acts against this ascent, the force of eruption is awesome.

If the lava is the sticky variety, it can clog up the vent resulting in a very dangerous situation. If the explosion cannot come out of the top of the volcano, the only alternative is to come out of the side. Then the enormous blast that would have shot into the atmosphere rockets across the landscape and destroys everything in its path. The unfortunate part is that magma that tends to be water-charged tends also to be the composition that is sticky. That means these lateral rather than vertical blasts are not so uncommon. Contamination of magma tends to make it both stickier and more water-laden further exacerbating the condition.

Water is not the only volatile in magma. Magma can carry a lot of sulfur which is released as an acidic aerosol during eruption. This can cause significant acid precipitation around volcanoes that has very deleterious effects on vegetation and the health of streams and other water bodies. Hawaii has serious problems with acid precipitation, and it is all volcanic rather than industrial as is commonly the case. Higher in the atmosphere, sulfur aerosols also can disperse incoming sunlight and cause significant global cooling. Global warming may be a great societal concern at present, but considering productivity of croplands, global cooling is worse.

Other volatiles like fluorine are typically in far too small a quantity to worry about. This is fortunate because in the atmosphere, it creates a far stronger acid than sulfur which can be quite deadly in even small concentrations. Carbon dioxide is also released, but considering that humans and animals exhale it and it is naturally occurring in the atmosphere, it should not cause a problem. This is only true if it is in low concentrations. At levels greater than 10%, it is deadly to all oxygen-breathing organisms. There are several other volatiles that are of equal concern, but they are just not as common.

CHAPTER 4

Types of Volcanoes

CHAPTER OUTLINE

4.1 What Is a Volcano? 49

4.2 Classification by Shape 49

4.3 Parts of a Volcano 50

4.4 Classification of Volcanoes by Threat 51

4.5 Decade Volcanoes 52

4.6 Evacuation of Volcanic Areas 54

4.7 Supervolcanoes 57

Earth's Fury: The Science of Natural Disasters, First Edition. Alexander Gates.
© 2022 John Wiley & Sons Ltd. Published 2022 by John Wiley & Sons Ltd.
Companion website: www.wiley.com/go/gates/earthsfury

Words You Should Know:

Cinder cone – A steep-sided, conical volcano that is composed completely of ash and other pyroclastic material and no lava.

Decade Volcanoes – A group of volcanoes designated as the most dangerous on the earth by virtue of their activity and their vicinity to population centers.

Ejecta – Solid and liquid volcanic material that is shot out of a volcano.

Fissure eruption – An eruption of lava and possibly pyroclastics from cracks in the ground.

Lava – Liquid rock on the surface of the earth.

Pyroclastics – Volcanic rocks that are deposited as solid particles rather than as liquid.

Shield volcano – A low rounded volcano in the general shape of an ancient circular shield.

Stratovolcano – A steep-sided, conical volcano composed of interlayered lava and ash so that it is stratified.

Supervolcano – A volcano with the capacity to produce an unprecedented eruption.

Volcanic explosivity index (VEI) – A scale for the power of volcanic eruptions from 1 to 8.

4.1 What Is a Volcano?

A volcano is an opening in the surface of the earth from which liquid rock, solid rock, and/or gas is expelled from a subsurface magmatic source. Magma is melted rock, but it is usually not just a rock found on the surface that is heated to the point of being a liquid. It always contains dissolved gases, some of which could be liquid under standard surface conditions, and it typically contains solids. Magma is driven toward the surface by pressure either through buoyancy from being less dense than surrounding rock or from being forced up from below by expansion of materials, or a combination of the two. The liquid rock is magma until it escapes to the surface through a vent where it becomes lava. The vent is the volcano.

Volcanoes do not always emit lava; they can emit gas, liquid, and/or solid material, or any combination of the three. Emissions can be ferocious eruptions or even explosions or just minor nonthreatening events. It is the degree of energy of the eruption and the location of the eruption that dictates whether it will be a natural disaster or go unnoticed. Volcanoes can be the most dangerous of natural disasters because not only can the event be devastating but the effects can wreak havoc on the planet for many years after the eruption ends.

4.2 Classification by Shape

Volcanoes can have a variety of shapes dictated primarily by the viscosity and stickiness of the lava, the amount of solid material, and the surface conditions. The stickiness and viscosity are largely a function of the chemical composition. Mafic lavas tend to be thin (low viscosity) and not very sticky, so they tend to form relatively flat and spread out features. Felsic lavas are very viscous and sticky, so they can form very steep-sided features. Steepness based on composition is generally gradational across the range of igneous compositions. Depending upon the lava type and emission, eruptions can produce significant amounts of pyroclastics. Solid material can greatly affect the steepness and shape of the volcano. Finally, the conditions into which the lava is emitted also affect the shape. Lava that is emitted underwater cools very quickly and forms steep volcanoes with distinctive shapes.

Under these constraints, volcanoes can have a variety of shapes (Figure 4.1). If lava pours out of linear cracks in the ground with no centralized surface feature, it is called a *fissure eruption*. The lava must be very fluid in these eruptions. The shapes of the flows can be greatly controlled by surface features such as river valleys. *Shield volcanoes* resemble ancient circular shields and form from fluid lavas being emitted from a single central vent. If the eruption contains more solid material, it forms a cone because more material is deposited closer to the vent and progressively less, the farther away. The *stratovolcano* is the classic cone volcano mountain and is composed of interlayered lava and pyroclastics forming strata. They can be snow-capped at most latitudes. If no lava is emitted from the central vent but only pyroclastics, a *cinder cone* will form. These are typically very symmetric except in windy areas where the emitted pyroclastics can be blown in a specific direction. Also known as a *tuff cone*,

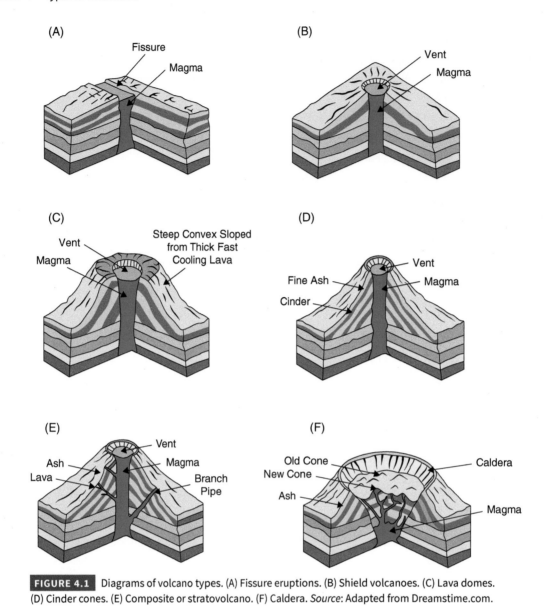

FIGURE 4.1 Diagrams of volcano types. (A) Fissure eruptions. (B) Shield volcanoes. (C) Lava domes. (D) Cinder cones. (E) Composite or stratovolcano. (F) Caldera. *Source*: Adapted from Dreamstime.com.

maars are shallow, flat-floored craters that are believed to have formed above diatremes as the result of a violent explosion of gas or steam. The craters are commonly filled with water forming lakes, and the rims of maars are low and composed of a mixture of fragments of volcanic rock and rocks from the walls of the diatreme. If the volcano erupts underneath a glacier, the ice forms a physical barrier to the flow and the shape is constrained. These volcanoes are called *tuyas* or *table volcanoes*. Submarine volcanoes are also strongly controlled by the temperature, pressure, and mechanical properties of the water they erupt into. In many cases, very delicate features can be preserved that might not survive on the surface. *Black smokers* are the chimney-like features of these deep-sea vents found at mid-ocean ridges. *Littoral cones* are small features that form when flowing lava hits a body of water and piles up.

4.3 Parts of a Volcano

There are some common features to all volcanoes, but many others are limited to specific volcano types. All volcanoes have a magma chamber beneath them that feeds magma into a conduit pipe (Figure 4.2). In turn, the conduit transmits the magma to the surface where it can erupt. In fissure eruptions, the conduit is a tabular crack in the bedrock through which magma ascends. These volcanoes may have a throat in some cases, but, more commonly, the lava just fountains from the vent or pours out onto the earth surface.

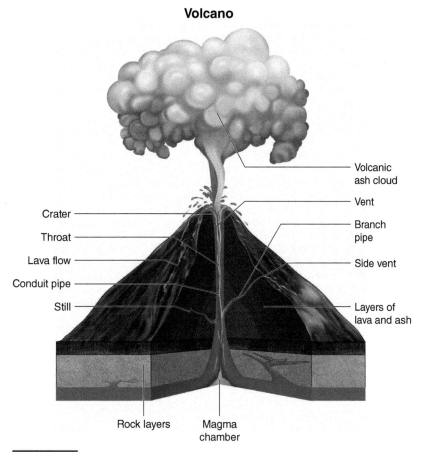

FIGURE 4.2 Diagram of the parts of a volcano as shown.

Shield volcanoes also have a magma chamber, but the conduit pipe is more circular in cross section as well as a vent. However, the vent can connect to a caldera at the top of the volcano that can fill with lava forming a lava lake. The Kīlauea shield volcano on Hawaii commonly forms lava lakes in the caldera. When the lake overflows or bursts through the sides, a lava flow is produced. The stratovolcano is a cone volcano with a magma chamber, pipe conduit, a throat at the vent, and a crater at the summit. The sides of the volcano are the flanks. There can be branch pipes from the main conduit that lead to flank vents and eruptions.

Slow extrusion of lava at the summit can produce a lava dome. They commonly form in the crater of a stratovolcano and can grow and collapse during an eruptive cycle. If a volcano explodes removing the cone and leaving a basal wreck, continued lava extrusion commonly produces a resurgent dome. A good example of this is Wizard Island in Crater Lake, Oregon.

4.4 Classification of Volcanoes by Threat

4.4.1 Volcanic Explosivity Index

The volcanic explosivity index (VEI) is a scale to rank the power of a volcanic eruption (Figure 4.3). The VEI ranges from 0 to 8 with increasing intensity. The factors that determine the number are: (i) the volume of ejecta produced during an eruption, (ii) the height of the eruptive column, and (iii) the duration of the eruption. Certain magma types and volcano types are more commonly associated with different VEIs. For example, basaltic fissure and shield volcanoes almost exclusively erupt with VEIs of 0 or 1. These are termed Icelandic and Hawaiian eruptions based on usual occurrences. Andesites produce stratovolcanoes or cinder cones having a full range of VEIs with a variety of eruption styles from the most intense such as Peléan eruptions to moderate eruptions such as Strombolian and Vulcanian types. These are named after volcanoes where the eruption type was observed and characterized. Vesuvian type is also named for a volcano, but Plinian types are named for the ancient Roman, Pliny the Younger who first described this type of eruption.

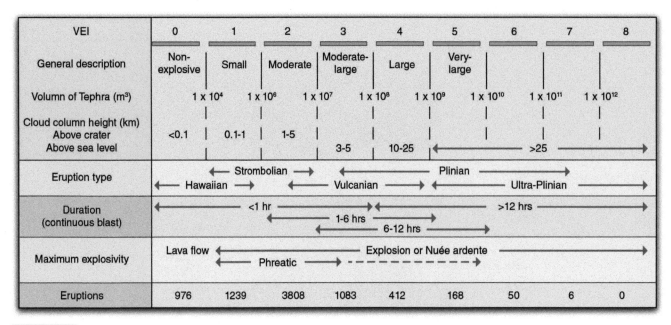

FIGURE 4.3 Table of volcanic explosivity index (VEI) showing the criteria for each rank, and the types of eruptions most closely associated with the various ranges of VEIs.

Rhyolite volcanoes are, by far, the least common. However, the thick and water-charged lava can produce very powerful eruptions. They tend to produce lava domes and thick ash deposits. They are so uncommon that they rarely cause disasters.

4.5 Decade Volcanoes

A major contribution of the professional organization International Association of Volcanology and Chemistry of the Earth's Interior (IAVCEI) of the International Decade for Natural Disaster Reduction (1990s) was to designate the Decade Volcanoes. The goal of designating Decade Volcanoes was to draw attention to a small group of active volcanoes in populated areas to encourage scientific research and public awareness to enhance understanding and literacy of volcanoes and the hazards they pose. These 16 volcanoes have been officially designated as Decade Volcanoes (Figure 4.4):

1. Avachinsky–Koryaksky, Kamchatka
2. Colima Volcano, Mexico
3. Mount Etna, Italy
4. Galeras Volcano, Columbia
5. Mauna Loa, Hawaii
6. Merapi Volcano, Indonesia
7. Nyiragongo Volcano, Republic of Congo
8. Mount Rainier, Washington
9. Sakurajima Volcano, Japan
10. Santa Maria/Santiaguito Volcano, Guatemala
11. Santorini Volcano, Greece
12. Taal Volcano, Philippines
13. Teide Volcano, Canary Islands
14. Ulawun Volcano, Papua New Guinea
15. Unzen Volcano, Japan
16. Mount Vesuvius, Italy

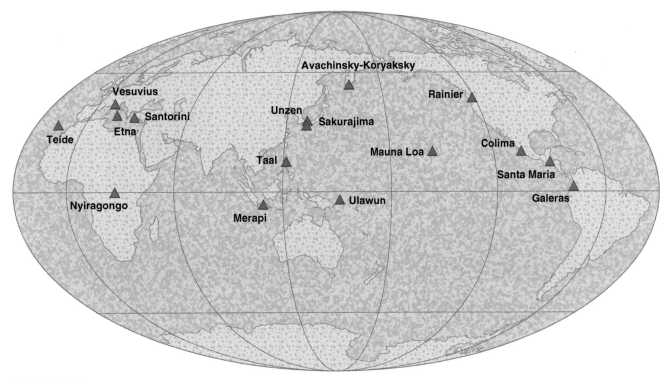

FIGURE 4.4 Map of the earth showing the locations of the 16 Decade Volcanoes.

These volcanoes are not only regularly active but they have the potential to produce very powerful eruptions. Further, they are close enough to highly populated areas that they have the potential to cause among the most catastrophic natural disasters ever. It is for this reason that these volcanoes are the most carefully monitored in the world. They have the most sophisticated monitoring systems possible including gas analysis, seismic monitoring, and laser surveying to monitor ballooning of the cone. The residents of these areas are also well educated on the signs of potential volcanic eruptions and how to evacuate in case of an eruption.

CASE STUDY 4.1 1600 BCE Thera Eruption

The Volcanic Eruption that Changed Western Civilization

Why would Santorini, the Greek island vacation destination, be included in the list of Decade Volcanoes? The answer is in the ancient history of the area. The Mediterranean Sea region saw a dramatic development of civilization between the third and second millennia BCE (Figure 4.5). Centered in Crete, the Minoan Civilization was perhaps the most advanced in the region. They developed advanced technology, art, and culture. The technology included plumbing, earthquake-resistant buildings, and double-hulled ships among others. They dominated the Mediterranean region of the time. However, there were other settlements around the region that also saw great advances in civilization. One such settlement was on the island of Thera. Current excavations show that Thera was equally advanced and a trading partner with the Minoans. The difference with Thera is that it was situated on a volcanic island with an active stratovolcano.

The conical stratovolcano likely erupted regularly but with minimal impact on the residents allowing development of the settlement. However, at some point, the intensity of eruptions increased considerably. In terms of the scientific evidence, it appears that the cone opened and seawater poured into the volcanic chamber producing a catastrophic phreatomagmatic explosion. The reverberation was probably heard from the British Isles to Persia as the entire volcanic cone exploded into debris. The explosion produced a huge tsunami that left fresh sediment on the top of a nearby island at an elevation of 230 ft (70 m). The wave spread out around the whole Mediterranean basin scraping up all the deposited sediments and sweeping them into a new deposit of mixed up sediment that geologists termed "homogenite." This deposit is up to 9 ft (2.7 m) thick in places and found throughout the basin. Only a small lifeless, crescent island, now called Santorini, remained from the once magnificent Thera (Figure 4.5). The exact timing of the eruption is unclear, but the best current estimate is 1650–1570 BCE.

The wave wreaked havoc on the coastal settlements all around the region, which altered the course of history and wound up in legends and possibly religions. The tsunami decimated the Minoan Civilization, which led to its downfall and the emergence of the Mycenaeans, which, in turn, led to the rise of the Golden Age of Greece. The legend of Thera is interpreted to have been remembered as the legend of Atlantis. The story of the Exodus also greatly resembles the eruption sequence of other stratovolcanoes. Ash emissions can fill the sky and cause all liquid to appear blood red as it reflects the filtered sun. Sulfur emissions can acidify surface water, killing all of the aquatic life, and encouraging insects to infest their rotting corpses. Finally, the sea retreats substantially for several minutes prior to a tsunami striking the shoreline in a wave that is capable of destroying even the largest of armies. All of these correlations with historic and biblical events have been suggested, but the timing of them has yet to be proven.

If a similar eruption occurred today, it would cause the worst natural disaster in history.

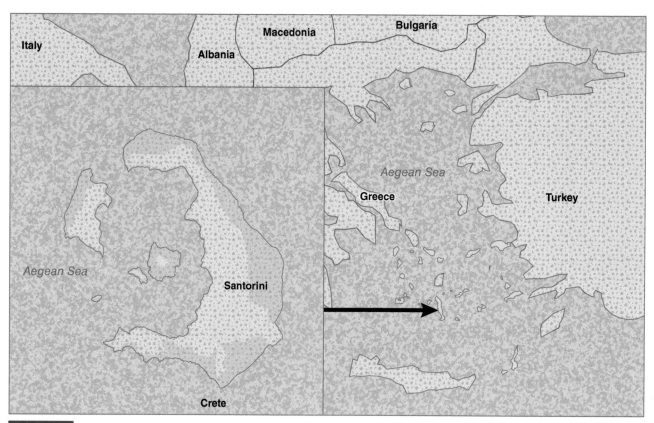

FIGURE 4.5 Map of the Eastern Mediterranean Sea Basin showing the location of Santorini/Thera, Greece. Inset contains a map of the island of Santorini showing the Thera caldera.

4.6 Evacuation of Volcanic Areas

Large eruptions and long-duration eruption cycles can destroy settlements and the local ecology, driving out the residents, if they are lucky enough to survive. However, the emissions can be toxic such as fluorine and sulfur compounds or hazardous by nature, such as ash which can damage the lungs and eyes. Lava flows and pyroclastic flows can also damage or destroy infrastructure and make cultivation impossible for an extended period. Even heavy ash flows or ashfalls can cause enough damage that settlements must be abandoned. Andesitic and rhyolitic volcanoes are especially dangerous. Areas around large volcanic explosions are typically abandoned by the survivors and the area is not reinhabited for many decades, if at all. When Mount Vesuvius erupted in 79 AD, it was in an inhabited area. It destroyed the cities of Pompeii and Herculaneum, never to be inhabited again. However, other cities were established in the damaged zone in decades to centuries later. More recently, the 1991 eruption of Mount Pinatubo heavily damaged the Clark Air Base and Subic Bay Naval Station, two largest US military bases in the Philippines. Both were abandoned and there are no plans to reoccupy them.

Areas with frequent but low impact lava flows can still maintain settlements with just occasional evacuations in extreme circumstances. Flows typically give plenty of warning to the residents who might be impacted so that they can be safely evacuated. Only those settlements that are in the direct path of the flows require evacuation and suffer damage. Only if there is significant emission of poisonous gases, the nearby settlements require evacuation, but these are rare and residents can return to their homes once the danger has passed. Examples of this are Hawaii and Iceland, in most cases. Lava flows are common in both areas, but because they are rarely explosive and restricted geographically, communities can exist around them with only occasional dangerous situations. Both of these areas are dominated by basaltic volcanism.

CASE STUDY 4.2 1995–2013 Soufrière Hills Eruption

Paradise Lost

The island of Montserrat is one in the chain of the Caribbean island arc which includes the islands of St. Kitts, Barbuda, and Antigua among others. It is a territory of the United Kingdom that was known as a paradise of white beaches, lush hillsides, and palm trees. As a result, Montserrat was a tourist destination in the 1960s through 1980s and even mentioned in the popular Beach Boys song "Kokomo." To add to the mystique, the ex-manager of the Beatles, George Martin, opened a recording studio in Montserrat in 1979, attracting a number of top recording stars to the island. Even more tourists flocked to the island hoping to catch a glimpse of a famous rock star.

The first problem for Montserrat came in 1989 when Hurricane Hugo passed right over the middle of the island. On the morning of Sunday, 22 September 1989, and for the next 14 hours, sustained winds of 145 mph (233 kmph), with gusts up to 180 mph (290 kmph), devastated the island. Nearly every building on the island was damaged and 20% were destroyed. All electricity and communications were lost, and the hilltop water tanks were blown over leaving no fresh water supply. More than 2500 people were homeless, and dozens of people died either directly from the storm or from waterborne diseases in ensuing months. However, this was not the disaster that destroyed this paradise. Efforts were started immediately to rebuild and revitalize the resort. This would be short-lived because the real disaster began on 18 July 1995, just six years later.

Montserrat, like all the other Caribbean islands, is a volcanic island. Being on an island arc, all of the volcanoes are stratovolcanoes and capable of violent eruptions. On Montserrat, the volcano is the 3300 ft (1050 m) high Soufrière Hills, which had been relatively quiet since the beginning of the twentieth century. This ended on 18 July 1995 when it suddenly became active. Over the next 18 years, it experienced semicontinuous activity in five main eruptive phases which ended the rebuilding efforts and caused most residents to abandon the island (Figure 4.6).

Early eruptions were primarily pyroclastic flows and lahars, or volcanic mudflows, and these predominated throughout the eruption sequence. Right from the beginning, the capital city of Plymouth was evacuated, and within few weeks it was covered in 5–10 ft (1.6–3 m) of debris from pyroclastic flows. Phreatic explosions began on 21 August 1995, which lasted for 18 weeks forming an andesite lava dome atop Soufrière Hills. This first phase of activity lasted until 1998 with the most powerful eruptions in 1997–1998 (Figure 4.6A). These were primarily caused by major dome collapses, two periods of explosive eruptions, and fountain-collapse pyroclastic flows at VEI = 1–3. An eruption on 25 June 1997 resulted in the deaths of 19 people from a major pyroclastic flow which destroyed the island's airport. The explosion produced enough ash to darken the sky across the island and to blanket Plymouth in a thick layer. Extensive and continuous seismic activity accompanied the eruptions from three zones, the Soufrière Hills volcano, from a ridge to the northeast, and beneath St. George's Hill.

The last eruption of the Soufrière Hills volcano ended in 2013. There were a minimum of three VEI = 3 eruptions during the periods of 1995, 2004, and 2005. Phase 1 of the eruptive cycle was the most energetic and it lasted from 1995 to 1998. Phase 2 was the most continuous and lasted from 1999 to 2003. Phase 3 was from 2005 to 2007 and Phase 4 and Phase 5 were short-lived lasting from 2008 to 2009 and 2009 to 2010, respectively. This does not mean that the later phases were not intense. For example, on Monday, 28 July 2008 at 11:27 p.m., an eruption produced pyroclastic flows that reached Plymouth and part of the east side of the lava dome collapsed, producing another pyroclastic flow in Tar River Valley. The ash column produced by this eruption reached a height of 40 000 ft (12 000 m).

Even in 2013, activity was still energetic enough to be a threat to the entire island. Heavy rains on 28 March 2013 generated large lahars, lasting several hours, and a pyroclastic flow was generated that traveled 0.95 mi (1.5 km). A phreatic explosion occurred on 18 July 2013 that caused ashfall across the island and a new vent developed in the summit crater on 28 July. Lahars and pyroclastic flows continued on a semi-regular basis (Figure 4.6B). On 19 August, both vents erupted ash and steam for 35 minutes producing another ashfall.

The lahars and pyroclastic flows from this 18-year eruptive cycle destroyed 18 towns and villages (Figure 4.6C). In addition, more than 15 communities were evacuated. An "exclusion zone" was established for the southern 2/3 of the island which was and is off-limits to everyone. Montserrat had a population of about 13 000 people prior to the eruption, but more than 8000 abandoned the island, mostly to relocate to the United Kingdom. This left a Montserrat population of less than 5900 people on the northernmost part of the island.

The United Kingdom and Montserrat provided aid to the people of Montserrat, but they deemed it insufficient and rioted. Additional aid followed, but the most notable source of funds was a large benefit concert held in London with the express purpose of helping Montserrat. The concert included Elton John, Eric Clapton, Sting, Paul McCartney, and Phil Collins among others. This event brought great notoriety to the plight of the fleeing people.

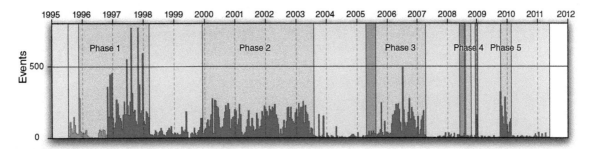

FIGURE 4.6 Timeline for eruptive cycles of Soufrière Hills volcano shown by seismic activity (after Cole et al., 2011).

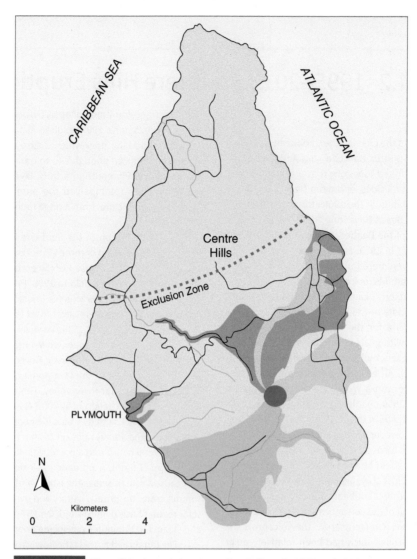

FIGURE 4.6A Map of Montserrat showing volcanic flows in 1997 and 1998.

FIGURE 4.6B Photograph of the Soufrière Hills volcano. *Source*: Courtesy of the British Geological Survey.

FIGURE 4.6C Photograph of a pyroclastic flow from the Soufrière Hills volcano. *Source:* Courtesy of the British Geological Survey.

4.7 Supervolcanoes

"Supervolcano" is not a real scientific term. It was coined by the producers of a British television program on powerful volcanoes. It is approximately defined as a volcano that resulted in exceptional impacts, largely through an explosion and the ejection of immense quantities of ash and other pyroclastic material. Supervolcanoes have the potential to change the earth's atmosphere for a temporary but significant period of time. Human civilization has been fortunate that a real supervolcano has not erupted during modern human history. Otherwise, civilization would probably have collapsed and the human species could have gone extinct. Even relatively small historical eruptions from volcanoes like Tambora, Indonesia and Laki, Iceland significantly disrupted modern civilization.

The most recent supervolcano eruption was from Toba, Indonesia. Toba is on the island of Sumatra in a chain of andesitic volcanoes. The last supervolcano eruption occurred about 74 000 years ago as recorded by the Young Toba Tuff. This eruption was a VE = 8 and produced about 700 million mi³ (2800 km³) of ejecta. In comparison, the 1980 Mount St. Helens eruption produced 0.24 mi³ (1 km³) and it made an exceptional impact in that part of Washington. It is likely that the Toba eruption caused a major impact on the worldwide climate and may have led to extinction events. The United States also has a potential supervolcano in the Yellowstone volcanic area.

CASE STUDY 4.3 Yellowstone Supervolcano

Major North American Threat

The beautiful geysers and thermal springs at the Yellowstone National Park in Wyoming cover an insidious past that has the potential of returning with catastrophic results. The thermal activity in the park results from magmatic activity at depth. This igneous activity is because the North American Plate is moving over a mantle plume at about 1.1 in. (2.7 cm) per year. There is a chain of progressive igneous activity extending from the current activity back across the Snake River Plain and Columbia River Plateau that began about 16.5 million years ago. These deposits define the track of the hot spot over time and there is evidence of other supervolcano eruptions in Idaho prior to reaching Yellowstone. For example, the Bruneau-Jarbidge Caldera formed 10–12 million years ago, and its supervolcano eruption deposited ash 1 ft (30 cm) deep, 1000 mi (1600 km) away in Nebraska where it killed herds of camel and rhinoceros among others preserved at Ashfall Fossil Beds State Historical Park.

Apparently, the Yellowstone area was the site of small rhyolitic eruptions in the early stages of volcanic activity about 8.7 million years ago. However, this small volcanic area would evolve into the site of three rhyolite supervolcano eruptions (Figure 4.7). The eruptions produced hot ash, pumice, and rock fragments that deposited and welded forming extensive sheets that are as thick as 1312 ft (400 m) in some areas. The first eruption occurred about 2.08 million years ago. The large magma chamber that had developed under the area suffered a catastrophic collapse of the roof. This created the Big Bend Ridge, Snake River, and Red Mountains caldera segments that are 121–153 mi

FIGURE 4.7 Photograph of the Yellowstone National Park. *Source*: Courtesy of USGS.

FIGURE 4.7A Photograph of Huckleberry Ridge flow. *Source*: Courtesy of USGS.

by 64–97 mi (75–95 km by 40–60 km) as parts of the huge Island Park Caldera which occupied an area of about 1000 mi² (2560 km²). The collapse generated a mammoth Plinian eruption forcing about 610 mi³ (2500 km³) of ejecta into the air. The ash spread across the United States covering an area of 5985 mi² (15 500 km²) and is now represented locally by the Huckleberry Ridge Tuff (Figure 4.7A).

The second eruption occurred about 1.3 million years ago and is the smallest of the three. This Plinian eruption produced the Henry's Fork Caldera which is about 10–17 mi (16–27.2 km) across. The eruption released about 70 mi³ (292 km³) of ejecta covering an area of 1042 mi² (2700 km²) and produced the Mesa Falls Tuff. Although sizable, this is too small to be considered a supervolcano.

The third and final eruption of this cycle occurred about 630 000 years ago (Figure 4.7B). This Plinian eruption formed the large Yellowstone Caldera which measures about 30 mi by 53 mi (48 km by 85 km). This supervolcano produced 240 mi³ (1000 km³) of rhyolitic ejecta that covered an area of 2896 mi² (7500 km²). This deposit is the Lava Creek Tuff and it has a maximum thickness of approximately 180–200 m (590–660 ft).

Naturally, the great concern is if and when Yellowstone will erupt again. Having such a dangerous volcano in the middle of the United States has resulted in considerable research on the area. At present, Yellowstone ranks as the 21st greatest volcanic threat in the United States with a threat score of 115 placing it at the top of the high threat category. Geophysical data in the area shows that the main magma chamber under Yellowstone is currently 50 mi (80 km) long and 24 mi (40 km) wide. This represents an upper felsic magma chamber that extends from 3.1 to 9.9 mi (5–16 km) in depth that is available for eruption. It is estimated that this magma is about 5–15% melt rock. However, there is a lower crustal reservoir of mafic magma of 2% melt that extends from 12 to more than 25 mi (20–40 km) in depth. The lower mafic magma reservoir is about 4.5 times as large as the upper rhyolite reservoir. It contains enough heat to maintain a sizable reservoir of felsic magma for a major eruption. The Yellowstone caldera floor rose almost 3 in. (7.6 cm) per year between 2004 and 2008. This was greater than three times the rate ever observed since measurements began in 1923. From 2004 to 2008, the caldera surface rose as much as 8 in. (20 cm) locally. In addition, there have been major earthquakes related to the tectonics of the area. For example, in 1959, the Hebgen Lake, Montana earthquake had a magnitude of 7.1 and caused significant disruption of the surface including landslides and scarps.

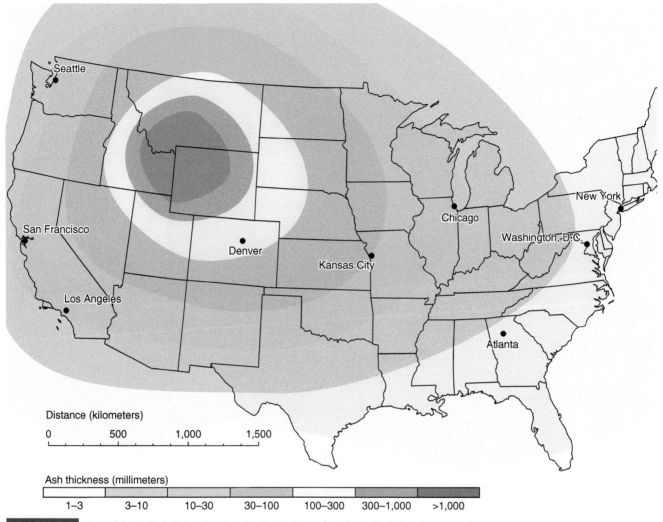

FIGURE 4.7B Map of the United States showing the distribution of ash from the Yellowstone eruption.

Reference

Cole, P., Bass, V., Christopher, T., Murrell, C., Odbert, H., Smith, P., Stewart, R., Stinton, A., Syers, R., and Williams, P. (2011). MVO scientific report for volcanic activity between 1 November 2010 and 30 April 2011, Open File Report OFR 11-01. Monsterrat Volcano Observatory (MVO).

CHAPTER 5

Volcanic Hazards

CHAPTER OUTLINE

5.1 Eruption of Stratovolcanoes 64

5.2 Eruption of Shield Volcanoes 87

Summit Eruption. An eruption from the main vent of a volcano at its peak. The explosion of eruption is caused by instantaneous expansion of water several thousand times its volume as it changes phase from liquid to gas. The phase change results from the pressure release as the magma rises in the vent. The explosion has so much energy that it can send the eruption cloud of gas and ash 20 miles straight up into the atmosphere.

Ashfall. Heavier ejecta tend to fall onto the volcano building up the cone but the eruption cloud drifts in the wind and can deposit ash and clasts of pumice great distances away. The thicker deposits are closer to the volcano and can be tens of meters thick.

Fumeroles. When a hot ashfall deposit occurs on an area with shallow groundwater or permafrost, the heat can cause the water to boil producing steam vents on the surface. Fumeroles can also be generated from magma beneath the groundwater causing steam vents at the surface.

Lahar. When hot ash lands on glacial ice at the top of a volcano, it melts and the two mix forming a thick boiling mass of mud. This mud flows down the valleys and engulfs everything it encounters. The pressure of the flow can crush homes, uproot trees and pull bridges off of their foundations. If the ash is deposited and then turned to mud by rain, it can be just as devastating but it is a mudflow rather than a lahar.

Lateral Eruption. If the summit vent becomes clogged, pressure can build up inside the volcano. Any release of pressure through the flank from a crack or landslide causes an eruption in a lateral direction that wreaks havoc on the surrounding area.

Lava Flow. As the gases are released from the magma and it reaches the surface, it becomes lava which flows down the slopes of the volcano. In an andesite or dacite volcano, the lava is 1000°C or more and burns everything in its path.

Pyroclastic Flow. Literally, "fire particles", pyroclastics are solid particles emitted by volcanoes. They flow down the side of a volcano mainly in valleys at 60 miles per hour in a cloud of 800°C gas. The cloud is so hot that it can vaporize anything in its path. The larger particles settle as the flow reaches the base of the volcano and it continues as primarily ash and gas.

Nuee Ardente. Literally, "fiery cloud" in French, a nuee ardente or pyroclastic surge is a cloud of gas and ash that rushes down the slopes of a volcano and out over the plains engulfing everything in its path. It is a kind of pyroclastic flow but it has a high gas to particle ratio and is not restricted to valleys.

ERUPTION OF A STRATOVOLCANO

Common to convergent margins, stratovolcanoes are among the most dangerous types with explosive eruptions and a number of hazards. The composition is typically andesite and dacite and includes both lava and ash.

Words You Should Know:

Fumaroles – Vents and craters in a pyroclastic flow from which steam from heated groundwater can escape.

Lahar – A boiling mudflow composed of fluidized ash that occurs during or soon after a volcanic eruption.

Lateral blast – A volcanic eruption or blast that is emitted from the flank of a volcano rather than the summit.

Lava flow – Pouring of lava out of the volcano flows downgradient as far as gravity drives it.

Nuée ardente – The hot gas and ash can separate from the coarse fraction of a pyroclastic flow. This "fiery cloud" can flow independently of the rest of the flow.

Phreatic explosion – When flowing lava comes into contact with water, it causes the water to flash boil. The instantaneous increase in volume is the explosion.

Phreatomagmatic explosion – If seawater or other surface water spills into base of an active volcano through cracks or other openings, it explodes through the instantaneous expansion of water converting to steam.

Pyroclastic flow – A flow of hot volcaniclastic debris and gas down the volcano slope as the result of an eruption.

Summit eruption – A volcanic eruption from the top of the volcano.

5.1 Eruption of Stratovolcanoes

CASE STUDY 5.1 1980 Eruption of Mount St. Helens

The Ultimate Lateral Blast

Most eruptions of stratovolcanoes are emitted from the vent at the summit of the mountain where the eruption column can shoot more than 25 km (15 mi) straight upward reaching the stratosphere. Considering the pull of gravity on the mass of the gas, rock, and ash in the column, the force of the explosion is unimaginable. The reason that it is so explosive is that as the wet ocean crust causes rocks to melt in a subduction zone like the Cascades, it produces water-rich magma. At 1000 °C or more, the water should be a gas, but the high pressure exerted on it by its depth in the earth keeps it in liquid state and mixed into the magma. When the magma is driven up the feeder pipe in the volcano, the pressure is released, and the water separates from it and instantaneously turns to steam. Each gallon of water turns into hundreds to thousands of gallons of steam in an instant and the expansion blows out of the summit vent. This is called a Plinian-type eruption.

The most appropriate analogy to this situation is a capped bottle of soda or champagne. As long as the cap is on the bottle, thus keeping the pressure higher, the carbon dioxide stays mixed into the liquid no matter how hard the bottle is shaken. Once the cap is removed, however, the pressure is released and the carbon dioxide changes to gas and is released, explosively in some cases.

In a lateral blast, the summit vent typically gets clogged by the sticky andesite magma (Figure 5.1). Pressure builds up in the volcano until a weak spot on the side of the mountain gives way. The eruption column that would have shot upward instead shoots sideways out of the opening in the mountain where gravity can speed it up as it descends the slopes. This was the case with Mount St. Helens in the Cascade Mountains of Washington state. The 1980 eruption of Mount St. Helens was the largest volcanic eruption in the continental United States in many centuries. Compared with the historical eruptions of Mounts Vesuvius, Tambora, or Pinatubo, however, at 0.24 mi^3 (1 km^3) of total ejecta, it was tiny. On the other hand, unlike these large eruptions which were summit eruptions, Mount St. Helens was the largest lateral eruption in recent history.

After a long period of inactivity or dormancy, Mount St. Helens suddenly came to life on 27 March 1980 when a summit eruption shot an eruption column 20 000 ft (6 km) into the air (Figure 5.1A). Scientists, the press, and numerous onlookers flocked to the area to observe the spectacular activity. Two eruptions on 8 April and 7 May kept the public attention fixed on the volcano as more activity was expected. The governor of Washington called in the National Guard to limit access to the area and keep reporters and onlookers at a safe distance.

Beginning on 17 April, attending volcanologists noticed that the north side of the volcano was starting to bulge as magma built up beneath it. The bulge grew continuously over the ensuing weeks

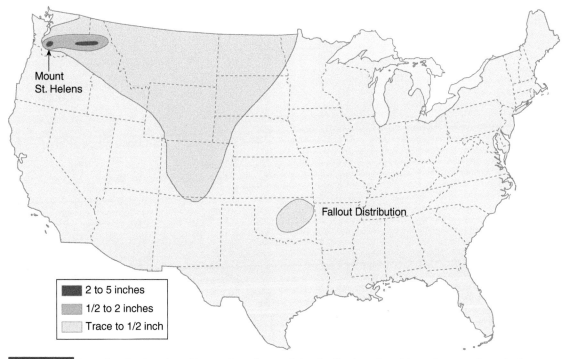

FIGURE 5.1 Map showing location of Mount St. Helens and the distribution of resulting ash across the United States.

and the steam eruptions subsided on the summit. The US Geological Survey set up the Coldwater I and Coldwater II continuously manned observation posts at about 6 mi (10 km) from the volcano to monitor this bulging. The volcanologists were on rotating shifts and used laser transits to measure the bulging through elevation changes.

On 17 May, volcanologist David Johnston substituted for the scheduled observer at Coldwater II because that person had commitments the next day. Filling in would be a deadly mistake. The following morning at 8:32 a.m., there was an earthquake of magnitude 5.1 under the mountain. The intense shaking triggered a massive landslide on the bulge from the weakened north face to the summit of the mountain. The landslide slid 0.6 mi³ (2.5 km³) of rock down the slope of the mountain. The sudden removal of rock released the pressure on the magma chamber beneath it, and like a shaken champagne bottle when the cap is pulled off, it exploded. The eruption was measured at a volcanic explosivity index (VEI) of 4–5 which is very strong. The massive eruption column shot laterally across the landscape at 220 mph (354 kmph) to as much as 670 mph (1078 kmph) depending upon location (Figure 5.1B). It raced along the surface mowing down mature forests and snapping huge trees like they were toothpicks. The great heat of the blast scorched everything in sight leaving a brown and desolate swath in an otherwise lush, green forest (Figure 5.1C). Unfortunately, it went straight over the Coldwater II observation post and killed David Johnston. His reported last words were a broadcast to the base station, "Vancouver! Vancouver! This is it!" His body was never found.

The blast continued to sweep over the landscape and still had enough power to knock down trees in 20 mi (32 km) away from the volcano (Figure 5.1D). As a result, and from ensuing volcanic hazards, a total of 57 people are confirmed to have been killed in the eruption. The only reason that the death toll was so low is that the area was relatively uninhabited and because of the restricted access. In an inhabited area, the death toll would have been in the tens of thousands or more. The eruption pulverized the rock on the north face of the mountain sending the ash all across neighboring states. It reduced the elevation of the summit by 1313 ft (400 m). The eruption column flattened 230 mi² (596 km²) of woodland, and killed more than 7000 game animals, 40 000 salmon, and 12 million salmon fingerlings (Figure 5.1E). More than 30 years later, the area had still not fully recovered. Mount St. Helens is a grim warning of the danger of allowing people to inhabit the areas around the other seemingly inactive Cascade volcanoes in the Northwestern United States.

FIGURE 5.1A Photograph of the summit eruption of Mount St. Helens. *Source*: Image courtesy of the USGS.

FIGURE 5.1B Photograph of the lateral blast shooting to the left out of the erupting Mount St. Helens. *Source*: Image courtesy of the USGS.

FIGURE 5.1C Photograph of broken trees from the lateral blast with a person for scale. *Source*: Image courtesy of the USGS.

FIGURE 5.1D Photograph of trees blown down by the lateral blast. *Source*: Image courtesy of the USGS.

FIGURE 5.1E Photograph of a lahar filling a river valley on Mount St. Helens. *Source*: Image courtesy of the USGS.

CASE STUDY 5.2 1993 Story of Galeras

Deadly Summit Eruption

The Galeras stratovolcano is a 16 000 ft (4877 m) mountain in the Andes of Southern Colombia. It is especially dangerous because the city of Pasto, Colombia, which has more than 450 000 inhabitants, sits directly on the eastern slope of Galeras, likely Colombia's most active volcano (Figure 5.2). Explosive eruptions since the geologically recent mid-Holocene time period have produced widespread pyroclastic flows that swept over the northern, eastern, and western flanks of the volcano. Galeras has been almost continuously active for the past one million years. Two major caldera-building eruptions occurred about 560 000 years ago, and between 40 000 and 150 000 years ago. The central volcanic cone has been the source of the numerous short and violent historical eruptions of small to moderate size. Historical eruptions have occurred 25 times since 1535 ranging from months to 66 years in duration, making it one of the most active volcanoes on Earth.

This regular violent activity coupled with the closeness to a sizable population center makes Galeras a very dangerous volcano. It is so dangerous that the International Association of Volcanology and Chemistry of the Earth's Interior chose Galeras as one of the 16 Decade Volcanoes. As part of the United-Nations-sponsored International Decade for Natural Disaster Reduction, the Decade Volcanoes project is designed to address a group of active volcanoes that pose particular threat to human populations (Figure 5.2A). The program aims to establish research and public-awareness activities to enhance understanding of these volcanoes and the hazards they pose. Decade Volcanoes are chosen based upon a history of large, destructive eruptions and proximity to populated areas where they may threaten tens or hundreds of thousands of people (Figure 5.2B). Mitigating eruption hazards at Decade Volcanoes is of critical importance.

A workshop was convened in January 1993 by the United Nations International Decade for Natural Disaster Reduction committee to review the research, monitoring, and disaster mitigation that had been done at the Decade Volcano, Galeras. Approximately 150 scientists assembled in Pasto to attend the workshop which had been organized by Professor Stanley Williams, a volcanologist from Arizona State University. On the morning of 14 January 1993, Williams led a group of 14 of his fellow volcanologists and several observers on a field trip into the Galeras caldera to take measurements and to test various eruption prediction methods. The caldera is a crater at the summit of a volcano from which the vast majority of volcanic eruptions take place. It is the surface expression of volcanic conduit or pipe which leads directly down to the magma chamber. Depending upon the volcano and time, the caldera can contain solid rock and/or lava and it commonly emits gases of various toxicity.

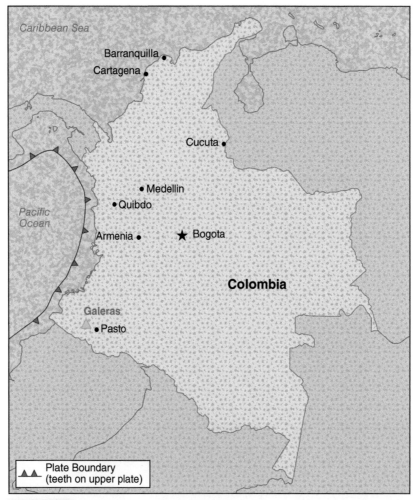

FIGURE 5.2 Map of Colombia and surrounding geography with the location of Mount Galeras and the city of Pasto.

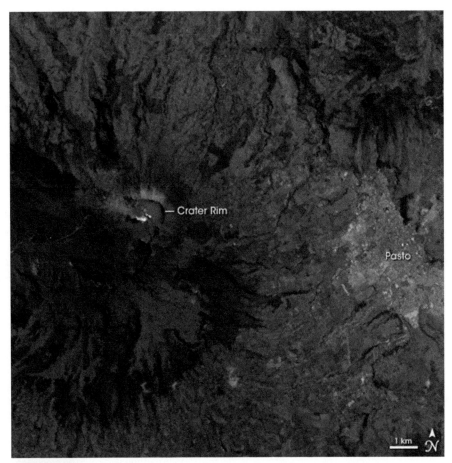

FIGURE 5.2A Satellite image of Mount Galeras and the city of Pasto showing the volcanic deposits around the mountain. *Source*: Image courtesy of NASA.

FIGURE 5.2B Photograph of a small summit eruption of the volcano Galeras. *Source*: Stf/EPA/Shutterstock.com.

Prior to this trip, Williams and his team of researchers had detected such gases escaping from Galeras. They also recorded 20 tornillo earthquakes as recently as 16 days before the trip. Tornillo earthquakes show a peculiar "screw shape" on seismograph records and are interpreted to result from magma moving through cracks in rocks underground. After this burst of activity that might have foretold of an eruption, however, the tremors died down and the volcano was emitting only low levels of gas. It was for this reason that the potential danger of an eruption had apparently passed, and the field trip was conducted.

This group, composed of some of the most knowledgeable scientists in the world about volcanic eruptions, hiked to the summit of Galeras. Many of the scientists and observers who entered the caldera did not even wear a helmet or take any safety gear. As the scientists completed their sampling of rock and gas and measuring temperatures, Galeras erupted (Figure 5.2C). Relative to other volcanic eruptions, it was minute, only 0.003% the size of the relatively small 1980 Mount St. Helens eruption, but it was deadly to the field party. The volcano exploded, spewing hot toxic gas and volcanic bombs at high velocity in all directions shattering bones and severely burning everyone in the caldera.

Six of the scientists and three observers were killed immediately by the blast, but Williams survived. Volcanic bombs from the explosion of the eruption crushed Williams's legs, fractured his skull, and burned him over much of his body. With great effort, unable to use his legs, he crawled out of the exploding caldera and part way down the mountain to hide behind a large boulder to shield him from the fiery ejecta, or solid volcanic emissions. The eruption abated as quickly as it began, and survivors called for help. The wounded, including Williams, were airlifted out to local hospitals. Later, they returned to their home countries where they underwent multiple surgeries for years in an attempt to repair the horrible injuries they had sustained. The ironic part of this disaster is that the dead and injured scientists were only in the volcano to develop methods to save people from volcanic eruptions in the first place.

Authorities later returned to the volcano to recover the remains of the dead and any damaged equipment. Autopsies were performed on the five bodies that were recovered and studies were done on the few pieces of equipment and clothing that were retrieved from the crater area. The main causes of death and injury were the force of the explosion at the volcanic vent and the bombardment by ejecta or pyroclastics within the first 15 minutes of the eruption. These volcanic ejecta ranged from bombs and blocks at up to more than 3 ft (0.9 m) in size which caused the death and destruction to pea-sized lapilli which fell last. There was no release of lava from this eruption, just gas and ejecta, two of the three possible emissions in a volcanic eruption.

This event was used to develop recommendations for the safety of volcanologists conducting research in volcanic craters. Hard hats are recommended to protect against blows to the head from ejecta during escape, and a lightweight, heat-resistant, and water-repellent coverall limits burns and reduces the risk of clothing being ignited from the ejecta. The coverall also protects against hypothermia due to the cold and windy conditions of a volcano summit.

Professor Williams published the book *Surviving Galeras* in April 2001 about his firsthand account of the 1993 eruption. It probably would have passed quickly into obscurity, but a controversy arose because a conflicting book was published at the same time. A journalist, Victoria Bruce, investigated the 1993 incident and found evidence that the field trip to the volcano should never have taken place. Her book *No Apparent Danger: The True Story of Volcanic Disaster at Galeras and Nevado del Ruiz* was also published in April 2001. Bruce claims that Williams was reckless and purposefully endangered the field party. She claims that Williams considered himself to be a geological "Indiana Jones" and took undue risks to prove it.

There was a 1991 report by a US Geological Survey geophysicist that tornillo earthquakes could be precursors of an eruption at Galeras. Two Colombian geophysicists claim to have warned Williams about the tornillos the night before the eruption, but, according to Bruce, Williams did not share the information with his colleagues. Otherwise, she claims, the disaster could have been avoided.

FIGURE 5.2C Photograph of a small summit eruption of the volcano Galeras. *Source*: Yogi Black / Alamy Stock Photo.

CASE STUDY 5.3 1902 Eruption of Mount Pelée

Nuée Ardente

Pyroclastic flows are masses of superheated gas and ejecta of various sizes that flow down the slopes of volcanoes at high velocity through a combination of the force of the eruption and the influence of gravity. Depending upon these two factors, pyroclastic flows can reach velocities of 100–150 mph (60–90 kmph) or more at temperatures in excess of 1500 °F (816 °C) and they can travel more than 6 mi (10 km). In many cases, they are the most deadly part of an eruption. Pyroclastic flows which are dense by virtue of their jumble of rocks and are typically confined to valleys. However, they are shrouded in a turbulent gas- and ash-rich cloud that can separate from the flow and travel a separate path. These clouds are pyroclastic surges and, unlike flows, they can travel over any obstacle and for much longer distances with even deadlier consequences. Some volcanologists simply lump all pyroclastic movement into the term pyroclastic density currents or PDCs rather than differentiating them.

The term *nuée ardente*, French for "glowing clouds," was coined for PDCs (pyroclastic surge) from the catastrophic 1902 eruption of Mount Pelée on the island of Martinique in the Caribbean Arc (Figure 5.3). Mount Pelée is a stratovolcano that overlooked the thriving city and port of St. Pierre at the base of its slope. St. Pierre had a population of 30 000 and the port was important for trade in the area. Tragically, an eruption of the volcano ended the existence of the town in a catastrophic manner, and to this day, the area has not been similarly resettled.

It was not that the residents did not have warnings of the impending doom, they just chose to ignore them or to believe the people who discounted them. After a long period of inactivity, Mount Pelée came to life on 23 April 1902 with fumarole activity, small ash eruptions, and earthquakes. On 25 April, a larger summit eruption was released from the Étang Sec caldera and ash dusted the area in a similar eruption on 26 April. All of this activity compelled a group of residents to climb the mountain on 27 April to make observations. They found a small debris cone emitting enough hot water to flood the crater in which it sat. Small eruptions continued for the next few days until 2 May when the activity increased dramatically in both intensity and duration. There were reports of farm animals dying of thirst and poisoning from ash, and there were even unsubstantiated reports of swarms of insects and snakes overrunning settlements as they fled from the volcano.

On 4 May, an ash cloud covered St. Pierre and the harbor causing many people to leave the city. On 5 May, a rim of the Étang Sec crater collapsed, sending a lahar of scalding mud down the Blanche River at

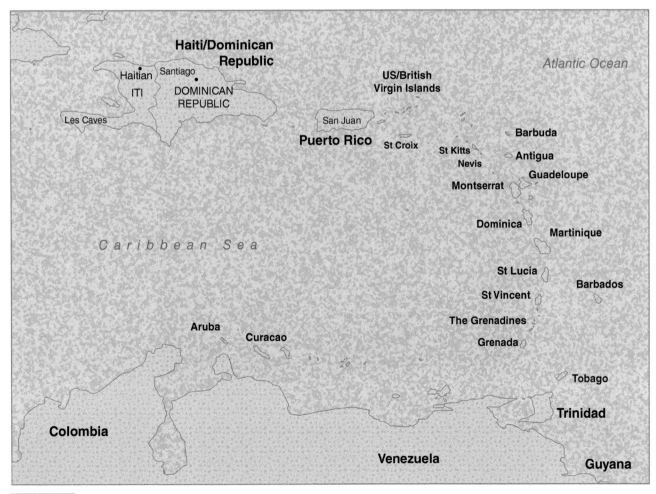

FIGURE 5.3 Map of the island of Martinique and surrounding area.

FIGURE 5.3A Series of photographs of the development of the *nuée ardente* generated in the 1902 Mount Pelée eruption. *Source*: US Library of Congress.

nearly 60 mph (100 km). It destroyed a rum distillery killing 23 workers and a sugar works where it caused many other casualties. The lahar then entered the sea and reportedly generated a 10 ft (3 m) tsunami that flooded the waterfront of St. Pierre. The wave, however, could have been the result of accompanying earthquake activity. Chaos ensued with many people fleeing the city and many others from outlying areas entering the city to seek shelter.

In response to the unrest, the governor appointed a "Volcano Commission" composed of doctors, pharmacists, and science teachers to determine if there was a danger to St. Pierre. Even with the volcanic activity increasing, the Volcano Commission met on the evening of 7 May and concluded, "There is nothing in the activity of Mt. Pelée that warrants a departure from St. Pierre" and "the safety of St. Pierre is completely assured." These conclusions were to be published early on the morning of 8 May.

On 8 May 1902, Ascension Day, at 8:02 a.m., Mt. Pelée underwent a major eruption with a VEI of 4. It was caused by an explosive collapse of a growing lava dome at the summit of the volcano in which the upper mountainside emitted a dense black pyroclastic flow that sped down the slopes (Figure 5.3A). A second black cloud erupted upward that darkened the sky over an area 50 mi (80 km) in radius. The velocity of the flow was more than 420 mph (670 km) with temperatures greater than 1967 °F (1075 °C). It flowed along the ground and sped down the slopes toward the city of St. Pierre, appearing black and heavy but glowing hot. It had lower section composed of flowing rock debris termed "glowing avalanche" (pyroclastic flow) and a dilute, gas-rich upper section termed "glowing cloud" (pyroclastic surge or *nuée ardente*). The flow separated as it sped downhill; the dense pyroclastic flow following the Blanche River Valley, while the lighter pyroclastic surge, depleted in heavy fragments, remaining turbulent and spreading out over the landscape. It flowed over the terraces and hills of the volcano and covered the entire city of St. Pierre with suffocating, scorching gas and ash.

The *nuée ardente* killed nearly the entire population of 30 000 people in St. Pierre, devastating an area of approximately 22 mi^2 (57 km^2) (Figure 5.3B). Turbulence in the cloud varied its effects locally. Some people were essentially carbonized upon impact, they were burned so badly, while others nearby were asphyxiated and not burned at all (Figure 5.3C). The only survivor within the city was a prisoner in a poorly ventilated underground cell, though there were other survivors on the outskirts of town (Figure 5.3D). The cloud continued out over the harbor killing nearly all of the sailors on the decks and destroying the top sections of the ships and, in some cases, the entire ship. The force of the flow created large waves on the harbor that swamped some ships and drowned some of the survivors who had jumped overboard for safety. Between the waves and the *nuée ardente*, not a single ship remained fully operational. By this time, the cloud had spread out so much that it took several minutes to pass out of the harbor (Figure 5.3E).

The story of this type of example of a *nuée ardente* does not even end here. There is a report that a ship hundreds of miles out to sea was affected by a very strange event later that day. First, it grew very hot. Then, sharks surrounded the ship and birds flocked all over the ship as the ocean water too became hot. The ocean became turbulent with huge waves crashing against the boat for several hours before ending abruptly. The sky grew hazy and then dark during the turbulence. The disturbance ended and everything returned to normal as if nothing had happened.

FIGURE 5.3B Photograph overlooking the destroyed city of St. Pierre after the 1902 *nuée ardente* passed through. *Source*: US Library of Congress.

FIGURE 5.3C Photograph of destroyed buildings in St. Pierre with a view up the slopes of Mt. Pelee after the *nuée ardente* passed through. *Source*: Image courtesy of the USGS.

9200—St. Pierre, Looking North—Mt. Pelee in the Distance, Martinique.

FIGURE 5.3D Photograph of the destroyed city of St. Pierre and Mt. Pelee. *Source*: US Library of Congress.

FIGURE 5.3E Photograph of destroyed buildings in the St. Pierre harbor. *Source* US Library of Congress.

CASE STUDY 5.4 1883 Krakatoa Eruption

Caldera Collapse at Sea

Perhaps the most famous volcanic eruption of all time is that of the island volcano Krakatoa in the Sunda Arc, Indonesia in 1883 and for many good reasons (Figure 5.4). The eruption was cataclysmic with many impressive features that have not been replicated by any volcano in recent history. There was plenty of fair warning of the main eruption with earthquake swarms and small eruptions well in advance. A significant eruption of ash was reported by a German warship on 20 May 1883 in which the eruption cloud reached 6 mi (9.6 km) high, and explosions could be heard in Batavia, 100 mi (160 km) away. On 16 June, eruptions renewed and covered the island in a thick black cloud for at least five days (Figure 5.4A). By 24 June, the cloud cleared, but the eruption seemed to be coming from a newly formed vent. Seismic and ash emission activity continued at such a pace that Dutch engineer Captain Ferzenaar visited the island on 11 August and reported that all vegetation had been stripped and a 1½ ft (0.5 m) layer of ash covered everything. He advised no further landings.

The eruption increased significantly on 25 August. By 2:00 p.m. on 26 August, there was a black cloud of ash 17 mi (27 km) high above the volcano in a Plinian-type eruption and nearly continuous explosions. Ships in the vicinity were thickly blanketed in ash and large pumice fragments and a small tsunami was reported between 6:00 and 7:00 p.m. some 25 mi (40 km) away in Java and Sumatra. On 27 August, the eruption grew even more violent. Explosions at 5:30 a.m., 6:44 a.m., and 10:02 a.m. were so loud that they could be heard in Perth, Australia, 1930 mi (3106 km) away and on the island of Rodrigues, 3000 mi (4828 km) away, where it was mistaken for cannon fire and the militia readied for an attack. In this Peléan phase of the eruption, huge pyroclastic flows were released down the volcano slopes (Figure 5.4B). As they hit the ocean, the pyroclastic surge or *nuée ardente* separated and sped frictionless like a hovercraft across the open water at temperatures of 1022 °F (550 °C) and speeds up to 200 mph (320 km) for 20 mi (32 km) or more before coming ashore and destroying everything in their paths. One such flow killed as

76 CHAPTER 5 Volcanic Hazards

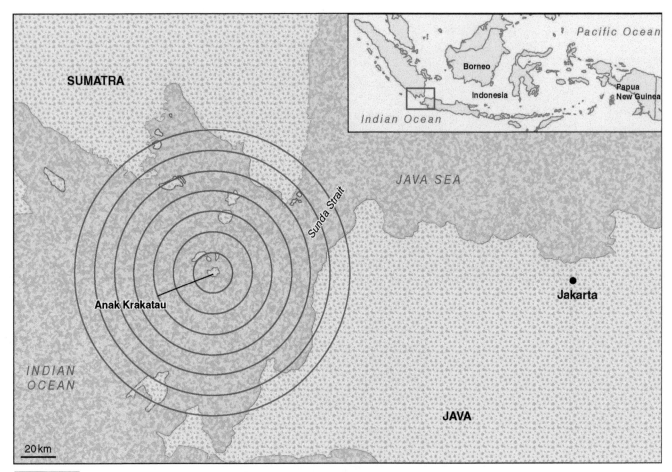

FIGURE 5.4 Map of Krakatoa and surrounding area with the inset map of the location in Indonesia.

many as 2000 people in Ketimbang. The dense pyroclastic flows hit the sea with such force that they generated tsunamis as much as 130 ft (40 m) high locally causing much more death and destruction. The material shed into the sea during this phase was spread in 4.3–5 mi^3 (18–21 km^3).

The final and loudest explosion was the projected caldera collapse that removed the entire volcano and left a crater in the sea (Figure 5.4C). It was thought to be a purely phreatomagmatic explosion from ocean water spilling into the magma chamber through cracks, but a more recent work suggests that a landslide caused a catastrophic release of pressure out of the side of the volcano with seawater involved in the pressurization and explosion. The result was an eruption column that shot 50 mi (80 km) into the air. The pressure wave that resulted traveled away from Krakatoa at 675 mph (1090 km). It ruptured eardrums of sailors on ships in the Sunda Strait, registered on every barometer on the planet, and continued to reverberate for five days, circling the earth seven times. The explosion is considered to be the loudest noise ever to have been made on Earth in recorded history.

Whether by the landslide into the sea, pyroclastic flow, or the explosion, huge tsunamis were generated by this final event. One wave is said to have pulled a 600-ton (544-mt) block of coral from the seafloor and deposited it onshore where it still resides. The town of Merak, Indonesia was destroyed by a tsunami reported to be more than 150 ft (46 m) high. There is speculation that the tsunamis may have been greater than 200 ft (61 m) high in the vicinity of the eruption and washed over the tops of uninhabited islands. Certainly, 100-ft (30 m) high tsunamis pounded the coast of Indonesia (Figure 5.4D). The steamship Berouw was carried 1 mi (1.6 km) inland on Sumatra and the whole crew was killed. Auckland was struck by a 6.5-ft (2-m) high wave 4825 mi (7767 km) away. Tsunamis were reported as far away as South Africa and bodies of victims were found floating across the Indian Ocean for months.

Ash darkened the skies 275 mi (442 km) from the volcano for at least three days during and after the eruption. Ash fell to the north on Singapore, 522 mi (840 km) away, to the southwest on Cocos Islands, 718 mi (1155 km) away, and on ships as far away as 3775 mi (6076 km) to the west. The ash circled the earth within two weeks where it produced colorful sunsets and adverse effects on the climate.

So much pumice was generated during the eruption that it clogged the Sunda Strait. In some places, it looked like solid ground and was sturdy enough for people to walk on. Relief ships were unable to land in many coastal communities for several weeks. With time, storms broke up the pumice rafts and spread them into the Java Sea and the Indian Ocean where ships encountered them for months after the eruption. One pumice raft floated 5075 mi (8170 km) to Durban, South Africa where it landed in September 1884. Other rafts reached Melanesia and remained floating and coherent for two years after the eruption.

FIGURE 5.4A Photograph/lithograph of the eruption of Krakatoa. *Source*: Parker & Coward, Britain/Wikimedia commons/Public Domain.

The total energy released by this eruption which had a VEI = 6 is estimated to be equivalent to 200 Mt of trinitrotoluene (TNT) or four times as powerful as the largest nuclear weapon ever built. Its effects totally destroyed 165 villages and damaged 132 others. The official death toll is 36 417 according to Dutch authorities with about 4500 from pyroclastic effects and the rest from tsunamis. Considering the devastation and uncertainty of death tolls associated with tsunamis, some estimate that it could be as high as 120 000 or more. It was probably the best reported natural disaster of its time owing to its broad effects in a well-traveled area and the recent introduction of underwater cables which allowed instant worldwide communication and reporting.

78 CHAPTER 5 Volcanic Hazards

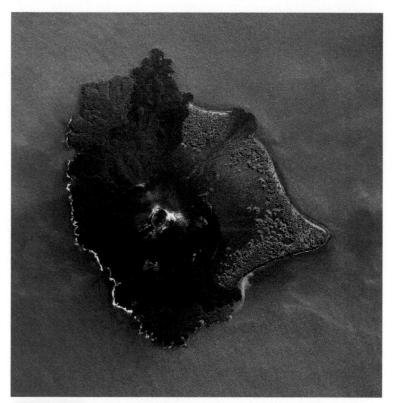

FIGURE 5.4B Satellite image of Anak Krakatoa, the resurgent dome in the middle of the crater. *Source*: Image courtesy of NASA.

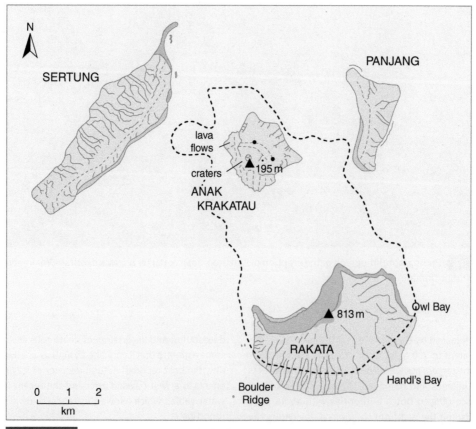

FIGURE 5.4C Map of the current islands around the Krakatoa eruption and the dashed lines show the former shape of the islands prior to the eruption.

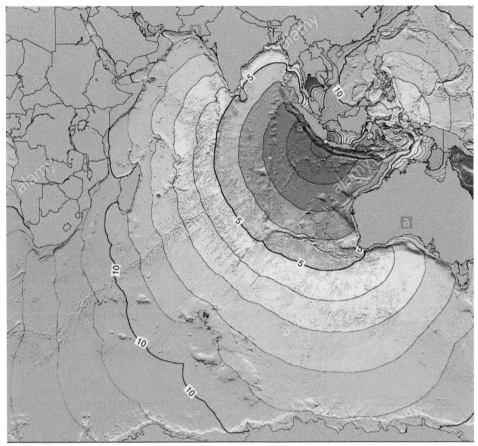

FIGURE 5.4D Map showing the distribution of the tsunami that was produced by the Krakatoa eruption. *Source*: Courtesy of NOAA.

CASE STUDY 5.5   1985 Nevado del Ruiz

A Deadly Lahar

Even though the volcano Nevado del Ruiz is only about 300 mi (500 km) from the equator in the jungles of Colombia, South America, the great height of more than 17 500 ft (5389 m) allows it to be snow-capped throughout the year (Figure 5.5). The summit area covered by snow and ice is about 9.7 mi^2 (25 km^2). In 1984, small earthquakes, sulfur emissions, and phreatic eruptions from the main crater, Arenas, began to occur at regular intervals. The intensity of the minor eruptions increased until 13 November 1985 at 3:06 p.m. when the initial blast of the main eruption struck.

At the outset of the eruption, the Colombian Geological Survey urged evacuations, but they were largely ignored. Within two hours of the beginning of the eruption, ash and pumice were raining down on the town of Armero, several miles downslope and on the Lagunillas River. The 28 700 residents, however, remained calm at the urging of the mayor through radio broadcasts and a local priest over a loudspeaker. The rain of ash and pumice ended by 5:00 p.m., but by 7:00 p.m., the Red Cross called for an evacuation of Armero and other towns, but the order never reached most of the people.

At 9:08 p.m., the eruption began in earnest when an eruption column shot 20 mi (30 km) into the air. The VEI of the eruption was 3 with 38.6 million tons (35 million mt) of material ejected. Lava flows and 706 300 ft^3 (20 000 m^3) of pyroclastic flows and ashfalls from the summit caldera covered the glacial ice at the peak of the volcano (Figure 5.5A). The hot ash melted the ice and mixed with the meltwater creating a mass of boiling mud or lahar. Unfortunately, the eruption was obscured to the people of Armero by a violent storm which caused power outages, so they had no forewarning of what was to come. Four major and numerous smaller lahars spread throughout the valleys radiating northeast of the summit at speeds of upwards of 37 mph (60 kmph). They dislodged rocks, removed vegetation including trees, and destroyed all houses and buildings in their path. The Cauca River Valley was the conduit for one lahar that engulfed the village of Chinchiná killing 1927 people and destroying 400 homes. Another lahar flowed down the Lagunillas River Valley, burst through a dam, and engulfed Armero, 46 mi (74 km) from the volcano in less than two hours after the eruption began (Figure 5.5B). The flow buried or swept away three-quarters of the town, killing more than 20 000 people. It is said that the mayor perished in the lahar while broadcasting assurances that there was no danger.

The lahar in Armero began with the arrival of a flash flood of water from melted glacial ice just before 11:30 p.m. that carried off cars and people. Three separate lahars then swept through the town in succession. The first lahar was 100 ft (30 m) deep and carried boulders and cobbles at a speed of 39 ft/s (12 m/s) for 10–20 minutes. The second

80 CHAPTER 5 Volcanic Hazards

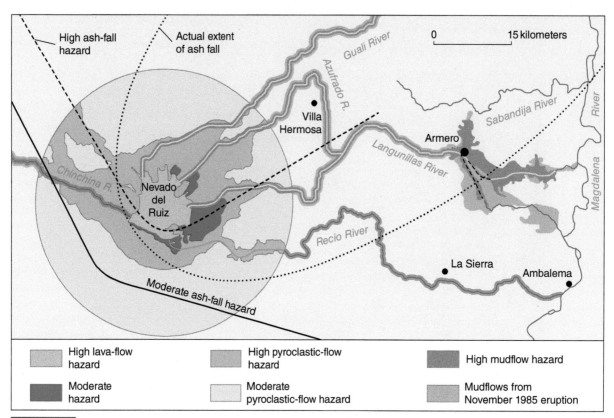

FIGURE 5.5 Map of the Nevado del Ruiz volcano and the hazards it produced. *Source*: Adapted from US Geological Survey.

FIGURE 5.5A Satellite image of Nevado del Ruiz volcano showing the lahar toward the upper left and the ice still intact to the right. *Source*: Image courtesy of NASA.

FIGURE 5.5B Photograph of the Nevado del Ruiz lahar. *Source*: Image courtesy of the USGS.

lahar lasted for 30 minutes and moved at a speed of 20 ft/s (6 m/s), and the third lahar extended the duration of the event to more than two hours. The mud engulfed 85% of the town. Buildings were crushed and collapsed, which killed numerous people (Figure 5.5C).

Rescuers were quickly mobilized and arrived in Armero by 14 November. The tragedy occurred at the same time as the 1985 Mexico City earthquake, so personnel and equipment were limited. Upon arrival, they were faced with a sea of thick, gray mud peppered with bodies and debris from cars, trees, and houses (Figure 5.5D). Survivors had to be dug out of the mud which had hardened into a cement-like consistency when it dried. In all, more than 23 000 people died in the tragedy, 5000 were injured, 20 000 were left homeless, and, in total, 230 000 were impacted in 13 villages. Some 27 000 acres (110 km^2) were laid waste and 15 000 animals were killed. The estimated cost of damage from the lahar is US$1 billion, which is about one-fifth of the gross national product of Colombia.

The account of Nevado del Ruiz did not end with the cleanup. Many people felt that the tragedy could have easily been averted. The volcano erupted a mere 69 years earlier and this was the third time in 400 years that it produced such extensive lahars. It produced steam eruptions and a swarm of earthquakes for 51 weeks prior to the eruption. The activity was so pronounced that a team of scientists and journalists visited the volcano as early as late February. In July, seismographs were placed around the volcano to monitor the earthquake swarm and the subsurface movement of magma. The United Nations contributed funds to Colombia to map the hazard risks around the volcano, the report for which was completed by 7 October, but not widely distributed. Even so, based upon this report, the Colombia National Bureau of Geology and Mines warned of lahars and identified Armero as a danger area. The Colombian government, however, dismissed the report because it was too alarming. A group of Italian volcanologists published a report on 22 October, warning of an impending eruption that was also ignored. A group of scientists even visited the volcano on 12 November because of the increase in activity, but their concerns were dismissed as well. It was for this reason that a banner was erected at one of the mass funerals that said, "The volcano didn't kill 22 000 people. The government killed them."

82 CHAPTER 5 Volcanic Hazards

FIGURE 5.5C Photograph of the Nevado del Ruiz lahar. *Source*: Langevin Jacques/Getty Images.

FIGURE 5.5D Photograph of the damage caused by the Nevado del Ruiz lahar. *Source*: Image courtesy of the USGS.

CASE STUDY 5.6 1912 Novarupta Eruption

Valley of 10 000 Smokes

The world's largest volcanic eruption of the twentieth century was from the Novarupta volcano near Mount Katmai, Alaska. This massive eruption occurred from 6 to 8 June 1912 and had a VEI = 6. It erupted for 60 hours emitting 3.1–3.6 mi³ (13–15 km³) of lava, 4.1 mi³ (17 km³) of air fall ejecta, and 2.6 mi³ (11 km³) of ash-flow. This is 30 times as much lava as Mount St. Helens and with 10 times as much force. The source of the magma for the eruption appears to have been from beneath Mount Katmai, 6.2 mi (10 km) away, which collapsed during the eruption accompanied by a series of 14 major earthquakes of magnitude 6–7.

Increased seismic activity was detected in the Mount Katmai area as early as late May 1912 and several large earthquakes were felt on the morning of 6 June. A huge Plinian eruption cloud at least 20 mi (32 km) high rose over Mount Katmai beginning with an immense explosion at about 1:00 p.m. on 6 June 1912. Within four hours, sulfurous ash was raining down on the village of Kodiak, Alaska 100 mi (160 km) southeast. Over the next three days, ash covered the entire northwestern North America including Vancouver in Canada and Seattle in Washington (Figure 5.6) where clothes on clotheslines were dissolved from it. The 3000-mi (4828 km) wide ash cloud then spread out and drifted covering much of the Northern Hemisphere. Emissions of it were detected at ground level in Wisconsin and Virginia over the next few days and in North Africa and Europe within the next two weeks.

In Alaska, the eruption was devastating. Trees all around the volcano were flattened and carbonized for a 40-mi² (104 km²) area. The sulfurous ash coated and killed vegetation over hundreds of square miles, which starved the already suffering larger animals to their death. Millions of birds were blinded and then killed by the thick ash that coated their feathers and prevented their flight. The ash filled streams, choked them, and turned them acidic. This killed all of the small organisms, which subsequently starved the fish. Alaska's salmon industry was devastated from 1915 to 1919, and took many more years to completely recover.

Although the details of this eruption are elaborate, the most remarkable part was the ash-flow deposits (Figure 5.6A). In addition to eight separate ashfall deposits, there were nine distinct ash-flow

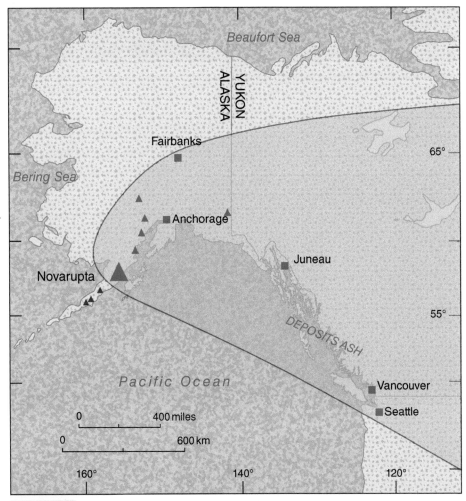

FIGURE 5.6 Map of the Katmai/Novarupta volcano in Alaska showing the direction of the ash cloud across Southern Canada and Northwest United States.

FIGURE 5.6A Current photograph of the ash field deposited by the 1912 Novarupta volcano. *Source*: Image courtesy of the USGS.

FIGURE 5.6B Photograph of the ash field, Valley of 10 000 Smokes from fumeroles deposited by the Novarupta volcano. *Source*: Image courtesy of the USGS.

FIGURE 5.6C Close-up photograph of fumeroles. *Source:* Image courtesy of the USGS.

deposits up to 700 ft thick in places. These rhyolite to dacite composition pyroclastic flows were so hot that they welded the underlying deposit during subsequent flows making the rock a welded ash-flow tuff. A large valley adjacent to the volcano had significant quantities of ground and surface water and even some permafrost that were overlain by these hot deposits, in some cases producing phreatic explosions and craters (Figure 5.6B). The result was that the underlying water was heated to boiling and percolated through the ash-flow deposits to the surface as highly mineralized solutions. The vents where the steam was released in fumaroles were highly mineralized (Figure 5.6C).

Botanist Robert Griggs explored this valley for the National Geographic Society in 1916 and was struck by the multitude of fumaroles. As a result, he named it the "Valley of 10 000 Smokes" (Figure 5.6D). The pictures and description of it were part of the reason that President Woodrow Wilson created the Katmai National Monument (now a national park) in 1918. In 1919, another survey found the temperature at the fumarole vents to still be 645 °C (Figure 5.6E). The temperatures slowly decreased until by 1930 when the once large fumaroles were just a few wisps of steam. The name "Valley of 10 000 Smokes" is still in use even though the maximum temperature is about 90 °C and the fumaroles have long since subsided.

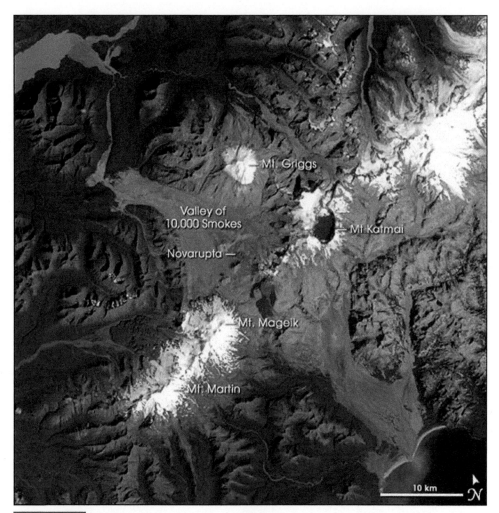

FIGURE 5.6D Satellite image of the Katmai and Novarupta volcanoes and Valley of 10 000 Smokes. The ash deposits appear in tan/brown flows. *Source*: Image courtesy of NASA.

FIGURE 5.6E Photograph of the Valley of 10 000 Smokes with fumeroles. *Source*: Image courtesy of the US Geological Survey.

5.2 Eruption of Shield Volcanoes

CASE STUDY 5.7 1783 Laki Fissure Eruptions

Poison Gas Emissions

On 8 June 1783, the Laki (or Lakagígar) volcano began its historic eruption (Figure 5.7). It was a fissure eruption that was 16.8-mi (27-km) long and included 130 craters along its length (Figure 5.7A). The eruption lasted eight months until 7 February 1784. The eruption, also known as the Skaftáreldar ("Skaftá Fires") or Síðueldur, was the single largest outpouring of lava in historical times at 3.5 mi^3 (14.7 km^3), and 0.2 mi^3 (0.91 km^3) of pyroclastic tephra was emitted as well (Figure 5.7B). This was the largest tephra emission in Iceland in 250 years.

There were 10 eruptive pulses during the first five months of activity, each with a 0.5–4-day explosive phase followed by a longer phase of lava fountaining. The early explosive activity was violent Strombolian to sub-Plinian type. The fire fountains were estimated to have reached maximum heights of approximately 2600–4600 ft (800–1400 m). Peak lava discharge during this phase is estimated at 176 573–211 888 ft^3 (5000–6600 m^3) per second, though some estimates place it as high as 295 000 ft^3 (8600 m^3) per second. The eruption was most vigorous in the first 1.5 months, followed by a slow and steady decline in activity over

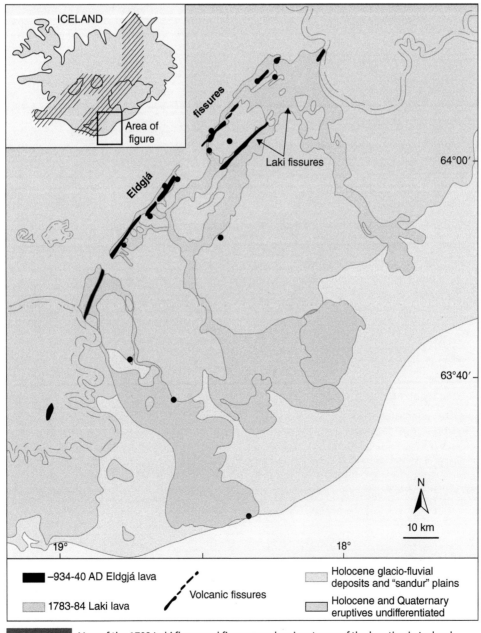

FIGURE 5.7 Map of the 1783 Laki flows and fissures and an inset map of the location in Iceland.

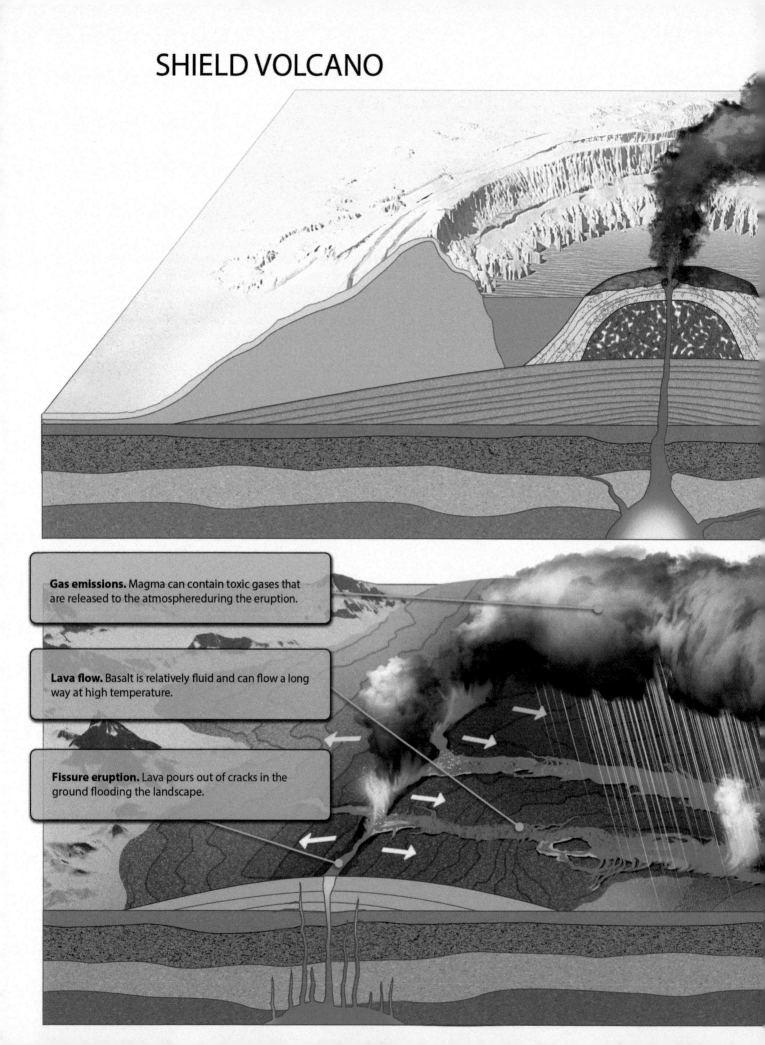

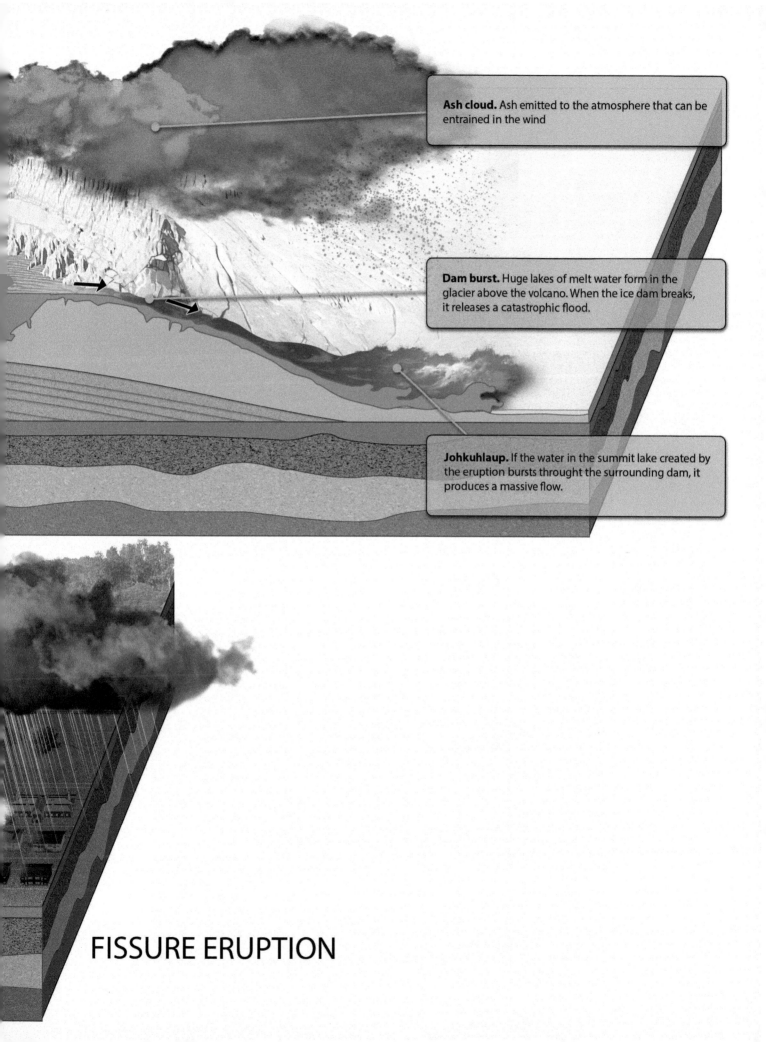

Ash cloud. Ash emitted to the atmosphere that can be entrained in the wind

Dam burst. Huge lakes of melt water form in the glacier above the volcano. When the ice dam breaks, it releases a catastrophic flood.

Johkuhlaup. If the water in the summit lake created by the eruption bursts throught the surrounding dam, it produces a massive flow.

FISSURE ERUPTION

FIGURE 5.7A Photograph of the Laki fissure system with volcanic cones. *Source*: Anne Schöpa/Imaggeo.

FIGURE 5.7B Photograph of a vent from the Laki volcanic field. *Source*: Image courtesy of the US Geological Survey.

the next three months. By the end of the fifth pulse in late July, about 60% of the magma had been erupted. By the end of the tenth pulse in October, about 93% of the volume had been erupted. The interval between successive eruptive phases increased with time. In the final 3.5 months, the eruption was a quiet emission of lava and gas.

The lava field covered an area of 218 mi^2 (565 km^2). The lava flowed 15 mi (24 km) southwestward from the volcano toward the sea (Figure 5.7C). The flow was 10 mi (16 km) at its widest point and had an average thickness of 100 ft (30 m). It filled two large valleys (Skaftá and Hverfisfljót) where rivers were dammed, and the thickness reached several hundred feet.

Despite the impressiveness of the eruption, the real impact of Laki was the emission of aerosols. Calculations indicate that eruption columns probably reached altitudes in excess of 8.1 mi (13 km) during the first few phases of the eruption and remained greater than 6.2 mi (10 km) high for the first three months. These eruption columns reached into the westerly jet stream which permitted emissions to be spread far into Europe and beyond. Not only was there ash but Laki also emitted an estimated 8 million tons (7.3 million mt) of hydrogen fluoride and 120 million tons (109 million mt) sulfur dioxide, 96% of which was emitted within the first five months. The sulfur was converted into sulfuric acid in the atmosphere and produced acid fog and rain.

The worst effects were in Iceland where ash settled over the entire island and a dry fog known as the "Laki haze" settled over the land for more than five months. Fallout of fluorine from this cloud poisoned the grass in grazing areas and as a result about 80% of the sheep, 50% of the cattle, and 50% of the horses in Iceland died. By the end of 1785, 20–25% of the human population (>10 000 people) had also died in the ensuing famine and fluorine poisoning. Ash fallout continued into Great Britain where the summer of 1783 was known as the "sand summer" as a result. The Laki haze continued to spread across Europe. By 26 June, almost all of Europe was covered, finally reaching Lisbon, St. Petersburg, and Moscow by 30 June. It also appeared over Tripoli, Libya by the end of June, and by July, it covered the skies all the way to the Altai Mountains in Central Asia, 4350 mi (7000 km) from the volcano. Later, the Laki haze occurred over Alaska and China, indicating that the sulfuric aerosol plume had spread across the entire Northern Hemisphere from 35° north latitude to the North Pole.

The acid fallout from the haze had a devastating effect on vegetation and human health. In Iceland, acid rain burned holes in leaves and left sores on the skin of animals and humans. Crops completely failed in Iceland, and even had low yields or failed throughout Europe as well. There were reports of fish kills in Scotland caused by acid deposition. People reported that inhaling the haze caused fatigue, shortness of breath, and palpitations of the heart. It was especially hard on the very old, the very young, and those with respiratory ailments.

The weather was also affected by the aerosol and ash emissions from the Laki eruption. July of 1783 was unusually hot in Western Europe, but during the rest of the summer and in most other areas, it was unusually cold. Rainfall over much of the Nile River watershed was well below normal producing a record low level for the river. The area around Altai Mountains experienced harsh overnight frosts in August. Severe drought occurred from India through the Yangtze River region in China and, in general, the summer was cold all over China. In Japan, unusually high precipitation and low late-summer temperatures resulted in widespread failure of the rice harvest and the most severe famine in its history.

The winter of 1783–1784 was one of the most severe in Europe and North America over the past 250 years, with long periods of sub-freezing temperatures. There are reports of extreme weather in Great Britain, France, and Scandinavia. By January, the straits between the Danish islands were frozen over and many other rivers were frozen. It was no better in North America. New England had its longest period of sub-zero temperatures and New Jersey had its highest accumulation of snow. In the Chesapeake Bay, the ice caused the longest known closure of the harbors and channels. Even the Mississippi River at New Orleans was filled with ice rafts between 13 and 19 February 1784.

It has been calculated that the winter mean temperature cooled sharply in 1783–1784 over Europe and Eastern United States by 5.4 °F (3 °C). The next two years were also colder than normal, but not as severe. The full recovery to normal temperatures is estimated to have taken four years.

Between the acid haze and precipitation and the severe weather, there was widespread famine and death both from direct health effects and starvation. It is estimated that more than 23 000 people died in Great Britain and well more than 16 000 died in France in 1783–1784. There was famine across Egypt, China, and Japan causing thousands of deaths. In all, it is estimated that as many as six million people died. The disruption in the food supply caused the economy in Western Europe to falter leading to widespread poverty and unrest. The lack of food and weak economy persisted until 1788 and is credited as a major contributing factor to the French Revolution in 1789.

FIGURE 5.7C Photograph of pillow basalts from the Laki volcanic field. *Source*: Onioram/Wikimedia Commons/CC by 4.0.

CASE STUDY 5.8 1996 Grímsvötn Eruption

Jökulhlaup

Grímsvötn is a very active subglacial volcano in Vatnajökull, Iceland. It erupts every decade or so and melts the glacier above it forming or expanding subglacial Lake Grímsvötn, which is contained by the ice dam that surrounds it (Figure 5.8). On occasion, the ice dam is breached or bursts and the lake drains catastrophically through the glacier forming a jökulhlaup, a meltwater flood. These jökulhlaups at Grímsvötn are the best examples in the world.

Although there have been many jökulhlaups at Grímsvötn, the 1996 event was extraordinary. A particularly energetic eruption of the volcano in the fall brought the glacial lake to its highest level ever observed at 4954 ft (1510 m) elevation (Figure 5.8A). The ice dam that held the water was lifted off of the glacier bed on the late evening of 4 November creating a breach. The coursing 0.77 mi^3 (3.2 km^3) of meltwater formed a tunnel under the ice through a pressure wave of 5–10 bars. The tunnel collapsed in the upper part leaving a canyon 3.7 mi (6 km) long, 0.6 mi (1 km) wide, and 328 ft (100 m) deep (Figure 5.8B). During the drainage, the surface of the lake subsided by 574 ft (175 m) and the area of the 820 ft (250 m) thick ice shelf shrank from 15.4 mi^2 (40 km^2) to less than 1.9 mi^2 (5 km^2).

In just under 10.5 hours after the dam burst, a wave of meltwater 9.8–16.4 ft (3–5 m) high coursed across the Skeiðarársandur outwash plain some 31 mi (50 km) away from the lake (Figure 5.8C). Although the meltwater started out at 46.4 °F (8 °C) in the lake, it was supercooled, carrying ice and freezing everywhere the flow slowed. The discharge of the flow peaked some 16 hours after the dam burst. It is estimated to have been 1 400 000–1 589 000 ft^3 (40 000–45 000 m^3) per second in the tunnel and an amazing 1 756 700 ft^3 (50 000 m^3) per second on the outwash plain. For comparison, the Amazon River, by far the largest river in the world (20% of all river drainage), has a discharge rate of 7 400 000 ft^3 (209 000 m^3) per second. The discharge of the Mississippi River is 200 000–700 000 ft^3 (7000–20 000 m^3) per second.

There was a low rumble for the 10.5 hours that the water worked its way to the edge of the glacier where the sound became a deafening roar. As the flood emerged from the 31-mi (50 km) long meltwater tunnel, it did catastrophic damage in the Skeiðarársandur outwash plain. The flood with a discharge of up to 2.5 times the maximum flow of the Mississippi River carried many large icebergs of up to 200 tons (181 mt). Upon viewing the damage, Prime Minister Davíð Oddsson is quoted as saying, "In just four hours this has knocked us back 20–30 years in terms of our road building endeavors." Many roads in the area were washed away as were all power lines and fiber optic telecommunications cables. One bridge over the Gýgja River that cost US$4.6 million was undermined and washed away. Several other bridges suffered a similar fate. As a precaution, fishing vessels off the south coast of Iceland and in the path of the flood were warned to remove their gear from the water and vacate the area for fear that the wave would continue into the ocean. In all, it is estimated that the 1996 Grímsvötn Jökulhlaup caused more than US$12 million worth of damage.

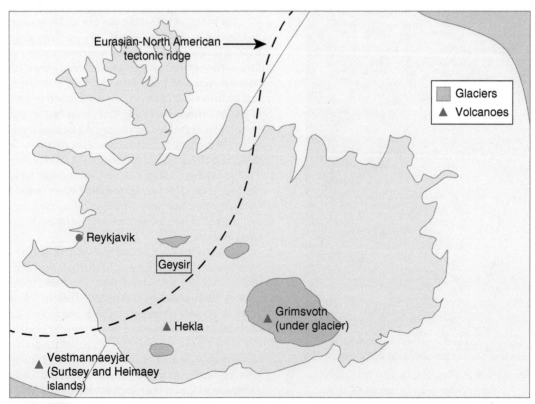

FIGURE 5.8 Map of Iceland showing the location of the Grímsvötn volcano under the glacier.

FIGURE 5.8A Photograph showing the Grímsvötn volcano erupting through a glacier. *Source*: Nordicphotos / Alamy Stock Photo.

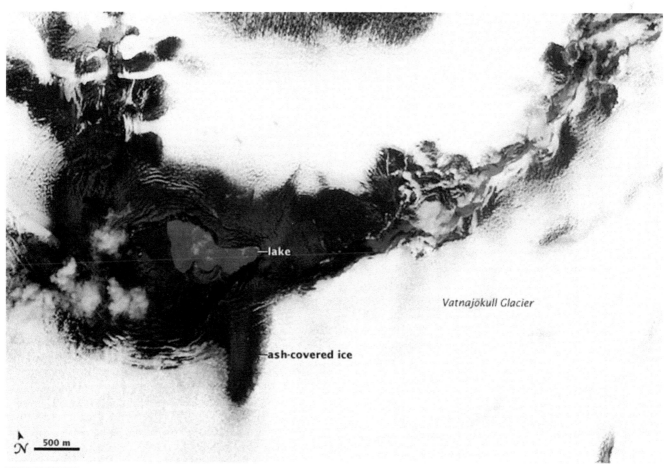

FIGURE 5.8B Satellite image of the Grímsvötn eruption showing the meltwater lake above the volcano, ash covered glacier, and meltwater channel of the jökulhlaup. *Source*: Image courtesy of NASA.

FIGURE 5.8C Photograph showing the flood of meltwater of the jökulhlaup from Grímsvötn. *Source*: Oddur Sigurðsson, Icelandic Meteorological Office.

CASE STUDY 5.9 1973 Eldfell Eruption

Controlling a Lava Flow

The 660-ft (200-m) high Eldfell volcano formed on the volcanic island of Heimaey on the south coast of Iceland (Figure 5.9). The area has experienced extensive historic volcanism, as with most of Iceland, as a result of its position on the divergent Mid-Atlantic Ridge. As with most basaltic volcanism, the eruptions are generally not explosive or particularly threatening to human settlements. The formation of Eldfell in 1973 was almost an exception to that rule.

Precursors to the eruption began at about 10 p.m. on 21 January 1973 with a swarm of weak but deep earthquakes around Heimaey. The number of earthquakes increased rapidly to 50 per hour between 1:00 and 3:00 a.m. on 22 January before subsiding until 11:00 a.m. when they stopped. At 11:00 p.m. and until 1:34 a.m. on 23 January, several shallower and stronger (up to M = 3) earthquakes occurred. At about 1:55 a.m., a 985-ft (300-m) long fissure developed on the east side of the island about 0.62 mi (1 km) from the center of the town of Heimaey.

The fissure grew to 1.24 mi (2 km) in length, crossing the entire island and began a spectacular fissure eruption with VEI = 0–1 (Figure 5.9A). More than 40 lava fountains were 164–492 ft (50–150 m) high along the entire length of the fissure which continued to lengthen to more than 1.9 mi (3 km). The main eruption then concentrated to a single vent of about 0.5 mi (0.8 km) that emitted lava and ejecta at a rate of about 3531 ft^3 (100 m^3) per second. Within two days, this now Strombolian eruption with VEI = 1–2+ had built a cone about 328 ft (100 m) high, and by 15 February, it was 722 ft (220 m) high. The eruption column, at times, reached nearly 5.6 mi (9 km) high.

Although this was a relatively powerful eruption for a basalt volcano, it was not the major problem for Heimaey. The real problems were the thick flow of lava that spread slowly to the east, northeast, and north and the ash that covered the town (Figure 5.9B). The lava flow was on average 131 ft (40 m) thick and up to 328 ft (100 m) thick in some places and burned and crushed everything in its path. The ejecta (tephra) blanketed the town up to 16.4 ft (5 m) thick, which was enough weight to collapse many buildings (Figure 5.9C).

Fortunately, for the people of Heimaey, a storm right before the eruption had forced the entire fishing fleet of 70 ships into the harbor. Fishing was the main industry of Heimaey. For that reason, when the eruption began, the Icelandic Civil Defense Organization sounded all of the sirens in town and the residents boarded the fishing boats. The first evacuation boats left the harbor by 2:30 a.m., a mere half hour after the start of the eruption. The airstrip also remained open, and the rest of the residents were flown out. Within six hours, the entire civilian population of 5300 people was safely on the mainland. Only 250 emergency personnel and volunteer workers trying to clear the ash from houses remained behind.

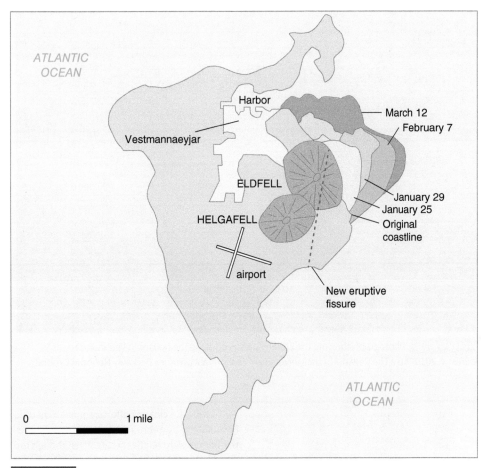

FIGURE 5.9 Map of Heimaey showing the location of the Eldfell vents and flows.

FIGURE 5.9A Photograph showing the Eldfell volcano erupting on the outskirts of Vestmannaeyjar.
Source: Image courtesy of the USGS.

FIGURE 5.9B Photograph showing the Eldfell flows overtaking the houses in Vestmannaeyjar. *Source*: © 2010-2019 The Regents of the University of California, Lawrence Berkeley National Laboratory.

FIGURE 5.9C Photograph showing the Eldfell flows overtaking and crushing a house in Vestmannaeyjar. *Source*: Image courtesy of the USGS.

Even with clearing efforts, some 400 houses collapsed or burned. Fortunately, by February, the tephra emissions abated and most of the homes were relatively safe. Even though the extrusion of lava fell steadily from the initial 3531 ft³ (100 m³) per second to 2119 ft³ (60 m³) per second by 8 February and just 353 ft³ (10 m³) per second by mid-March, the more ominous problem was the lava flows (Figure 5.9D). A submarine flow to the north of town severed the electric cables and water supply from the mainland. Then the flow began to encroach on the harbor. This was a serious situation because it would destroy the fishing industry which provided 25% of Iceland's annual catch and was the town's only source of income. If this happened, the evacuation would be permanent.

The Iceland government decided to attempt an experimental procedure of spraying water on the front of the lava flow to harden it and impede its forward motion. This method had been attempted only in Hawaii and on Mount Etna, Italy on a much smaller scale and with very limited success.

A test of the procedure took place on 6 February with a small volume of seawater pumped at 26.4 gal (100 l) per second on the front of a flow. All of the water flashed to steam and it took a while to harden the lava, but the flow slowed significantly (Figure 5.9E). Encouraged by the success, operations began in earnest and water was strategically sprayed on the flows. An incident of a rapidly advancing flow on 1 March led workers to bring in the dredging boat Sandey which pumped 106 gal (400 l) per second on the flow. A network of pipes was laid atop the lava spraying seawater over an area of 129 167 ft² (12 000 m²) at a time up to 427 ft (130 m) into the flow, but it was difficult work. All of the supports melted or caught fire, there were toxic CO gas emissions, and the flow was constantly moving.

Even with all of these efforts, by the end of March, one-fifth of the town had been engulfed by lava and the flow was overtaking it. At that point, 32 oil pumps were brought in from the United States, with pumping capacities of 264 gal (1000 l) per second apiece (Figure 5.9F). This

FIGURE 5.9D Photograph showing the Eldfell flows moving down streets in Vestmannaeyjar. *Source*: Image courtesy of the USGS.

FIGURE 5.9E Photograph showing efforts to slow the Eldfell flow shown in Figure 5.9D using seawater sprayed on the flow. *Source*: Image courtesy of the USGS.

greatly increased the volume of seawater that could be sprayed, and the flow soon slowed and finally stopped with the fishing harbor still intact. The eruption continued into mid-April at a rate of 177 ft^3 (5 m^3) per second and there was even renewed activity in late May to July, but no more problems arose. The cost of the entire operation was $1 447 742.

After the eruption ended, lava flows had created more than 1 mi^2 (2.5 km^2) of new land on the island and the ejecta was used to extend the runway at the airport and as fill material upon which 200 new houses were built. The heat from the cooling lava flows was used to drive several 40 MW steam turbine generators, thereby providing electricity for the entire island in addition to hot water. The new land now acts as a breakwater for the harbor improving its quality. More than 80% of the population returned by 1975 and Heimaey now has 33% of Iceland's fishing industry.

FIGURE 5.9F Photograph showing efforts to slow the Eldfell flow using seawater sprayed on the flow from tugboats. *Source*: Image courtesy of the USGS.

CASE STUDY 5.10 1986 Lake Nyos Disaster

Gas Poisoning

Lake Nyos lies in Northwest Cameroon within the Cameroon Volcanic Line on the west coast of Africa (Figure 5.10). The source of the volcanism on this line may be a mantle plume or a failed rift zone. Lake Nyos fills a deep depression high on the flank of an inactive volcano in the Oku Volcanic Plain. The depression is a nearly circular maar volcano that formed as an explosion crater when a lava flow contacted groundwater and produced a phreatic explosion about 400 years ago. The lake is approximately 1.1 mi (1.8 km) across and 682 ft (208 m) deep (Figure 5.10A). About 50 mi (80 km) beneath the lake, an active magma chamber releases gases that percolate up into the deep waters of the lake. As a result, the lower part of the lake is saturated with CO_2, held in solution by the pressure of the overlying water. The cold bottom waters are denser than the shallower warmer water, which helps to hold the CO_2 in place, but also allows it to accumulate to high concentrations.

The tragedy at Lake Nyos began on 21 August 1986. Heavy rains typical of the rainy season in the area had abated by 9:30 p.m. and the air was cool and calm. Life around the lake went on as usual until a series of rumbling sounds lasting 15–20 seconds each drew many people out of their homes to see what was happening. Some people experienced symptoms of warmth and smelled rotten eggs or gunpowder before falling unconscious. For others, there were no symptoms; they just fell unconscious, many never to wake. One observer on higher ground reported a bubbling sound and saw a white cloud rise from the lake. A large wave was also reported on the lake. It would later be determined that the wave struck the southern shore with an 82 ft (25 m) height, topped a spillway by 20 ft (6 m), and produced a jet that shot 262 ft (80 m) up a wall. A white dust was also found on the western shore and the lake level had dropped by about 1 m.

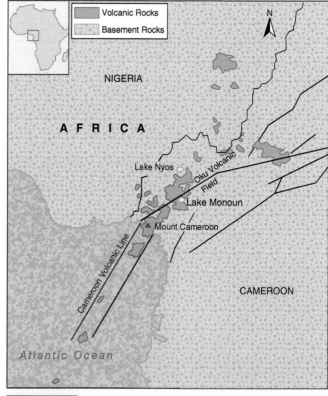

FIGURE 5.10 Map showing the location of Nyos in the Cameroon volcanic line in Cameroon. Inset map shows location in Africa.

Survivors of the tragedy awoke 6–36 hours later, weak and confused. They were the lucky ones. More than 1700 people perished, and 3000 cattle died along the lake shore and up to 6 mi (10 km) north of the lake. In all, livestock losses were more than 3900 cattle, 3300 fowl, 550 goats, and 350 sheep (Figure 5.10B). Wild animals, birds, amphibians, reptiles, and insects were also reported killed in great numbers and were scarce for several days, but plant life remained unaffected. The pattern was that at lower elevations everything was dead and at higher elevations there were more survivors. In the hills above the lake, there were no victims.

By Friday morning, 22 August, people from other villages began arriving to help with recovery and burial. On Saturday, a helicopter pilot viewed the disaster and reported it to the world (Figure 5.10C). It took two days after the governor of the region was notified of the tragedy for medical teams to arrive. By then, upwards of 4000 people were fleeing the area and many bodies had already been buried. For this reason, the death toll of 1700 can only be approximated. The doctors found that 845 of the survivors that they treated had symptoms compatible with exposure to an asphyxiating gas. Up to 20% had lesions determined to be caused by lying in a comatose state for a long period of time.

FIGURE 5.10A Panoramic photograph of a stirred-up Lake Nyos as a result of the events leading to the disaster with part of the impacted village to the left. *Source*: Image courtesy of the USGS.

FIGURE 5.10B Photograph of cattle that were killed by the gas cloud from Lake Nyos. *Source*: Peter Turnley/Getty Images.

FIGURE 5.10C Photograph of cattle killed by the gas cloud from Lake Nyos inside the village. *Source*: Image courtesy of the USGS.

A team of scientists studied the disaster and determined that some event caused the CO_2 gas to be rapidly released from the deeper waters. It could have been an earthquake, a small volcanic eruption, cooling of the lake surface, or the most likely scenario, a landslide because they identified a fresh landscape scarp (Figure 5.10D). Estimates of the amount of gas released vary wildly from 110 231 to 330 693 tons (100 000–300 000 mt) to 1.76 million tons (1.6 million mt) and volumes of 0.16–0.29 mi^3 (0.68–1.2 km^3). It rose to the surface at estimated speeds of up to 62 mi (100 km) per hour, breached the surface, and then traveled across the lake and up the valleys at speeds of 12.4–31.1 mph (20–50 kmph). The 164-ft (50-m) thick blanket hugged the surface because CO_2 is 1.5 times as dense as air. CO_2 becomes toxic when the concentration in air reaches 10% and did so at lower elevations, but was more dilute at high elevations causing people to lose consciousness. The cloud is estimated to have extended up to 14.3 mi (23 km) from the lakeshore.

This was not the first CO_2-poisoning incident. Just two years earlier, at the much smaller Lake Monoun, Cameroon, there was a similar poisoning incident in which 37 people died. Estimates were that such events happened every 10–30 years, but could really happen anytime. As a result, a system of siphoning tubes provided by the United States was installed in the lake in 2001 and 2011 to pump bottom waters to the surface, thus releasing the CO_2 in a slow and controlled manner. The problem now is that one of the walls of the lake has weakened and threatens to break and drain part of the water. This would cause a catastrophic release of gas and another tragedy.

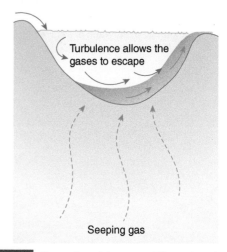

FIGURE 5.10D Illustration of the gases diffusing into the bottom of Lake Nyos and being catastrophically released to overwhelm the shore.

CHAPTER 6

Causes of Earthquakes

CHAPTER OUTLINE

6.1 Stress and Strain 103

6.2 Faults 104

6.3 Intraplate Versus Plate Margins 105

6.4 Earthquakes from Igneous Activity 110

Earth's Fury: The Science of Natural Disasters, First Edition. Alexander Gates.
© 2022 John Wiley & Sons Ltd. Published 2022 by John Wiley & Sons Ltd.
Companion website: www.wiley.com/go/gates/earthsfury

Words You Should Know:

Brittle deformation – Deformation involving the loss of cohesion and sharp breakage of rock similar to breaking glass.
Dip-slip faults – Faults across which the movement is in a dip sense or vertically up or down the fault plane. They can be normal, reverse, thrust, or vertical faults.
Ductile deformation – Permanent and distributed deformation of a material with no loss of cohesion, such as the deformation of clay, gum, or putty.
Elastic strain – Strain in material that only exists as long as stress is applied, such as a spring or a rubber band. Seismic waves impose elastic strain.
Intraplate faults – Faults that are within plates rather than along plate boundaries.
Megathrusts – Subduction zone plate contacts are also called megathrusts because they have thrust fault geometry.
Strain – The change in shape or other state of a material in response to stress.
Stress – The force per unit area exerted on a material such as rocks and faults.
Strike-slip faults – Faults across which movement is in a lateral sense. Also known as transcurrent or wrench faults, strike-slip faults can be left-lateral or right-lateral.

6.1 Stress and Strain

Earthquakes are among the most destructive of natural disasters as single events and certainly the most unexpected. The only way they can be less destructive or will ever be predictable is to understand their generation and processes. The most destructive earthquakes are generated by faults in rock and especially crystalline rock. The first step in earthquake damage reduction and possible prediction is to understand the mechanics of the faults that generate them.

Unfortunately, the words "stress" and "strain" are often confused in everyday language. Stress is the force that is imposed to cause deformation, or strain, in a rock including movement on a fault. Stress is defined as the force applied per unit area of a material. If the amount of stress is the same on all sides of the material, then it is pressure, such as air pressure or water pressure. To cause strain in rocks, stress must be different in at least two directions. This is known as differential stress. Differential stress requires a maximum stress direction (maximum force) in which objects are shortened and a minimum stress (least force) direction in which objects elongate (Figure 6.1).

This elongation and/or shortening of material as well as shear comprise the strain in a rock. Faults accommodate the elongation and shortening, in many cases. Areas of general compressional stress, such as convergent margins, are dominated by faults that accommodate shortening. Areas of general tension or extension, such as divergent margins, are dominated by faults that extend or lengthen the crust. Transform margins are dominated by faults that move landmasses laterally in shear, but they can also contain significant numbers of faults that lengthen or shorten the crust as well.

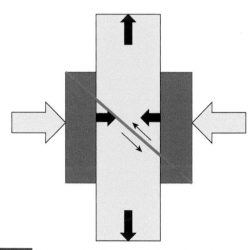

FIGURE 6.1 Diagram showing the relationships among compression (yellow arrows), extension and shear (red line and small black arrows), and extension versus elongation (black arrows) on an initial block (blue) deforming to a final block (yellow).

Strain or deformation can be divided into three types based upon the material response as brittle, ductile, or elastic. Brittle deformation is like the breaking of a bottle, pencil, or piece of chalk. Enough stress is applied until the strength of the object is reached and then it breaks in a single event. With a pencil, your fingers apply enough stress until it snaps. At that point, the stress has been released from the pencil by being, in part, converted to the energy in the sound waves that are emitted when it breaks (Figure 6.2). There is no more stress in the pencil on either of its broken parts, but there is a plane or zone of breakage or strain. The other two parts of the pencil are exactly the same as they were before the breakage, so, individually, they record no strain. They could be glued back together and, besides the glued line of breakage, the pencil would look exactly the same as it did prior to breaking it. This means that there was a loss of cohesion during the brittle deformation.

Brittle deformation is the way rocks behave primarily in shallow faults that produce earthquakes. In the vast majority of cases, tectonic forces supply the stress to the rocks. As the stress builds up, at some point, it exceeds the strength of the rocks and they break. This breakage is the earthquake, and it releases waves, just like the

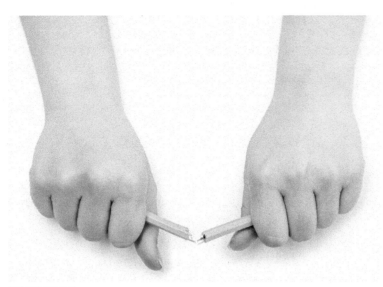

FIGURE 6.2 The breaking of a pencil in brittle deformation showing the release of sound waves. *Source*: WLADIMIR BULGAR/Getty images.

sound waves from the pencil that can be felt and even heard, in some cases. The waves are the energy built up by the stress being converted to and released as kinetic energy. The stress across the fault plane drops to levels below the strength of the rock and the situation becomes mechanically stable for the time being. However, tectonic forces are relentless and begin to build up again immediately. This buildup and release of stress is called stick-slip behavior and it is the reason that active faults produce earthquakes over and over again.

When the fault moves in one place, it is no longer helping to support the tectonic stress that has built up along the entire fault. This means that the stress at other points along the fault, and especially right next to the area that moved, experiences a rapid increase. In many cases, the stress exceeds the strength of the rocks in these areas as well and they break, generating earthquakes as well. This can continue until the entire fault has moved or at least until enough stress has been released that it no longer exceeds the strength of the rock in the rest of the fault. The earthquakes following the first movement are called aftershocks unless the first movement triggers an even bigger earthquake. Then it is termed a foreshock. The largest earthquake in the sequence is called the main shock and these designations can only be determined after the whole sequence is well underway or has ended.

Elastic deformation can be modeled by a rubber band. It only changes shape, or is strained, when a person stretches it. The person provides the stress with their fingers. As soon as the rubber band is let go, it returns to its original shape with no record of what just happened to it. For that reason, rocks and other materials contain no permanent record of elastic deformation. However, the seismic waves released by an earthquake impose elastic deformation on most things they pass through. If people experience a strong earthquake, they may see the ground surface move up and down or side to side as the waves pass through, but it returns to normal once the waves are gone.

Ductile deformation is permanent, like brittle, but there is no loss of cohesion and it is distributed and gradual. Clay, putty, or gum undergo ductile deformation if stressed, stretching or contracting across a large part without breaking. No earthquakes are produced by ductile deformation. Typically, rocks that are under high temperature and pressure undergo ductile deformation, so they are almost exclusively very deep in the earth. The asthenosphere undergoes ductile deformation in its flow that drives the plates, and the very deep extension of faults undergoes ductile deformation.

6.2 Faults

Faults are breaks in the earth's crust across which movement has occurred; otherwise they are just fractures. Faults can be generally divided into those with largely vertical movement, called dip-slip faults, and those with primarily lateral movement, called strike-slip faults. For this reason, dip-slip faults are better viewed in a vertical cross section view of the earth, whereas strike-slip faults are better viewed on a map or surface view of the earth. In most dip-slip faults, the fault plane is a sloping surface. The footwall is the rock beneath this fault plane. The rock above the fault plane is the hanging wall. The relative motion of the hanging wall to the footwall determines the kind of fault, normal or reverse.

The designation of dip-slip movement is relative to gravity. In normal faults, the hanging wall slides down the fault plane as gravity would dictate. Normal faults accommodate extension of the crust and are most common in divergent margins on continental crust. In reverse faults, the hanging wall is pushed up the slope of the fault plane, in the reverse direction of gravity. If the slope of the fault is shallow, 30° or less, reverse faults are called thrust faults. Reverse and thrust faults accommodate contraction of the

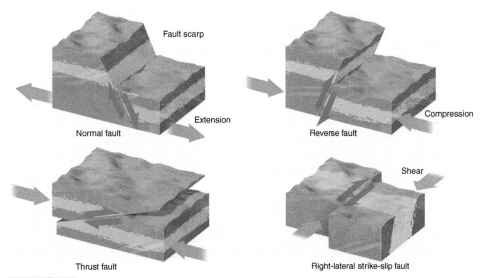

FIGURE 6.3 Block diagrams of several main fault types including normal, reverse, thrust and right-lateral strike slip faults as indicated.

crust and are most common in convergent margins of all types. If the dip-slip fault is vertical, there is no hanging wall or footwall, and the normal or reverse designation cannot be applied.

Strike-slip faults are usually very steep at the surface. They appear as lines on maps. Movement is designated by the relative lateral motion on each side of the fault. If a person looks across the fault, the opposite side can either move to their left or their right. If it moves to the left, the fault is a left-lateral or sinistral strike-slip fault. If it moves right, the fault is a right-lateral or dextral strike-slip fault.

These motions are end members on the spectrum of faults. Faults can have combinations of strike-slip and dip-slip motions and are called *oblique-slip faults*. The components of the fault, reverse, normal, right-lateral, or left-lateral, should be included in the naming of the oblique-slip fault.

6.3   Intraplate Versus Plate Margins

Plate margins are areas of concentrated stress. The continuous forcing together, pulling apart, or sliding against of massive tectonic plates produces numerous highly active faults. That is the reason that these areas produce so many earthquakes and are so dangerous. However, faults and earthquakes can also occur in other areas as well. If a geologic feature is located within a plate, it is termed intraplate. Both earthquakes and volcanoes occur in intraplate settings, but they are uncommon. Hawaii is an example of a volcano at an intraplate location. In some cases, intraplate earthquakes can be more destructive than plate margin varieties. The strongest historical earthquake in the continental United States was in New Madrid, Missouri, and was an intraplate earthquake. Because intraplate earthquakes do not recur at regular intervals, a long enough time lapse can separate the events that people may forget the risk and be unprepared. They can also occur virtually anywhere and at any time, without warning.

CASE STUDY 6.1   1811–1812 New Madrid Earthquakes

Intense Intraplate Seismic Activity

A vast majority of earthquakes occur along plate boundaries. For the continental United States, the entire West Coast is in this category with the San Andreas Transform in the south and the Cascadia Subduction Zone in the north. Intraplate seismicity includes all earthquakes that are not associated with plate boundaries. These earthquakes are nearly impossible to predict with any degree of certainty and are almost exclusively small. However, in rare cases, they can be very large and even exceed the magnitude and rate of those at plate margins. This was the case in 1811 and 1812 in New Madrid, Missouri when a group of very powerful earthquakes caused great damage (Figure 6.4).

The greatest concern is whether the disaster can happen again considering the vast population in the area. If it does, it will most certainly produce the greatest natural disaster in US history.

In 1811 and 1812, the Midwest was not very settled and the population density of people who built permanent structures was sparse. Therefore, although powerful, the New Madrid earthquakes did only a moderate amount of damage. The area of damage with shaking intensity greater than or equal to VII on the Modified Mercalli Scale which is considered very strong or worse, was a huge 231 661 mi^2 (600 000 km^2). The area where shaking intensity was moderate but alarming or greater than or equal to V on the Modified Mercalli Scale was 1 million mi^2 (2.5 million km^2), and the area of perceptibility was 2 million mi^2 (5 000 000 km^2).

106 CHAPTER 6 Causes of Earthquakes

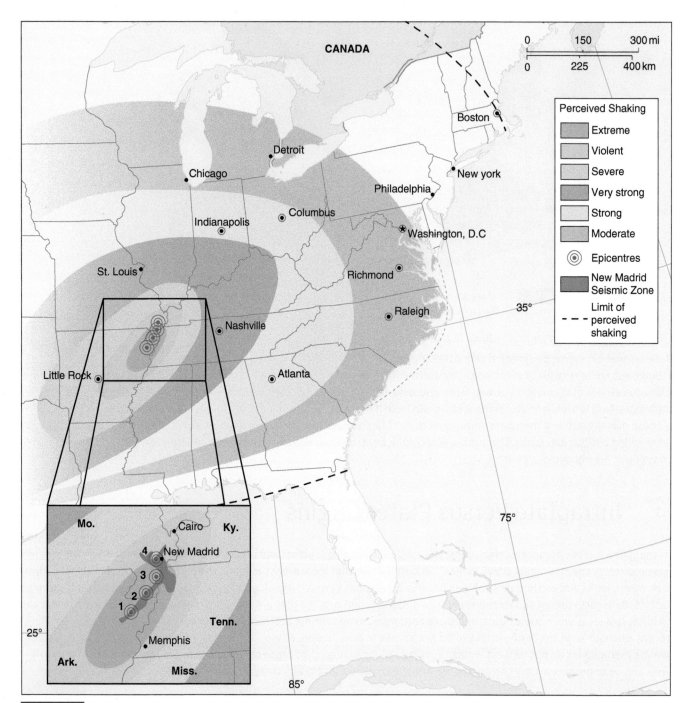

FIGURE 6.4 Map of the Eastern United States showing the location of the New Madrid earthquake and impacted area. *Source*: Adapted from New Madrid earthquakes of 1811–12. Encyclopædia Britannica, Inc.

The first event was on 16 December 1811 at 2:15 a.m. with an epicenter in Northeast Arkansas. The estimated magnitude was 7.5. The intense shaking awakened people in New York City, Washington, D.C., and Charleston, South Carolina. The area that would become the city of Memphis, Tennessee experienced a shaking intensity of IX on the Modified Mercalli Scale. It is said that sand and water from liquefaction were shot tens of feet into the air in fountains and heavily damaged the town of Little Missouri. An earthquake-generated seiche wave propagated up the Mississippi River. However, the shaking caused generally slight damage to man-made structures in the region, because of the sparse population.

The second event was really a strong aftershock, and it occurred the next morning at 7:15 a.m. on 16 December with the epicenter again in Northeast Arkansas. The magnitude of this earthquake is estimated at 7.0, and it too was felt on the East Coast of the United States.

What is considered the second main shock struck on 23 January 1812 at 9:15 a.m. with the epicenter in New Madrid, Missouri. The estimated magnitude is 7.3. The damage to the epicentral area included general ground warping, water and sand fountains, large ground fissures, extensive landslides, and caving of stream banks. This earthquake appears to have been caused by a rupture on the New Madrid North Fault. This may have increased stress on the Reelfoot Fault.

The final main shock occurred on 7 February 1812 at 3:45 a.m. also with an epicenter in New Madrid, Missouri. The magnitude of this event was about 7.5 and it occurred on the Reelfoot Fault of Missouri and Tennessee. The town of New Madrid was completely destroyed, and as far away as St. Louis, many houses were severely damaged and chimneys toppled. Uplift along parts of this reverse fault formed waterfalls on the Mississippi River at Kentucky Bend, and waves that propagated upstream making it look like the river was flowing backward. It also formed the Reelfoot Lake by obstructing streams in Lake County, Tennessee (Figure 6.4A).

These, however, were not the only events. In total, between 16 December 1811 and 15 March 1812, there were over 200 moderate to large aftershocks reported in the region. Seven to ten of the aftershocks had estimated magnitudes greater than 6.0, and there were about 100 between magnitude 5.0 and 5.9.

There were many reports of surface effects from the earthquakes. They included widespread landslides on steeper bluffs and hillsides, large areas of surface uplift, and still larger areas of subsidence, many of which were flooded. Large waves on the Mississippi River sunk a number of boats and washed others high onto the shore (Figure 6.4B).

FIGURE 6.4A Photograph of submerged trees in Reelfoot Lake as a result of subsidence produced by movement on the Reelfoot Fault. *Source*: Courtesy of the US Geological Survey.

FIGURE 6.4B Print of upheaval on a river resulting from the New Madrid earthquake. *Source*: Sarin Images/GRANGER.

However, the most extensive effects of the 1811–1812 earthquakes are the large sand blows, which were formed by eruption of water and sand onto the surface (Figure 6.4C). This resulted from earthquake-driven liquefaction and the sand blows occurred over an extremely large area of 4015 mi^2 (10 400 km^2). Liquefaction and lateral spreading over the surface were severe and widespread. The effects of liquefaction extended 124 mi (200 km) northeast of the New Madrid Seismic Zone, 150 mi (240 km) to the north-northwest near St. Louis, Missouri, and 155 mi (250 km) to the south near the mouth of the Arkansas River.

The powerful New Madrid earthquakes were even claimed to have been felt by President James Madison at the White House, nearly 900 mi (1450 km) away. Most of the homes of the town of New Madrid left in apocalyptic ruin. The entire Midwest was still in flux at the time. In June of 1812, the Missouri Territory was formed by Congress from the Louisiana Purchase Territory. Soon after, the War of 1812 began which occupied the country until January of 1814. At that point, the Territorial Governor William Clark, of the Lewis and Clark Expedition, finally was able to address the damage from the earthquakes.

Clark had influence with the White House and on 17 February 1815, Congress approved $50 000 for the New Madrid Relief Act. This was the first disaster relief bill in the history of the United States. Anyone who lost land during the earthquake was eligible to receive a certificate redeemable for between 160 and 640 acres of public land in Missouri. Unfortunately, administering the act was filled with corruption as only 20 of the 516 certificates issued by Congress were received by actual New Madrid residents. Even Governor Clark attempted to profit by sending agents to make deals with unknowing victims. Most certificates went to people in St. Louis and all the money used to buy the land became worthless. Lawsuits challenging the claims lasted until 1862.

The story of New Madrid, however, is not over. In November 2008, the US Federal Emergency Management Agency (FEMA) warned that an earthquake in the New Madrid Seismic Zone similar to the 1811–1812 activity could produce highest economic loss from a natural disaster in history of the United States. They predicted catastrophic damage across Alabama, Arkansas, Illinois, Indiana, Kentucky, Mississippi, Missouri, and particularly Tennessee. They speculated that a 7.7-magnitude quake would damage tens of thousands of structures affecting water distribution, transportation systems, and other vital infrastructure. Estimates of death tolls in the 250 000 range have been predicted by some. Adding to this danger, the US Geological Survey has estimated that the risk of a catastrophe similar to the 1811–1812 New Madrid earthquakes in the next 50 years is a very high 7–10%. The risk of a smaller but still devastating magnitude 6.0 earthquake in the next 50 years is a frightening 25–40%.

FIGURE 6.4C Photograph of a liquefaction channel leading to a now gone mud volcano. *Source*: Courtesy of the US Geological Survey.

Riverbanks caved and collapsed, sandbars and points of islands washed away, and some of the whole islands disappeared. The area with the most pronounced of these surface features was about 30 000–50 000 mi^2 (78 000–129 000 km^2), extending from Cairo, Illinois to Memphis, Tennessee, and from Crowley's Ridge, Arkansas to Chickasaw Bluffs, Tennessee.

The uplift and subsidence were notable and much still exist today. An area of subsidence that formed during the 7 February 1812 earthquake was at the Reelfoot Lake where subsidence ranged from 5 to 20 ft (1.5–6 m) (Figure 6.4A). Other areas subsided up to 16.5 ft (5 m), although 5–8.3 ft (1.5–2.5 m) was more common. The elongate 40-mi (64-km) by 0.6-mi (1-km) Lake St. Francis, Arkansas was formed by subsidence during both prehistoric and the 1811–1812 earthquakes.

Highly destructive intraplate earthquakes are rare compared with the common ones at plate boundaries. They are especially common in subduction zones where the subducting oceanic plate looks and operates like an immense thrust fault. For this reason, they are commonly referred to as megathrusts. These faults can exhibit large amounts of movement in a single event on very long ruptures while producing large-magnitude earthquakes. The 2011 Tōhoku, Japan, 2010 Maule, Chile, and 2004 Indian Ocean earthquakes were the latest major megathrust-type events. Kamchatka, Chile, and Alaska had such events in 1952, 1960, and 1964, respectively, and Cascadia off the Northwest Coast of the United States produced a megathrust-type event (estimated 9.0 magnitude) in 1700. Because these events are submarine and thrust in nature, there is a strong potential for them to produce very large tsunamis.

CASE STUDY 6.2 1700 Cascadia Earthquake

American Megathrust

The shallow portion of subduction zones exhibits the same geometry and mechanics as a thrust fault. However, the fault plane in a subduction zone is much larger and the rupture zones during earthquakes can be huge. For this reason, they are regularly termed megathrusts. The biggest recorded earthquakes such as the 1960 Valdivia, Chile earthquake (M = 9.5) and the 1964 Good Friday, Alaska (M = 9.2) were megathrusts as were the more recent extreme earthquakes such as the 2004 Banda Aceh, Indonesia (M = 9.0) and 2011 Tōhoku, Japan (M = 9.0). Smaller subduction zones such as Cascadia are not usually considered to be capable of producing such huge earthquakes, but this may not be the case. However, it took some detective work to figure that out.

The story begins with an "orphan" tsunami in Japan meticulously recorded from multiple locations describing high waves on 27 and 28 January 1700. It was called an orphan because no local earthquake could be identified that might have triggered it. The waves ranged from 3.3 to 17.5 ft (1–5.4 m) and killed several people, washed away more than 20 buildings, flooded rice paddies, and damaged salt kilns. Startled and wet villagers awoke and hastily scrambled to high ground. For years, the wave was a mystery until, in 1996, Japanese researchers looked toward Cascadia for a source (Figure 6.5). It would have taken a tsunami about 10 hours to travel from Cascadia to Japan. If that was the location of earthquake that generated the orphan tsunami, it would have occurred at about 9 p.m. on 26 January 1700.

American researchers began looking for evidence of the event in Washington to Northern California. There are a number of oral traditions among indigenous coastal peoples from British Columbia to Northern California that describe the shaking and tsunami-like flooding from around the time period. Archaeological research found evidence of several coastal Native American villages having been flooded and abandoned around 1700. Scientists discovered from the outermost tree rings of red cedar trees in Oregon and Washington that they were killed by lowering of large coastal forests into the ocean tidal zone (Figure 6.5A). As determined by radiocarbon dating, the trees died in the winter of 1699–1700 when the coasts of Northern California, Oregon, and Washington apparently dropped by 3.3–6.6 ft (1–2 m), flooding them with seawater. Sediment layers in the locations of the old forests demonstrate sedimentation patterns that are consistent with earthquakes and tsunamis around this time. Core samples from the nearby seafloor and debris samples from earthquake-induced landslides in the Pacific Northwest also support the timing and shaking.

The severe drop in the vast areas of forests that subsided along the coast as much as 6.6 ft (2 m) indicates a large event. The production of a tsunami that crossed the entire Pacific Ocean and struck the

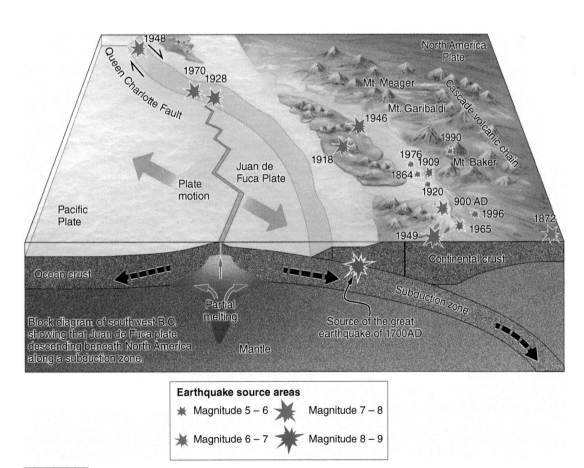

FIGURE 6.5 Tectonic map of the Northwestern United States showing the location of the Cascadia Subduction Zone.
Source: After Geological Survey of Canada.

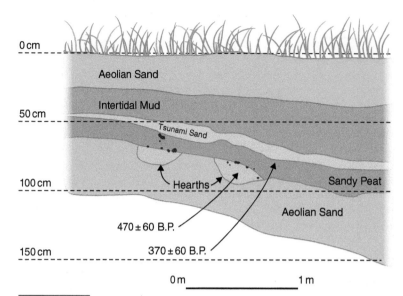

FIGURE 6.5A Diagram showing the stratigraphy of the Oregon Coast and the age constraints on the tsunami deposits. *Source:* After Minor & Grant, 1996. Earthquake-Induced Subsidence and Burial of Late Holocene Archaeological Sites, Northern Oregon Coast. American Antiquity, 61(4), 772–781. doi:10.2307/282017.

coast of Japan with heights as much as 17.7 ft (5.4 m) requires a very large event. The plentiful landslide scarps and extensive thick submarine deposits also require a large earthquake. These effects are coupled with a historic catastrophe in about 1700 when all communities on Pachena Bay were wiped out on the mountainside up to approximately 75 ft (23 m) above sea level, likely the result of a tsunami of that height. Based on this evidence, experts conclude that the Cascadia earthquake occurred on the Cascadia Subduction Zone megathrust on 26 January 1700 with an estimated magnitude of 8.7–9.2.

The megathrust earthquake was on the Juan de Fuca Plate subducting under the North American Plate from Mid-Vancouver Island, south along the Pacific Coast, to Northern California. To produce an earthquake of this magnitude, the length of the fault rupture was about 682 mi (1100 km), and the width of the rupture was likely on the order of 62 mi (100 km). The modeled average amount of slip on the rupture surface is about 66 ft (20 m), but other models estimate as much as 132 ft (40 m) of slip at least locally. It is further interpreted that severe shaking could have lasted for five minutes or more. Sequences of severe aftershocks would have lasted for many months.

There is no doubt that there was a very major earthquake in 1700. The problem is that this earthquake is on a subduction zone where the buildup of stress is constant and it has now been building up for more than 320 years. Modeling of the recurrence intervals on this zone is that magnitude 8+ earthquakes can occur every 300 years and magnitude 9 earthquakes may take as much as 500 years. Based upon these calculations, there is a huge 37% chance that a magnitude 8.2+ earthquake will occur within the next 50 years, and even a 10–15% chance that the entire 682-mi (1100-km) long Cascadia megathrust will rupture producing a magnitude 9+ earthquake over the same period. It is disconcerting that some reference works begin these predictions with the phrase "if it hasn't happened already."

The problem is that a lot of people now live in this part of the country and they have developed a large infrastructure. Evacuation maps were constructed based on the modeling of a Cascadia tsunami striking the US mainland. They found that the at-risk population based on census data for areas within 0.6 mi (1 km) of the ocean or waters connected to the ocean exceeded 150 000 year-round residents in the year 2000. By now, it is significantly greater.

These models predict that the next Cascadia earthquake and tsunami will impact millions of people, property, infrastructure, and the environment. One model predicts that the number of deaths will exceed 10 000, and more than 30 000 people will be injured. Economic losses will exceed $70 billion for Washington, Oregon, and California alone. This would be the worst natural disaster in the history of the United States.

6.4 Earthquakes from Igneous Activity

The movement of magma below the earth surface can produce earthquakes. Magma can force its way upward toward the surface. As it moves, it cracks the bedrock ahead of it, so it can squeeze into them allowing it to continue to ascend. Each of the cracks creates a small earthquake. Taken together, groups of these small earthquakes are called earthquake swarms. Such swarms in volcanic areas are considered a precursor to an eruption. Some of these small earthquakes have distinctive arrival patterns on a seismograph. The waves show a decaying pattern that looks like a small screw and, as such, they are called tornillos.

Even stronger earthquakes can be generated by evacuating magma. Magma chambers support the crust above them as they push upward. If the magma retreats or is removed from the magma chamber, the crust above can collapse producing an earthquake. Most earthquakes in Hawaii are produced in this manner. In most cases, these earthquakes are small, but, occasionally, larger earthquakes of magnitude 5 or more can result.

CHAPTER 7

Earthquakes 101

CHAPTER OUTLINE

7.1 Earthquake Basics 113

7.2 Earthquake Waves 113

7.3 Foreshocks, Main Shocks, and Aftershocks 116

7.4 Magnitude and Intensity 116

7.5 Determining the Epicenter and Movement on Faults 119

7.6 Earthquake Risk Maps and ShakeMaps 122

Earth's Fury: The Science of Natural Disasters, First Edition. Alexander Gates.
© 2022 John Wiley & Sons Ltd. Published 2022 by John Wiley & Sons Ltd.
Companion website: www.wiley.com/go/gates/earthsfury

Words You Should Know:

Body waves – Seismic waves that travel through the volume of the earth in three dimensions including P-waves and S-waves.

Fault-plane solution – A system that determines the character of an earthquake source based upon analysis of seismograph arrivals.

Intensity – The shaking imposed by surface waves on the earth and impacted communities.

Love wave – Surface waves with side-to-side motion.

Magnitude – A logarithmic scale that measures the power of an earthquake or the energy released as determined by seismograph arrivals.

P-wave – Primary waves are the first to arrive at the seismograph. Also known as compressional waves.

Rayleigh wave – Surface waves with vertical and rolling motion like an ocean wave.

Seismograph – A device to measure the power and timing of seismic waves.

Surface waves – Seismic waves that travel along the surface of the earth, including Love and Rayleigh waves.

S-wave – Secondary waves are the second arrivals at a seismograph. Also known as shear waves.

7.1 Earthquake Basics

When rock breaks along a fault or around intruding magma, it releases vibrational waves that can be destructive. This event is an earthquake. There are a number of different types of waves, each with different characteristics. The waves travel in all directions and produce varying amounts of damage. There are many factors that control the power with which these waves affect structures and communities. Further, the study of these waves as recorded on seismograph arrivals can determine the source and location of the earthquake as well as the power of the event. The earthquake basics control types and extent of the earthquake dangers and disasters. For that reason, they should be understood before considering the examples of disasters.

7.2 Earthquake Waves

When an earthquake occurs, the stress that has built up on the fault is converted to seismic energy in the form of waves as the rock breaks. The point of breakage on the fault is the earthquake focus or hypocenter. The point on the surface above the focus is the epicenter. The seismic waves radiate outward from the focus in all directions. The best analogy for imagining the radiation of seismic waves is throwing a rock on a pond. When the rock hits the water, rings of waves radiate away growing larger with time (Figure 7.1).

FIGURE 7.1 The concentric expanding ripples on a pond after a rock is thrown in.
Source: Cheryl Casey/Dreamstime.com.

These waves on a pond are only at the water surface, but in earthquakes, some waves also travel in spheres rather than rings and continue throughout the entire volume of the earth.

There are a number of wave types released by an earthquake. They can be divided into two basic types, body waves and surface waves. Surface waves are like the rings on a pond; they radiate outward from the epicenter on the earth's surface and are by far the most damaging. There are two types of surface waves, Rayleigh waves and Love waves. Rayleigh waves move across the earth's surface like waves on the ocean. The surface rolls in an upright circular motion opposite to the forward motion of the wave as it moves through. The surface, however, returns to its original position after the wave passes (Figure 7.2A). The damaging motion is primarily vertical. Love waves cause the surface to move side-to-side as the waves propagate forward (Figure 7.2B), so the damage is primarily lateral.

By contrast, body waves travel throughout the volume of the earth instead of only along the surface. There are two types of body waves, P-waves and S-waves. P is for primary because these waves are the first to arrive at the seismograph and S is for secondary because they arrive second. P-waves are compressional and move by compressing material in the direction of travel as they pass through (Figure 7.3). Stretching a slinky toy (spring) out and plucking it to generate a wave along its length illus-

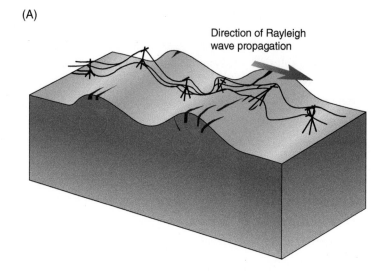

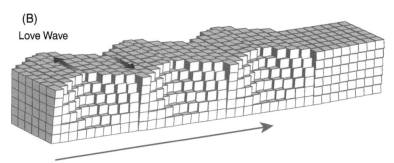

FIGURE 7.2 (A) Block diagram showing the propagation of a Rayleigh wave through the crust. (B) Block diagram showing the propagation of a Love wave through the crust.

trate a P-wave. S-waves cause rock and soil to move up and down perpendicular to the direction of travel (Figure 7.3). This motion is analogous to a stretched out jump rope where the person on one end whips the rope. The wave that travels along the rope shows the motion of an S-wave. The side-to-side motion means the motion includes a shear component.

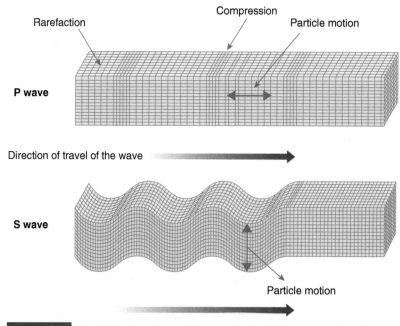

FIGURE 7.3 Block diagrams showing the propagation of a P-wave (top) and an S-wave (bottom) through the crust.

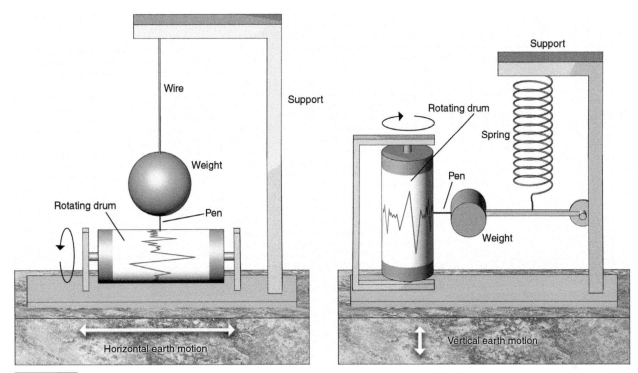

FIGURE 7.4 Schematic diagrams of simple seismographs showing vertical motion vibrations (right) and horizontal motion vibrations (left).

All waves pass through the earth at a velocity dictated by the type of wave and the material that it is passing through. In the same material, P-waves are the fastest, S-waves are second fastest, and surface waves are the slowest. In general, waves move fastest in crystalline rock where crystals are knitted tightly together. Sedimentary rocks are slower, and unconsolidated sediments and soils are even slower. Fluids, like water, are still slower and they do not transmit S-waves.

All waves are detected on a seismograph, also called a seismometer, or a geophone. The arrivals discussed in the preceding text are the seismic waves arriving at a seismograph. The earliest seismograph was a small statue in China in the second century AD in which balls fell from the mouths of dragons into the mouths of frogs when seismic waves passed through. There were other variably effective seismographs or seismoscopes over the centuries, but they were not reliable until the late nineteenth and early twentieth centuries.

The seismograph is fixed to the ground to detect the waves and vibrations, but they must be compared to a fixed location. They must also record continuously. The early seismographs had rotating drums attached to the ground that were marked by a pen suspended by a wire, spring, and shock absorber to keep it isolated from the waves (Figure 7.4). As technology advanced, the sensing devices were replaced with electronics and three rotating drums were used to track motions in all three directions because seismic waves can shake the ground in many ways (Figure 7.5). Later, seismographs became completely digital and the rotating drum was eliminated, so that monitoring of the waves would just be recorded directly to computer memory.

The paper on the rotating drum would have to be changed every 24 hours because the pen would progress across the entire page in that time. The removed paper records are called seismograms as are the digital readouts of the seismic monitoring. This is basically a graph of amount of vibration versus time. When an earthquake occurs, depending upon distance, power, and material through which waves travel, the seismograph detects some combination of the seismic waves. If all waves are detected, the first arrival is the P-wave, the second arrival is the S-wave, and the third arrival is the surface wave, and they can be easily discerned on the seismogram (Figure 7.6). The surface

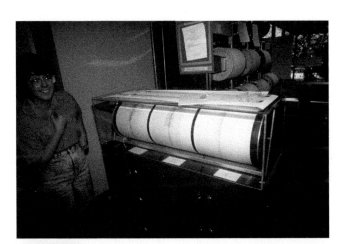

FIGURE 7.5 Photograph of a three-drum seismic system showing the monitoring of vibrations in three dimensions. *Source*: J.K. Nakato – US Geological Survey.

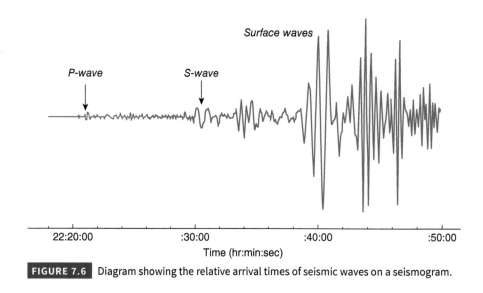

FIGURE 7.6 Diagram showing the relative arrival times of seismic waves on a seismogram.

waves attenuate quickly depending on the material they pass through, so they are only seen near the earthquake and there is a shadow zone in the earth where S-waves do not appear. Some seismograms, therefore, only include P-waves.

7.3 Foreshocks, Main Shocks, and Aftershocks

Earthquakes typically occur in clusters or sequences. The reason for the multiple earthquakes is the fault mechanics that produces them. When a fault slips to produce an earthquake, the stress is released on the area that slipped, but it increases on other areas of the fault, especially adjacent to the slipped area. These adjacent areas also slip, producing earthquakes as a result. This slip, in turn, increases the stress on other parts of the fault. Once the local stress on the fault does not exceed the strength of the rock, the sequence of earthquakes ends. However, by the end of the sequence, a large part of the fault may have slipped, if not the entire fault as well as a number of subsidiary faults.

This sequence of earthquakes can be divided into three categories, foreshocks, main shock, and aftershocks. These may need to be assigned after much of the sequence is past. The main shock is the most powerful earthquake in the sequence. If there are two main shocks, then it is a doublet earthquake or main shock. Once the main shock is identified, then every earthquake before it is a foreshock and every earthquake after it is an aftershock. If the first earthquake in a sequence has a magnitude of 5, the classification is unknown. If all following earthquakes have magnitudes less than 5, then they were the main shocks. If the next earthquake has a magnitude greater than 5, then the first earthquake was a foreshock.

In some cases, there is only one earthquake which is the main shock. This is common for smaller earthquakes. With larger earthquakes, aftershocks can persist for a year or more. In the early stage of the aftershock sequence, they tend to be strong and in rapid succession. With time, the earthquakes get weaker and less frequent. In some cases, another strong earthquake occurs in this sequence and the aftershocks increase again. This new strong earthquake is considered to be a separate event and a new main shock.

7.4 Magnitude and Intensity

The magnitude and intensity of an earthquake are two measures of its power and destructiveness. An equation to quantify earthquakes based on the information in the seismogram was developed by Charles Richter in 1935. He defined the magnitude of an earthquake by the waves observed on a seismogram using terms from a typical sine wave from physics (Figure 7.7).

In very general terms, the formula is a log function of the amplitude of the wave divided by the period of the wave, adjusted for distance from the earthquake (Figure 7.8). The log term means that each higher number in magnitude has 10 times as much power. That means that a magnitude 6 earthquake is 10 times as powerful as a magnitude 5 earthquake, and 100 times as powerful as a magnitude 4 earthquake. The amplitude is the amount of movement the wave moved the ground surface, but it is divided by period or the time it takes for the wave to pass through.

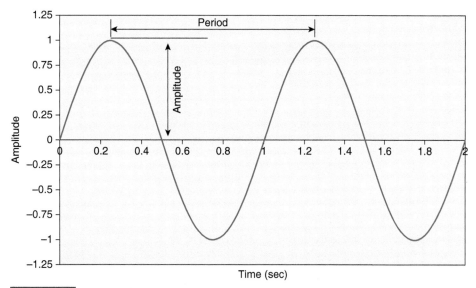

FIGURE 7.7 Diagram of a sine wave showing the quantities of amplitude and period which are used in the calculation of magnitude.

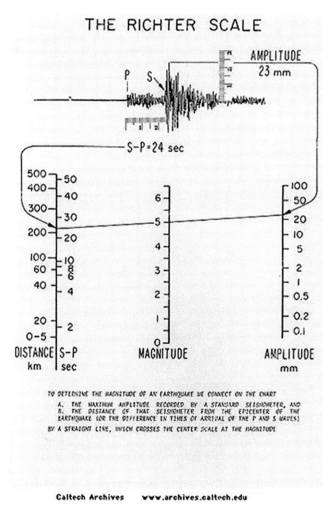

FIGURE 7.8 Diagram showing how Richter magnitude is determined based upon wave amplitude from a seismogram, and distance from the earthquake determined by the delay between the arrivals of P- and S-waves. *Source*: The Richter Scale, Caltech Archieves, California Institute of Technology. Licensed under CC BY-4.0.

The seismograph registers a lower magnitude of the earthquake the farther away it is from the focus. The reason for this can be illustrated with Figure 7.1. When the rock hits the pond, it creates a ring of wave on the water with all the energy it will ever have. The circumference of the ring is small at that point, so energy per length of circumference is high. It has a high density of energy, and, therefore, the wave is high. But as the ring grows, the circumference grows rapidly. Now the energy density is much smaller as is the wave height because the energy is the same or less and it is spread out across a much longer length. Due to this and the surface friction, the energy of the surface waves drops off quickly and the damage area is constrained.

This phenomenon is even more pronounced with body waves. They spread out in a sphere, which attenuates the energy even quicker. However, there is less friction in bedrock than at the surface, so they can persist for long distances. This is the reason that the distance factor must be included in the Richter magnitude calculation. The difference in travel speed of the three wave types means that the time delay between them is a reflection of the distance. This time delay can even be felt. The surface waves do the most damage. They feel like the rattling that occurs if a big truck is driving near your house. The windows can rattle as can small objects on shelves. Close to the earthquake, the waves are so close together that you can feel the surface waves at the outset. If the rattling starts first, it means you are close to the epicenter and it is urgent for you to seek shelter. By contrast, body waves tend to make the surface sway slowly. If you feel them first, it means that you are at a distance from the epicenter and you will have more time to seek shelter if you need it at all.

Earthquake magnitude previously referred to Richter magnitude, but now it refers to moment magnitude (Table 7.1). Moment magnitude of an earthquake is calculated by measuring the seismic moment. This is the measure of the power of an earthquake based on the leverage of forces across the area of rupture of the fault plane. Seismologists integrate data from multiple seismic

TABLE 7.1 Magnitude scale for earthquakes and their characteristics.

Magnitude	Label	Frequency (eqs/yr)	Equivalent power (trinitrotoluene, TNT)	Possible impacts
0–2.9	Micro	>100 000	≤4000 lb (1800 kg)	Not felt, detected on sensitive instruments
3–3.9	Minor	100 000–12 000	4000–12 300 lb (1800–5600 kg)	Felt by many, no damage
4–4.9	Light	2000–12 000	12 300–4 million lb (5600–1.8 million kg)	Felt by all, breakage of minor objects
5–5.9	Moderate	2000–200	4–123 million lb (1.8–53 million kg)	Slight structural damage possible
6–6.9	Strong	200–20	0.123–4 billion lb (0.056–1.8 billion kg)	Moderate damage, loss of life
7–7.9	Major	20–3	4–123 billion lb (1.8–56 billion kg)	Severe damage over large areas, heavy loss of life
>8	Great	<3	>123 billion lb (56 billion kg)	Extreme destruction and loss of life

stations to determine a single, more representative magnitude measure. Table 7.1 illustrates the relations among magnitudes, frequency, and equivalent power.

Another measure of the power of an earthquake is the intensity (Table 7.2). Surface waves cause shaking that can cause severe damage. The damage can be qualitatively evaluated to determine the intensity of the shaking of an earthquake. A scale was developed by the Italian scientist Mercalli and it had 10 levels. In 1902, the scale was expanded to 12 levels. In 1931, it was redefined and renamed to the Modified Mercalli Scale which is still used today. It is the most used of the qualitative scales. The scale is ranked in Roman numerals determined by the factors in Table 7.2.

TABLE 7.2 Earthquake intensity factors of the Modified Mercalli Scale.

Mercalli intensity	Description	Witness observations
I	Not felt	Felt by very few people, barely noticeable.
II	Weak	Felt by a few people, especially on upper floors.
III	Weak	Noticeable indoors, especially on upper floors, but may not be recognized as an earthquake.
IV	Light	Felt by many indoors, few outdoors. May feel like heavy truck passing by.
V	Moderate	Felt by almost everyone, some people awakened. Small objects moved. Trees and poles may shake.
VI	Strong	Felt by everyone. Difficult to stand. Some heavy furniture moved, some plaster falls. Chimneys may be slightly damaged.
VII	Very strong	Slight to moderate damage in well-built, ordinary structures. Considerable damage to poorly built structures. Some walls may fall.
VIII	Severe	Little damage in specially built structures. Considerable damage to ordinary buildings, severe damage to poorly built structures. Some walls collapse.
IX	Violent	Considerable damage to specially built structures, buildings shifted off foundations. Ground cracked noticeably. Wholesale destruction. Landslides.
X	Extreme	Most masonry and frame structures and their foundations destroyed. Ground badly cracked. Landslides. Wholesale destruction.
XI	Catastrophic	Total damage. Few, if any, structures standing. Bridges destroyed. Wide cracks in ground. Waves seen on ground.
XII	Enormous	Total damage. Waves seen on ground. Objects thrown up into air.

The Modified Mercalli Scale is a purely observational scale that is specific to an area for a specific seismic event. Attempting to compare intensity measurements from different areas or different times has several major limitations. It requires a human settlement and witnesses of the event to assign intensities. Further, building quality and codes vary by area. Poorly constructed buildings can be more strongly impacted by seismic waves than well-constructed buildings, so a single earthquake may impose intensity VII in one area but intensity X in the same town. Various types of rock, sediments, and soils transmit waves differently. Areas underlain by wet sediments, soil, or fill tend to amplify surface waves resulting in worse destruction. Distance from the focus also impacts the amount of shaking; the greater the distance, the weaker the shaking.

Magnitude measures the power of an earthquake regardless of all contributing factors in comparison with intensity which measures impact on humans and damage to their settlements. Although both quantities are applied to events, their relationship is not as dependent as might be expected. Small earthquakes might produce high intensity numbers and enormous earthquakes can occur in unpopulated areas, so produce low intensity.

7.5 Determining the Epicenter and Movement on Faults

Seismograms from multiple seismographs can be used together to determine the location of the focus and even the type of fault movement. Locating the focus or epicenter of a specific earthquake can be accomplished using a technique called triangulation. This method requires three seismographs at different geographic positions around the earthquake. If the distance is great enough, each seismogram will show separate P- and S-wave arrivals. If the P-wave and S-wave velocities in the crust are known between the focus and seismograph, the distance between them can be calculated. This is accomplished by way of plotting the time delay between arrivals of the P- and S-waves on a P- and S-wave travel-time curve for the area. These time delays correspond to distances that the waves traveled (Figure 7.9). That means the earthquake could have occurred anywhere along a circle with a radius equal to the distance determined on the travel-time curve. This clearly is not a tight enough constraint to determine the active fault. However, if the same exercise is done using three seismographs, a single location for the earthquake can be determined (Figure 7.10).

The example shown in Figure 7.10 shows seismograph stations that are distant from the earthquake making the circles large. In this situation, the margin of error is large. This makes it difficult to determine depth of focus, so only the epicenter can be determined. Further, in an area with multiple close-spaced faults, even the active fault may be difficult to identify. However, if there are multiple stations in the vicinity of the earthquake, the resolution of triangulation can be much finer. In that case, specific active faults can be identified (Figure 7.11) and even depth of focus is possible. These simplistic solutions illustrate the basis of even current methods. However, modern analysis involves rapid computer solutions of all of the seismograms in the epicentral and even distant areas.

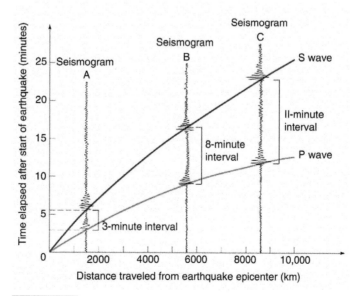

FIGURE 7.9 Diagram showing travel-time curves for P- and S-waves for an area. The delay between P- and S-waves determines the distance of the three seismographs from the earthquake.

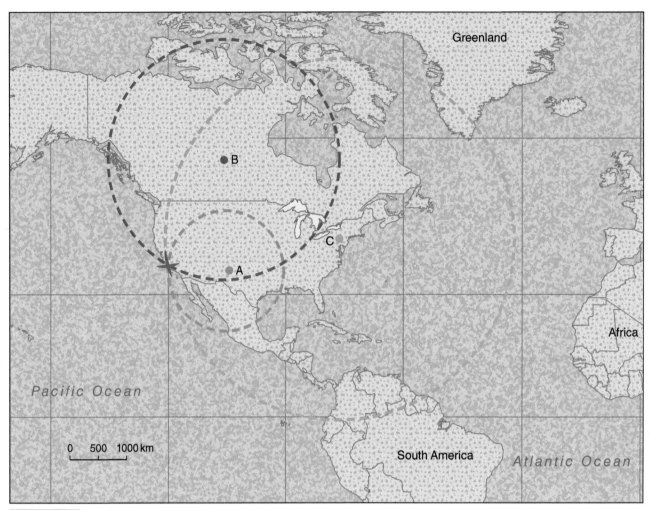

FIGURE 7.10 Map of North America with the circles of travel distance drawn around each seismograph like that determined in Figure 7.9 showing the earthquake epicenter to be in Southern California.

Fault-plane solution or focal mechanism analysis of multiple stereograms around the epicenter of an earthquake can even determine the movement on a fault that produced the earthquake. This technique relies on the spatial distribution of first motions on stereograms. The first motion is the direction of ground movement upon arrival of the P-wave at the seismograph and is recorded on seismograms of the earthquake. The seismograph needle either moves up or down and the direction it moves yields information on how the fault moved (Figure 7.12). In simple terms, a line can be drawn perpendicular to the fault line at the epicenter called the auxiliary plane. The crossing of the fault and auxiliary plane forms four geographic quadrants around the earthquake epicenter. One of two first motions, up or down, is recorded depending upon which quadrant the seismograph is located. The upward first motion of the needle reflects compression from that quadrant. The wave in that quadrant pushes the rock toward the seismograph. Downward first motion in a quadrant reflects dilation as the wave pulls the rock away from the seismograph.

These first motion studies are the bases to construct fault-plane solutions, which determine the source mechanism of an earthquake. These are circular diagrams that look like beach balls at various determinative angles (Figure 7.13). The angles of beach balls show the motion on the fault that produced the earthquake. If the beach ball has an "X" (on the end of the ball), the fault moved in a strike-slip sense and the filled versus white quadrants determine if it was left-lateral or right-lateral. If the beach ball shows the sides, so contains arc lines as opposed to "X," the fault moved in a dip-slip sense. The filled quadrants determine whether the fault was reverse or normal. If the "X" is off-center, the fault had oblique-slip motion. Figure 7.13 shows block diagrams of the various faults and the compressed versus extended areas in each type. It also shows the corresponding fault-plane solution for each type including compressed versus extended areas.

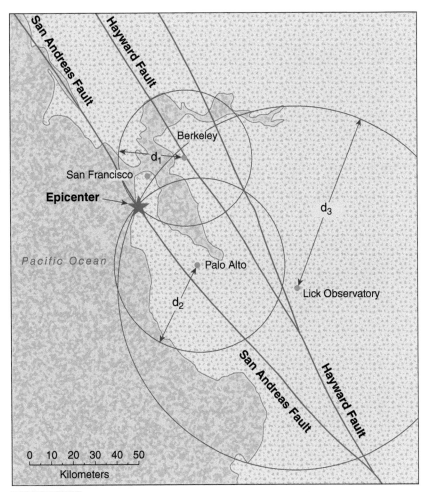

FIGURE 7.11 Much more focused triangulation exercise showing that with the right choices of seismographs, the active faults can be identified even in complexly faulted areas. *Source*: Adapted from figure by Stephen A. Nelson, https://www2.tulane.edu/~sanelson/eens1110/earthint.htm.

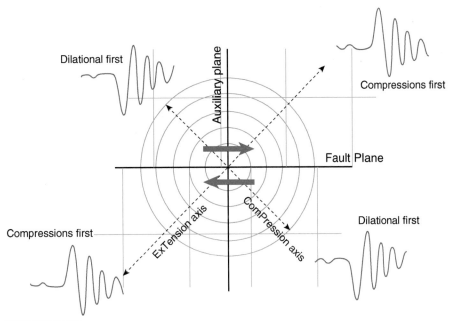

FIGURE 7.12 Diagram of a fault plane and its auxiliary plane dividing the area into quadrants of initial downward and upward needle movements on the seismograph. These reflect areas of dilation and compression, respectively.

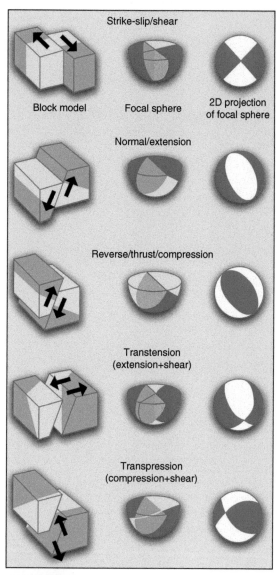

FIGURE 7.13 Diagram showing block diagrams of five fault-earthquake scenarios and the resultant fault-plane solutions on the beach balls. The red areas on the block diagrams show areas of compression, whereas the gray areas show dilation. These are directly transferrable to the corresponding areas on the fault-plane solutions.

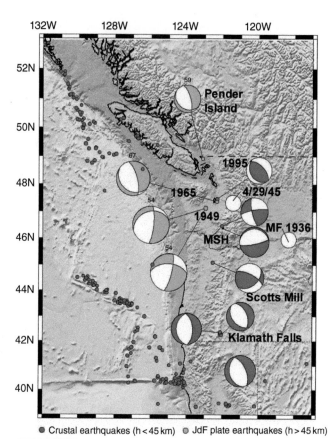

FIGURE 7.14 Map of the Northwest Coast of the United States showing fault-plane solutions for a number of earthquakes in the area that are both shallow (crustal) to the west and deep and related to the subduction zone to the east. Use Figure 7.13 to determine the type of fault movement that created each of the earthquakes. *Source:* Image courtesy of USGS.

Once the focal mechanisms are determined for a sequence of earthquakes in an area, they can be used to characterize the events. Maps can be constructed with each focal mechanism beach ball plotted for each earthquake of the sequence (Figure 7.14). By considering the spatial relations and movement sense of the earthquakes in a sequence, the fault mechanics of the entire tectonic event can be deciphered. This is a very powerful tool in understanding the behavior of the crust during deformation.

7.6 Earthquake Risk Maps and ShakeMaps

To address the seismic risk that many Americans are exposed to in a number of areas of the country, the US government established the Advanced National Seismic System (ANSS). This program provides accurate and timely information on seismic events including the impact on buildings and other infrastructure. The ANSS is a component of the National Earthquake Hazards Reduction Program (NEHRP). The ANSS is designed to encompass 7000 seismic stations with dense concentrations at 26 designated high-risk urban centers across the United States. Functions of ANSS include real-time monitoring of seismicity, thorough analysis of seismic events, and automatic broadcasts of seismic hazards in real time.

Another program to analyze seismicity in the United States is ShakeMap, which is a tool for earthquake evaluation that provides near-real-time maps of ground motion and shaking intensity. When an earthquake strikes an area, emergency management personnel must quickly decide where to focus their resources and efforts to best address the disaster. An overwhelming amount of information is received and can only be processed in a cursory manner. Because ground shaking causes the most damage to man-made structures, knowing where the ground shakes the most and to what degree determines the areas of highest damage. ShakeMap is a map that shows the distribution of relative shaking intensity and thus the areas of highest risk for damage. They are constructed using geologic base maps and experience from previous earthquakes. As the seismic waves are still impacting an area, rescue teams can mobilize to best reduce the damage and shave precious minutes off response time to save more lives.

CHAPTER 8

Earthquake Hazards

Earth's Fury: The Science of Natural Disasters, First Edition. Alexander Gates.
© 2022 John Wiley & Sons Ltd. Published 2022 by John Wiley & Sons Ltd.
Companion website: www.wiley.com/go/gates/earthsfury

Words You Should Know:

> **Doublet earthquake** – An earthquake sequence in which there are two main shocks.
> **Fire storms** – Surface waves can cause fire-bearing objects and devices to fall over and ignite everything around them, leading to widespread blazes. When the fire becomes intense and involves strong updrafts, it can become a storm.
> **Hammering** – Shaking can cause adjacent tall buildings to sway and collide with each other multiple times, destroying both.
> **Liquefaction** – Shaking drives groundwater toward the surface in certain deposits that liquefies the soil and sediment in the shallow subsurface.
> **Natural frequency** – Tall buildings can have a natural frequency where seismic waves can cause them to ring like a tuning fork and break apart.
> **Slope failure** – Any number of mass movements from avalanches to slump resulting from shaking caused by seismic waves.

The surface waves generated by an earthquake are capable of extreme destruction. Although there are other possible destructive effects of earthquakes, such as tsunamis, this chapter focuses on the hazardous results of shaking. There are several phenomena that are produced by earthquake waves which are illustrated here by case studies of actual events. The hazards with the most variability are slope failure and mass movements. For this reason, there are several examples of extreme slope failure. The science is included in the stories.

CASE STUDY 8.1 1985 Mexico City Earthquake

Soil Amplification, Hammering and Natural Frequency

The geology underlying an area impacts how the surface waves will be amplified or subdued and ultimately affect the buildings and infrastructure during an earthquake. Whereas solid bedrock presents a challenge to seismic wave transmission, some softer deposits shake like gelatin when surface waves pass through. This is called soil amplification of seismic waves and it is a danger. An earthquake which appears to be causing little damage can pass into soft sediments and suddenly cause great damage. This was the case for the Mexico City earthquake. On 19 September 1985, a devastating earthquake of magnitude 8.0 struck with an epicenter just offshore of Michoacán, Mexico in the Pacific Ocean (Figure 8.1). The duration of the event was between three and four minutes. It was generated by the subduction of the Cocos Ocean Plate beneath the North American Plate in an ocean–continent collision at the Middle America Trench. Movement along the fault was about 10 ft (3 m) in reverse motion (megathrust) with a focus at a shallow 12-mi (20 km) depth.

The waves traveled 220 mi (350 km) eastward before reaching the capital of Mexico City approximately two minutes later. A large part of the city is located on the ancient Lake Texcoco which was drained by the Spanish. The sediments in the middle of this lakebed are primarily water-saturated silts and volcanic clays. When the surface waves passed through them, the waves were amplified, especially as they passed under the buildings of the downtown area. The Rayleigh waves caused the base of buildings to rise faster than the top, and middle floors were crushed (Figure 8.1A). Love waves caused buildings to be sheared and topple over sideways. Adjacent tall buildings experienced hammering in which they swayed wildly, colliding at the higher floors and leaving both buildings badly damaged or destroyed. The passing waves were at the natural frequency of the lakebed sediments, which caused some tall buildings to undergo harmonic vibrations which shattered them (Figure 8.1B). Unless damped, objects vibrate at a specific natural frequency if struck by a force or impulse. If a string on a guitar is plucked or a tuning fork struck, the noise is the natural frequency. The problem arises if there is constructive interaction of the natural frequency of a building with the frequency of earthquake waves. A series of vibrational nodes form which can cause the building to shake itself apart. In addition, groundwater was driven to the surface by the shaking causing liquefaction which undermined the foundation of a number of buildings (Figure 8.1C). The downtown area of Mexico City was demolished as a result of these compounding effects with 412 buildings completely destroyed and 3124 seriously damaged.

The death toll of this event is officially estimated at 10 000 people, but only 5000 bodies were recovered. However, the actual death toll could be closer to 30 000 and one group even estimates it at 45 000 with the missing taken into account. Even at the higher numbers, this is a very low number for a city of 14 million people struck by a magnitude 8 earthquake and the damage it inflicted. The reason that the death toll was not higher is because of the timing of the event. Basically, not many people were in the downtown area because it was too early in the morning. If it had been a short time later and the downtown was populated with commuters and/or workers, the death toll could have been in the hundreds of thousands.

CHAPTER 8 Earthquake Hazards

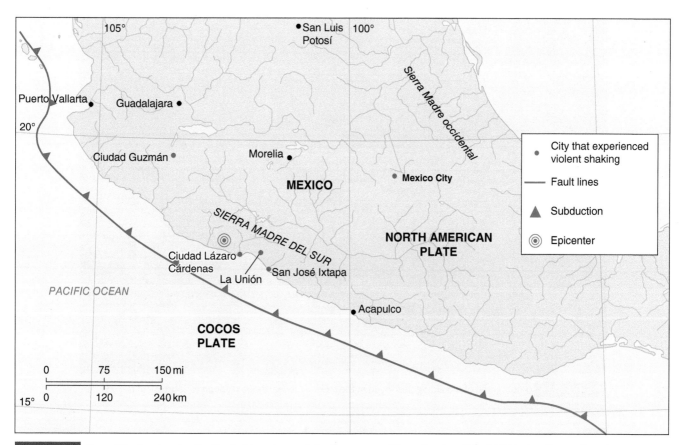

FIGURE 8.1 Map of Mexico showing the Central American subduction zone and the epicenter of the 1985 Mexico City earthquake. *Source*: Adapted from Mexico City earthquake of 1985. Encyclopædia Britannica, Inc.

FIGURE 8.1A Photograph of a building with collapsed middle floors in the Mexico City earthquake.
Source: Courtesy of the US Geological Survey.

FIGURE 8.1B Photograph of a building destroyed in the Mexico City earthquake. *Source*: Courtesy of the US Geological Survey.

FIGURE 8.1C Photograph of a building with the foundation destroyed in the Mexico City earthquake. *Source*: Courtesy of the US Geological Survey.

The problem is that the next time the area might not be as lucky. In fact, the population occupying this dangerous lakebed area has now swelled to 32 million. On the other hand, the building codes have been made much more stringent to address the amplification of surface waves. An early warning system has been installed that senses large earthquakes in the subduction zone and sends the information to Mexico City, so it can be evacuated. Safety and evacuation drills are commonplace at schools and throughout the city to act on those warnings. It will be seen if these precautions are enough.

CASE STUDY 8.2 1964 Niigata Earthquake, Japan

A Case of Extreme Liquefaction

The residents and officials of Niigata, Japan were well aware that the area was earthquake-prone and major events could strike. They took precautions by constructing new apartment buildings that were engineered to be earthquake-resistant to ensure their safety. Structural reinforcement of these buildings improved rigidity which made them able to withstand significant shaking from strong surface waves. The buildings were quickly occupied by residents who felt that they were protected. However, everyone had overlooked the process of liquefaction.

A major earthquake struck the area at 1:01 p.m. on 16 June 1964 with a powerful magnitude of 7.5–7.6. The epicenter was just 30 mi (50 km) north of Niigata below the Sea of Japan near the island of Awashima (Figure 8.2). The focus was at a depth of 35.4 mi (57 km) on a thrust fault on the eastern margin of the Japan Sea Mobile Belt extending from Sakhalin to Niigata. This fault is part of a developing trench along the Eastern Sea of Japan. This trench is part of a subduction zone of ocean crust beneath the Eurasian Plate. The earthquake-producing fault lies N15°E and dips 11° to the east.

The surface waves generated a shaking intensity of VIII on the Modified Mercalli Scale, which ranks as severe. The earthquake also generated tsunamis that reached a maximum height of more than 13 ft (4 m) on the coast near the epicenter and a height of 6 ft (1.8 m) in Niigata. However, the most notable effect was liquefaction.

Liquefaction is a phenomenon in which the shaking of the shallow earth by surface waves forces groundwater upward toward the surface. Water can actually be shot out of the ground in fountains in strong earthquakes and it can entrain so much mud and fine sediment that it produces mud volcanoes above these vents or pipes. However, it more commonly involves the saturating of soils and surface sediments. Saturation is when the pore spaces between soil and sediment particles are completely filled with water which is injected into them under high pressure. This causes the strength and stiffness of the soil and sediment to be greatly reduced so that it can no longer support buildings and other structures. As a result, it takes on the consistency

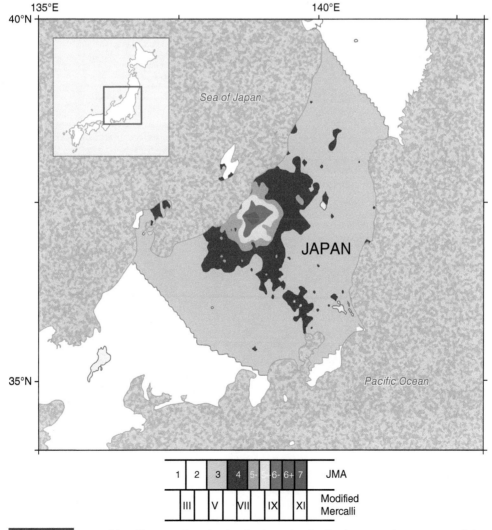

FIGURE 8.2 Map of the Niigata earthquake showing intensity distribution. Inset shows Japan and the location of the map. *Source*: Adapted from Mori and Somerville (2006).

of quicksand or quick mud like a fluid and the upward pressure of the water can cause it to flow out onto the surface.

The 1964 earthquake caused severe damage along the Shinano River in Niigata city as a result of soil liquefaction. This area consists of loose sandy soils in the top 33–50 ft (10–15 m) and a water table around 6.5 ft (2 m) deep. The ground surface in this area is essentially flat. Yet, even on these very gentle slopes of less than 1% grade, the lateral flow of the surface exceeded 13 ft (4 m) from the liquefaction in some areas. In addition, the surface waves caused the soft, saturated soil and sediment to shake violently. The peak surface acceleration in these soft sediments was a strong 0.16 g.

There were two main structural features that were affected in this area. A major bridge over the river collapsed about 70 seconds after the beginning of the earthquake (Figure 8.2A). The liquefaction-induced failure of the footings of the bridge locally and the lateral spreading pushed the bridge foundation. The slight shift of the superstructure caused the roadway segments to rotate and drop into the river.

The other problem was the newly constructed apartment buildings. The Kawagishi-cho Complex contained eight, four-story buildings that were approximately 98 ft (30 m) by 26 ft (8 m). They were built on reclaimed land that was largely unconsolidated fill over more unconsolidated sediments of a recent delta. The earthquake caused severe soil settlement under the complex. Although these buildings were designed and constructed to withstand violent shaking from an earthquake, they were built with shallow foundations. When the soils beneath their foundations turned to "quicksand," these reinforced concrete buildings sank into the earth and tilted at angles of up to 80° from vertical (Figure 8.2B). These expensive buildings had to be destroyed. Since then, earthquake-resistant and most other construction includes

FIGURE 8.2A Photograph of collapsed bridge in the Niigata earthquake. *Source*: Courtesy of the US Geological Survey.

FIGURE 8.2B Photograph of earthquake-resistant buildings tilted over by liquefaction in the Niigata earthquake. *Source*: Courtesy of the US Geological Survey.

deep pilings installed into the sediments underneath multistoried buildings. This has largely solved the problem.

The number of victims from the earthquake was fortunately small with 36 dead or missing and 385 injured. The damage to buildings, on the other hand, was significant. In all, there were 3534 houses destroyed, and 11 000 houses damaged from the liquefaction, tsunami, and the surface waves.

CASE STUDY 8.3 1923 Great Kantō Earthquake

Fire Storms

Some of the hazards associated with earthquake have changed as technology and society have changed depending upon the area of the world. One of the most prominent changes is the threat of fire. Before about 75–100 years ago, depending upon location, fire was the main source of cooking and heating, first with wood and later with the addition of coal. Shaking from the surface waves could knock over stoves and furnaces which was devastating especially for wood houses. In many natural disasters, it was the fire and not the seismic waves that caused the most damage.

One of the most horrific examples of the destruction caused by fire was the Great Kantō earthquake of Japan. At 11:59 a.m. on 1 September 1923, a massive earthquake struck the Sagami Bay, approximately 30 mi (66 km) southeast of Yokohama and 60 mi (100 km) from Central Tokyo (Figure 8.3). The focus was at an estimated depth of 14 mi (23 km), which is relatively shallow, so more damaging. The source was a megathrust of the Philippine Sea Plate subducting beneath Japan and the rupture area is estimated to have been 60 mi (100 km) by 60 mi (100 km).

When the main shock struck, it was so powerful that it badly damaged the local seismograph, so the magnitude can only be estimated at between 7.9 and 8.2. Reliable seismographs had only recently been installed. The duration of the shock can also only be estimated at 4 minutes, but the reported range is 48 seconds to 10 minutes. The surface wave intensity is better determined at XI or extreme on the Modified Mercalli Scale. This extreme shaking in such a populated area was damaging enough, but because the epicenter was under water, a large tsunami was generated that struck the shore areas just a few minutes later with waves as high as 30 ft (9 m). The uplift and depression from the earthquake altered the bathymetry of Sagami Bay.

The surface waves, slope failures, and tsunamis did enough damage to make this a major disaster on its own (Figure 8.3A). However, because it struck at noon, many people were cooking lunch on open fires which

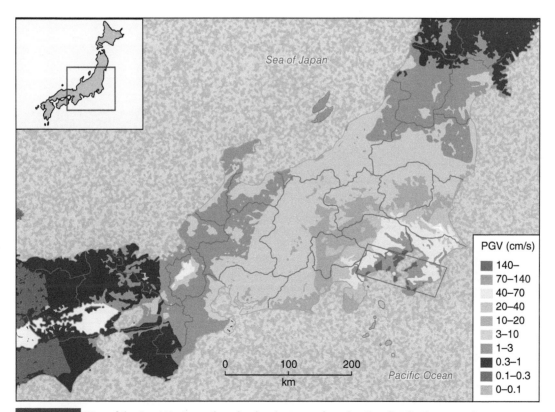

FIGURE 8.3 Map of the Great Kantō earthquake showing ground acceleration distribution. Inset shows Japan and the location of the map. *Source:* Adapted from Wakamatsu and Matsuoka (2006).

FIGURE 8.3A Photograph of Tokyo destroyed in the Great Kantō earthquake. *Source*: Courtesy of the US Geological Survey.

were overturned by the shock. Fires quickly engulfed both Tokyo and Yokohama. The earthquake had severed the water lines and fire brigades could not pass through the debris-covered streets allowing the developing inferno to burn out of control (Figure 8.3B). Many people were killed outright by the earthquake and tsunami, but the refugees scrambled to avoid the fire and many perished in the flames and damage caused by them. In Tokyo, the fire was so hot that firestorms developed and even a 300-ft (95 m) high fire tornado formed that was strong enough to pick people and large objects high into the air. Survivors quickly gathered in an army depot and city park, away from the burning buildings for safety, but a firestorm swept down on them killing 35 000–40 000 of them and leaving only a few hundred badly burned survivors.

FIGURE 8.3B Photograph of Tokyo destroyed in the Great Kantō earthquake. *Source*: Courtesy of the US Geological Survey.

The death toll for the disaster is estimated between 105 000 and 143 000 people. It left both cities in ruins. The strong shaking damaged an estimated 75% of the buildings in Tokyo and the fire incinerated 67%. This destruction left 1.9 million people homeless.

The disaster also had many social repercussions. At this point in the history of Tokyo, the middle class, who lived in the topographically lower part of the city, had begun to gain power and influence from the rich who lived in the elevated areas (Figure 8.3C). However, the damage was much greater in the lower city and all political headway was lost (Figure 8.3D). In addition, there was a growing community of Koreans in Tokyo at the time. Many of the Japanese residents strongly opposed their presence in Tokyo. They took advantage of the chaos to persecute the Koreans including execution and lynching of several hundred residents. The disaster led to a much less-open society that was much more pro-Japanese and strict. This change in attitude ultimately carried into Japan's nationalism and their aggressiveness in World War II.

FIGURE 8.3C Photograph of buildings in Tokyo destroyed in the Great Kantō earthquake. *Source*: Courtesy of the US Geological Survey.

FIGURE 8.3D Photograph of the city of Tokyo destroyed in the Great Kantō earthquake. *Source*: Courtesy of the US Geological Survey.

CASE STUDY 8.4 1976 Tangshan Earthquake, China

Doublet Main Shock

The most dangerous kind of earthquake event is if there are two or more major main shocks. This was the case with perhaps the deadliest earthquake in history. The two Tangshan earthquakes struck the very highly populated area of Northeastern China, just 85 mi (137 km) east of the densely populated city of Beijing (Figure 8.4). The first earthquake occurred at 3:43 a.m. on 28 July 1976 with a moment magnitude of 7.6. The duration of the event was about two minutes. The focus of this main shock was a shallow 5.5 mi (12 km) directly beneath the city of Tangshan, so the surface waves were at near maximum power. The damage from shaking reached a maximum intensity of XI on the Modified Mercalli Scale, which is extreme. A red and white glow of earthquake light was observed a distant 200 mi (322 km) from the epicenter. The source of such earthquake light is still debatable, but it is typically not seen for such great distances or at such intensity.

The rupture that produced the first event was the northeast-oriented Tangshan Fault with right-lateral strike-slip movement. The stress that drove this movement was caused by escape tectonics driven by the ongoing Himalayan collision to the southwest. Resulting surface offset extended for 3.6 mi (8 km) through the center of the city. Significant subsidence occurred along this trend that caused extensive damage to the transportation networks (Figure 8.4A). This initial shock damaged or destroyed nearly 85% of the houses in the city, sending survivors into the streets in the middle of the night.

Residents began returning to their homes during the following day to assess damage and recover personal items when, at 6:45 p.m., the second event struck. That earthquake measured a surface magnitude of 7.4 (7.0 overall). The epicenter was a close 32 mi (70 km) to the east-northeast of the first event with a shallow focus at a depth of 7.6 mi (16.7 km). The northwest-oriented fault that generated this event crossed the Tangshan Fault as a conjugate left-lateral strike-slip fault. Shaking was nearly as intense as the first main shock and many of the buildings that remained from the first event were destroyed by the second main shock causing even more casualties (Figure 8.4B).

The two closely spaced major earthquakes caused such panic that people were afraid to sleep indoors and as many as six million people slept outdoors in temporary shelters for more than two weeks.

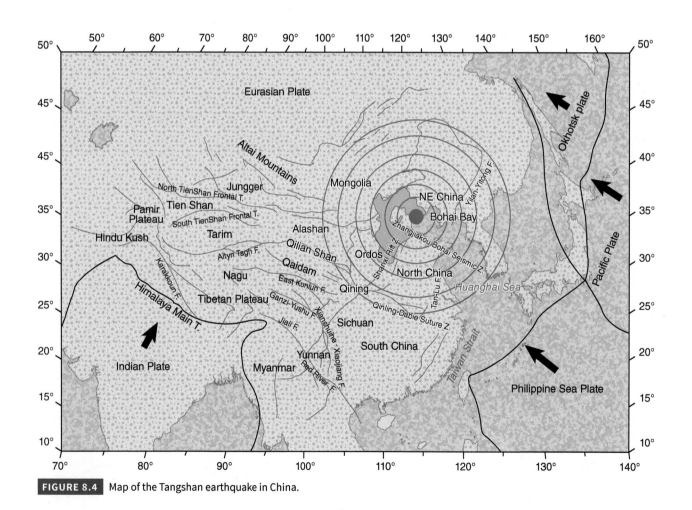

FIGURE 8.4 Map of the Tangshan earthquake in China.

FIGURE 8.4A Photograph of a bridge destroyed in the Tangshan earthquake. *Source*: Courtesy of the US Geological Survey.

FIGURE 8.4B Photograph of buildings destroyed in the Tangshan earthquake. *Source*: Courtesy of the US Geological Survey.

The devastation to the city was extreme (Figure 8.4C) with loss of utilities and transportation and the collapse of many of the coal mines in the area. Because this earthquake occurred during the height of the Cold War, details of the disaster were not widely disseminated. The communist Chinese government refused to release accurate information for fear that it would show weakness and encourage their enemies to take advantage of them. However, they could not completely deny the disaster. They released an official death toll of 242 000, but later studies of the event suggested a death toll closer to 655 000. If this later number is correct, it may have been the greatest natural disaster in history.

FIGURE 8.4C Photograph of destruction in the Tangshan earthquake. *Source*: Courtesy of the US Geological Survey.

CASE STUDY 8.5 1960 Agadir Earthquake, Morocco

Danger of Long Recurrence Times

Urban areas in developed countries that are subjected to regular seismic events tend to be better prepared. The weaker buildings become unstable with regular shaking and are replaced. Building codes tend to be updated to incorporate the latest earthquake hazard reduction technology, so newer buildings tend to be more resistant to shaking. This means that cities in areas with seismic potential but with infrequent activity can have particularly devastating earthquakes even in low-magnitude events. If the city is in a less-developed country, the situation can even be worse.

This was the case in the small city of Agadir, Morocco on 29 February 1960 at 11:40 p.m. when a strong earthquake struck (Figure 8.5). The tremor had only a moderate magnitude of 5.8 from a focal source at a shallow depth of 6.8 mi (15 km). The epicenter was directly beneath the city, which is the worst location for damage. The duration of the tremor was less than 15 seconds. Shaking had an intensity of X on the Modified Mercalli Scale, which is considered extreme. The earthquake produced a tsunami that damaged the coastal area.

The entire old city of Agadir was destroyed as were 70% of the modern buildings (Figure 8.5A). The old buildings were constructed from compressed local clays which offered no resistance to shaking. The new buildings were built from 1945 to 1955 of reinforced concrete, but the strength was variable and also commonly badly damaged (Figure 8.5B). This made the streets impassable (Figure 8.5C). The water mains were ruptured as well making firefighting efforts ineffective and allowing fires to burn unchecked. Sewers ruptured allowing rats to emerge from them and spread throughout the city (Figure 8.5D). The event occurred on the third day of Ramadan and badly damaged the local mosques.

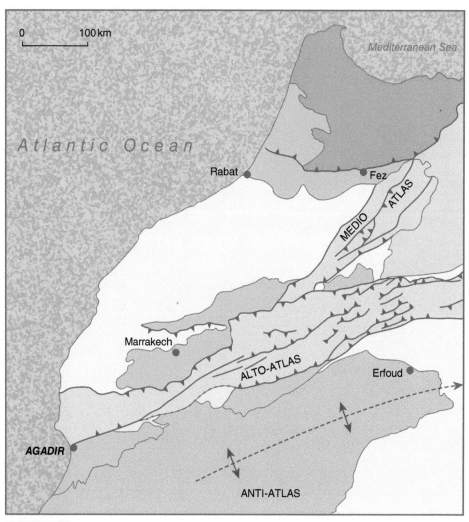

FIGURE 8.5 Geologic map of Morocco showing the location of Agadir.

FIGURE 8.5A Photograph of the destruction to the city of Agadir. *Source*: AP Images.

FIGURE 8.5B Photograph of damage to buildings by the 1960 Agadir earthquake. *Source*: Kerry Dunstone/agefotostock/Alamy Stock Photo.

Even though the earthquake was not large, the death toll was at least 12 000 and as much as 15 000. In addition, 12 000 residents were injured out of a total population of 33 000. This means that less than one-third of the population of the city was not injured or killed. This is a huge percentage for any event. There were so many dead that they had to be buried in a mass grave using a front-end loader. Consider that a shallow earthquake of magnitude 5.8 struck near Richmond, Virginia in 2011 and no one was killed. The reason that the Agadir earthquake was so devastating is that the area is infrequently active. The last major earthquake in the area was in 1751, 209 years earlier. The epicenter was directly in the city and the focus was shallow, meaning maximum impact from surface waves. Subsequent studies, however, concluded that poor building practices were the main culprit. The infrequent activity meant that there was no need for better building codes to that point.

FIGURE 8.5C Photograph of massive damage to buildings from the Agadir earthquake. *Source*: ZUMA/ZUMA Press, Inc./Alamy Stock Photo.

FIGURE 8.5D Photograph of damage to the Saada Hotel during the Agadir earthquake. *Source*: Keystone-France/Getty Images.

CASE STUDY 8.6 1920 Gansu Earthquake, China

Slope Failure

Local geology and cultural practices can turn a dangerous earthquake into a catastrophe. This was the case with the Gansu or Haiyuan earthquake, China. This massive earthquake struck the Haiyuan area of Ningxia on 16 December 1920 at 10:07 p.m. (Figure 8.6). The estimated magnitude of the event ranges from 7.8 to 8.7, but, recently, the lower number is favored. The maximum intensity at the epicenter was XII on the Modified Mercalli Scale, the highest classification (total destruction). Shaking from surface waves affected more than 25 000 mi^2 (64 750 km^2) from the Yellow Sea in the east to the Qinghai (Tsinghai) Province in the west and from Inner Mongolia in the north to central Sichuan Province to the south.

The main shock was produced by movement on the 621-mi (1000-km)-long, west-northwest-oriented Haiyuan fault. The fault underwent left-lateral strike-slip movement along the northeast edge

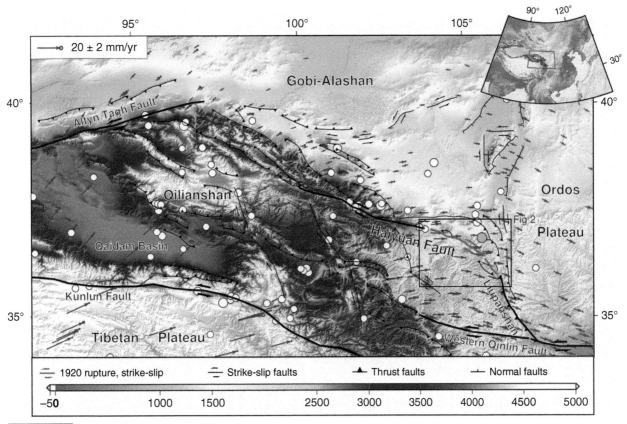

FIGURE 8.6 Map of Central China showing the location of the 1920 Gansu earthquake. Inset shows the location relative to the Himalayan continental collision. *Source*: Kulikova, G. et al. 2020/John Wiley & Sons/Creative Commons CC BY.

of the Tibet–Qinghai Plateau. These faults accommodate escape tectonics generated by the ongoing Himalayan collision between India and Eurasia. The main shock produced a rupture that was 147 mi (237 km) long in four segments. The maximum displacement along the fault was about 33 ft (10 m). More than 125 mi (200 km) of surface faulting was also documented.

An area that was strongly impacted by the Gansu earthquake is underlain by thick loess deposits (Figure 8.6A). Loess is windblown

FIGURE 8.6A Photograph of loess bluffs and plateau near Luochuan, China. Largest loess deposit in the world. *Source*: Courtesy of JY International Cultural Communications Co., Ltd.

dust from deserts and arid regions that accumulates on slopes in the downwind direction. The dust is eroded by wind and carried in suspension until it encounters an area of lower wind velocity. Large hills and mountains create a wind shadow that allows deposition. Where it accumulates, it can be quite thick, but not particularly stable as the result of the fine particle size. China has particularly thick loess deposits blown in from the Gobi Desert to the west. The thick deposits are excellent for burrowing and building homes into the walls as well as cultivation. Large populations can inhabit these carved and engineered hillslopes, but they have poor resistance to shaking.

The Gansu earthquake occurred in an area with large settlement in loess slopes. The severe shaking from the earthquake caused major mass movements in the area. More than 1000 landslides are estimated to have occurred. Some dammed rivers produced landslide or earthquake lakes that flooded areas and later burst producing downstream floods. Some shifted the course of rivers. Some landslides buried entire towns. As a result, the death and destruction caused by the Gansu earthquake was astounding (Figure 8.6B). The estimated death toll for this event varies, depending on whether the aftermath of disease and severe cold weather are taken into account. The lowest estimate for the earthquake itself is 180 000 deaths with 20 000 more from the aftermath. There are other estimates of 235 502 and 240 000 deaths, but the most recent investigation yields a death toll of 273 400 people in 77 counties. This staggering number makes it the third worst earthquake in China and the fourth most devastating earthquake in history. Fortunately, the recurrence interval for events comparable to the Gansu earthquake is between 800 and 1600 years.

FIGURE 8.6B Photograph of a temple destroyed by the Gansu earthquake. *Source*: Jingning County Archives.

CASE STUDY 8.7 1964 Good Friday Earthquake, Alaska

Ground Failure

Strong earthquakes can cause ground failure which involves surface movement as the result of shaking from the passing of surface waves. Ground failure includes landslides, slumping, fissures, lateral spreading, and any other comparable mass wasting. It does not include rock falls and rock slides or any other steep slope process. Ground failure can be responsible for the majority of the damage to buildings, bridges, and roads. Although the Good Friday earthquake had many severe earthquake hazards, it is well known for the ground failure that it caused (Figure 8.7).

The Good Friday earthquake was the second most powerful seismic event ever recorded. On 27 March 1964 (Good Friday) at 5:36 p.m., a devastating earthquake struck Prince William Sound, 78 mi (125 km) east of Anchorage and 40 mi (64 km) west of Valdez, Alaska (Figure 8.7A). The magnitude of the earthquake was 9.2, the highest ever in the Northern Hemisphere. The earthquake was generated by the megathrust of the Aleutian Subduction Zone in which the Pacific Oceanic Plate is being driven northward beneath the North American Plate. The rupture length on the fault was 600 mi (970 km) long and it lifted the seafloor as much as 60 ft (18 m) in some areas. The depth of the focus was a moderate 16 mi (25 km). The lift of the seafloor generated a

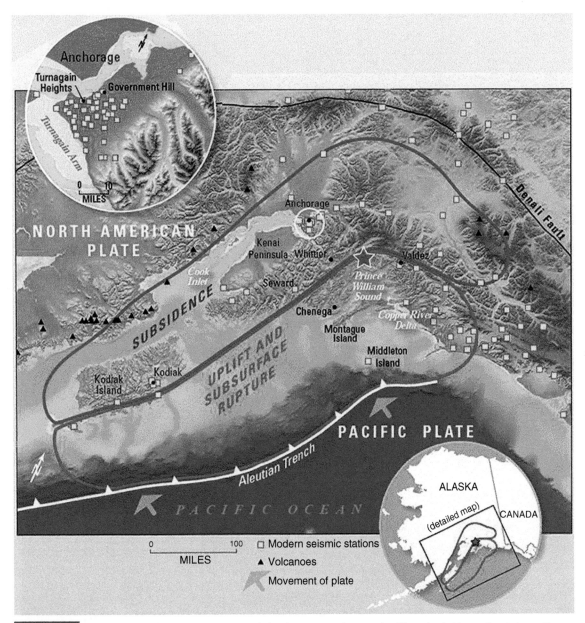

FIGURE 8.7 Tectonic map of the Aleutian Arc in subduction zone and areas of uplift and subsidence. Inset shows the political boundaries of the area. *Source*: Courtesy of the US Geological Survey.

major tsunami that struck throughout the Pacific Ocean Basin and was the final justification for installing the Pacific Tsunami Warning System (PTWS) to monitor activity, warn residents, and initiate evacuations. It was the only early warning system until after 2004 when a second system was installed in the Indian Ocean. The tsunami was at its highest in the region around Alaska with waves reaching heights of 220 ft (67 m). These were catastrophic. However, damaging waves reached as far as Peru.

The power that this earthquake delivered to the epicentral region was enormous. Ground acceleration was as high as 0.18 g and shaking lasted four to five minutes. This severe shaking for such a long duration produced a surface-shaking intensity of XI on the Modified Mercalli Scale, which is considered extreme. The damage it did to the surface topography and buildings and infrastructure of Anchorage was devastating (Figure 8.7B). The downtown area was built on sand deposits which developed fissures and slumped in the shaking (Figure 8.7C). Sections of streets dropped as much as 10 ft (3 m) and the public school was sheared in half, dropping a full floor across the head of a slump (Figure 8.7D). In one area, the ground failure and slumping were so extreme that the surface broke into blocks and each rotated as much as 45° in all different directions. Houses, trees, and telephone poles pointed in all directions in a chaotic jumble (Figure 8.7E).

The death toll from this event was just 131 in Alaska owing to the sparse population of the area. Damage was estimated at US$730 million in 2018 dollars, but was US$116 million at the time. There is no question that this active subduction zone will continue to produce large earthquakes, but the Good Friday event is considered a once-in-500-year earthquake, so recurrence in the near future is unlikely.

FIGURE 8.7A Photograph of a roads destroyed by slumping during the 1964 Good Friday earthquake. *Source*: Courtesy of the US Geological Survey.

FIGURE 8.7B Photograph of houses jumbled at the head of a slump in the 1964 Good Friday earthquake. *Source*: Courtesy of the US Geological Survey.

FIGURE 8.7C Photograph of a section of downtown Anchorage having undergone surface collapse. *Source*: Courtesy of the US Geological Survey.

FIGURE 8.7D Photograph of a school broken in half by slumping during the 1964 Good Friday earthquake. *Source*: Courtesy of the U.S. Geological Survey.

FIGURE 8.7E Photograph of a houses destroyed by slumping during the 1964 Good Friday earthquake. *Source*: Courtesy of the US Geological Survey.

CASE STUDY 8.8 2008 Sichuan Earthquake

Poor Construction and Slope Failure

A powerful earthquake struck the Sichuan region of China on 12 May 2008 at 2:28 p.m. The epicenter was located 55 mi (90 km) west-northwest of Chengdu, the capital of Sichuan, in Wenchuan County (Figure 8.8). The focal depth was 11.8 mi (19 km) and the moment magnitude was a powerful 7.9. The maximum intensity of the shaking was XI on the Modified Mercalli Scale, which was classified as "very disastrous." Shaking lasted about 120 seconds and most of the energy was released in the first 80 seconds.

There were seven major aftershocks within three hours of the main shock with magnitudes between 4.0 and 6.0. By 6 November 2008, aftershocks totaled 42 719 with 246 of magnitudes 4.0–4.9, 34 of 5.0–5.9, and 8 of 6.0–6.4.

The Sichuan or Wenchuan earthquake resulted from movement on a reverse fault along the northwestern edge of the Sichuan Basin. The slip was on the Longmenshan Fault or a related fault. The estimated displacement for this slip is up to 29 ft (9 m) along a 150 mi (240 km) long by 12 mi (20 km) deep fault segment. The stress that caused the movement is generated by the convergence of India with Asia and uplift of the Tibetan Plateau. The northeast trending faults result from escape tectonics across the area.

The bulk of the damage in the area was caused by intense shaking of steep unstable slopes. The shaking caused catastrophic rockfalls, avalanches, and landslides (Figure 8.8A) that severely damaged or destroyed mountain roads and railway lines and buried numerous buildings in the Beichuan–Wenchuan area. Piles of debris and destruction of the roads cut off access to the region from search and rescue efforts for several days. A single landslide in Qingchuan buried more than 700 people. These landslides dammed a number of rivers creating 34 "landslide" or "earthquake" lakes. Tremendous amounts of water filled the lakes behind the landslide dams at very high rates. The result was rapidly building water pressure on the debris dams that would eventually breach them. This drastic situation had the potential to catastrophically impact more than 700 000 people downstream of the dams. The Chinese government reacted quickly to the disaster and pending catastrophe. Chinese army troops were brought in with heavy machinery to quickly drain the earthquake lakes in a controlled release of water to prevent the disaster. 50 000 army troops and armed police were dispatched on 12 May by the area military command for rescue and relief operations in Wenchuan County. Several days later, at the peak of operations, there were 135 000 Chinese army troops and medics involved across 58 counties and cities. The aircraft involved in the operation including air force, army, and civil aviation totaled more than 150, making it China's largest ever noncombat airlift operation.

Damage to buildings and infrastructure was extreme (Figure 8.8B) and included nearly 2500 dams, more than 32 933 mi (53 000 km) of roads, and 29 825 mi (48 000 km) of water pipelines. Surface faulting along a 0.95-mi (1.5-km) length was observed as cracks and fractures in the city of Qingchuan and the surrounding mountains. The cities of Beichuan, Dujiangyan, Wulong, and Yingxiu had more than 80% of the buildings collapsed (Figure 8.8C). Approximately 12.5 million domestic animals perished, and a large portion of agricultural fields and processing facilities were destroyed. The economic loss for the disaster is estimated at US$86 to $150 billion in 2008 dollars.

In addition to the damage, an estimated 87 587 people were killed, 374 177 were injured, and 18 392 were missing and presumed dead making this among the deadliest earthquakes. More than 45.5 million people were affected in the area surrounding the epicenter. More than 15 million people were evacuated and more than 5 million and as many as 11 million were left homeless. It was particularly devastating

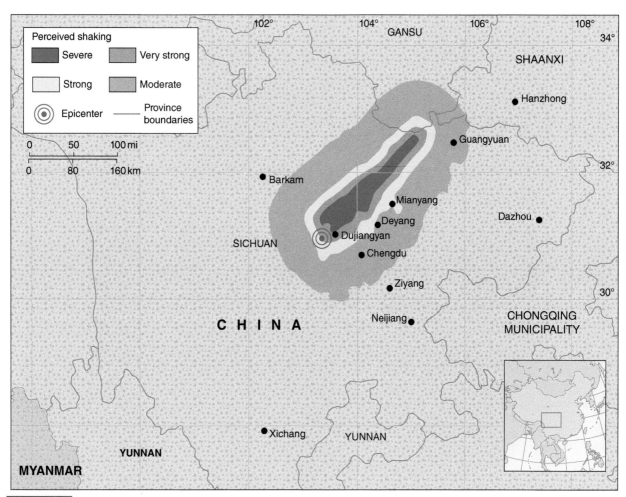

FIGURE 8.8 Map of Central China showing the distribution of shaking intensity during the 2008 Sichuan earthquake. Inset shows the location of the map in China. *Source*: Adapted from Encyclopaedia Britannica, Inc.

FIGURE 8.8A Photograph of rockslides generated during the 2008 Sichuan earthquake. *Source*: Courtesy of the US Geological Survey.

FIGURE 8.8B Photograph of buildings destroyed during the 2008 Sichuan earthquake. *Source*: Courtesy of the US Geological Survey.

FIGURE 8.8C Photograph of building foundations damaged during the 2008 Sichuan earthquake. *Source*: Courtesy of the US Geological Survey.

for schools and their students with more than 19 065 schoolchildren dead. The cause of their deaths was the collapse of more than 7000 school buildings (Figure 8.8D). In Mianyang City alone, seven schools collapsed and at least 1700 students and school staff were buried. As many as 1300 students and teachers perished in the collapse of the single Beichuan Middle School. The schools were highly vulnerable to shaking because construction standards were not adhered to in the buildings. The residents of the affected areas now refer to the schools as "tofu-dregs schoolhouses" to describe the poor construction. The high number of children killed was especially devastating for the impacted families in the region because China's single-child-limit laws left them childless.

Several Chinese and international experts theorized that the high pressure of the deep water in the reservoir behind the Zipingpu Dam in Sichuan may have influenced the timing of the earthquake. The reservoir is just 3.5 mi (5.6 km) from the epicenter and the high fluid pressure on the rock beneath it may have weakened its resistance to rupture and hastened the earthquake. This theory, however, has not been verified, but if it proves true, dam construction in seismically active areas will need reconsideration.

FIGURE 8.8D Photograph of buildings demolished during the 2008 Sichuan earthquake. *Source*: Courtesy of the US Geological Survey.

CASE STUDY 8.9 1970 Ancash Earthquake

Earthquake-Generated Avalanche

One of the main sources of damage from earthquakes is mass movement of rock and soil shaken loose by the surface waves. These are most commonly earthflows and other moderate speed slope failures that cause local structural damage. Others can cause debris flows and higher speed failures. However, the 1970 Ancash earthquake produced perhaps the highest speed mass movement ever in history.

The Ancash earthquake or Great Peruvian earthquake struck on Sunday, 31 May 1970 at 3:23 p.m. The epicenter was 22 mi (35 km) in the Pacific Ocean off the coast of Casma and Chimbote, Peru (Figure 8.9). The magnitude was a powerful 7.9 and shaking intensity was a strong VIII on the Modified Mercalli Scale. The shaking lasted about 45 seconds and primarily affected the Ancash and La Libertad regions. The depth of the focus was a deep 28 mi (45 km) which should have made it a megathrust earthquake on the subducting Nazca Ocean Plate, but it was generated by a compensating normal fault.

The main shock affected an area that was about 40 000 mi^2 (103 600 km^2). It is estimated that it destroyed roughly 200 000 homes and buildings leaving some 800 000 people homeless (Figure 8.9A). It is also estimated that the earthquake caused the death of 70 000 people and caused 50 000 injuries. However, it was a small part of this disaster that included a very rare event.

The shaking from the surface waves of the Ancash earthquake dislodged a slab of ice-covered rock about 2600 ft (800 m) wide from the western face of Nevados Huascarán, the highest peak of the Peruvian Andes (Figure 8.9B). Huascarán rises to an elevation of 22 205 ft (6768 m) and the dislodged slab was removed from the north peak, at an elevation between 18 400 and 20 300 ft (5600–6200 m). The slab

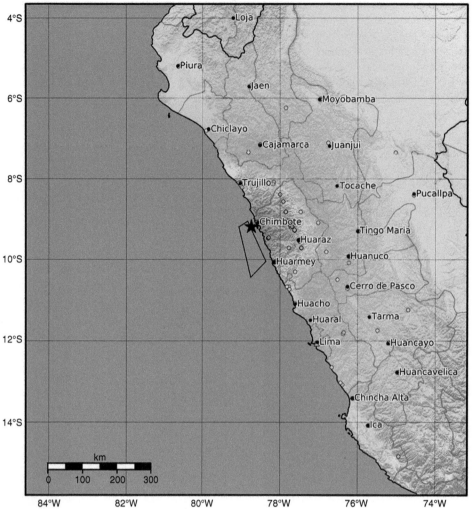

FIGURE 8.9 Geologic map of Peru and surrounding area showing the location of Chimbote. *Source*: Shakemap of the Ankash earthquake of 31 May 1970 in Peru, United States Geological Survey.

free-fell at least 2000 ft (600 m) before colliding with Glacier 511 and sliding nearly 1.9 mi (3 km) down its surface, accumulating ice and snow as it went. The initial volume of this ice-rock avalanche was well in excess of 33 million cubic yards (25 million m³). However, as the mass careened down the steep Llanganuco Valley, it accumulated large volumes of dirt, loose glacial moraine, and water, and uprooted trees and grew rapidly in volume. The ice melted during the journey turning it into a mudflow further down the mountain where it reached a maximum volume of as much as 130 million cubic yards (100 million m³).

The real story, however, was the speed and processes during travel. In addition to ice and debris, the avalanche contained boulders of 3 tons (4400 kg) or more (Figure 8.9C) that careened down the hillside, leaving huge craters in the ground (Figure 8.9D) and flattening all buildings that had not been already destroyed by the earthquake. The largest single boulder was found near Ranrahirca and weighed 31 000 000 lb (14 000 mt). The avalanche mass attained speeds of up to 270 mph (435 kmph), but some boulder projectiles were launched ahead of the main flow by impacting each other and may have exceeded speeds up to 680 mph (1100 kmph). These impacts hurled the boulders as high as 2.5 mi (4 km) into the air. Residents would have heard a distant roar and in seconds a mass of huge speeding boulders would have pulverized their homes and all structures in their path. A strong air blast preceded the flow, followed by the boulders and debris and finally a semi-liquid wave. The wind blast was strong enough to knock people over and, in some places, to topple trees. The speed of the flow left no time to evacuate.

Even the mudflow moved at higher-than-expected velocities (Figure 8.9E). It was described by survivors as "wavelike." It may have achieved unusually high speeds because a mixture of snow and ice trapped air underneath the mudflow that allowed it to essentially float over the ground in air-cushioned flow. As the flow moved down the mountain, the warmer temperatures and heat generated by impacting boulders melted the snow and ice. The impacting boulders pulverized their surfaces into a rock flour. That and the soil entrained by the flow turned the mass into a thick, mud-rich flow 10 ft (3.3 m) or more in thickness but flowing at much slower speeds.

As would be expected, this avalanche and mudflow caused death and destruction throughout its path which began in the Llanganuco Valley. By the time the flow struck the town of Yungay, it was 8.1 mi (13 km) from its avalanche origin and just 1 minute and 42 seconds

FIGURE 8.9A Photograph of a lone statue left in the rubble generated by the Ancash earthquake. *Source*: Courtesy of the US Geological Survey.

FIGURE 8.9B Photograph showing the origin of the rock and ice fall on Nevados Huascarán and the travel path of the avalanche and debris flow down the river valley. *Source*: Courtesy of the US Geological Survey.

after the Ancash earthquake. Although there was the 660 ft (200 m) Cerro de Aira hill in front of Yungay, the flow struck the ridge at such a high speed, 110 mph (170 km), that part of it sped over the crest, destroying Shillkop, Aira, and Ongo before descending on Yungay. The flow was still at high speed when it flattened Yungay, killing virtually all residents and burying the town under about 16 ft (5 m) of debris.

The bulk of the mudflow continued down the Río Llanganuco Valley. The towns of Incayoc and Huashau, at the base of Huascarán, were struck first. Then it devastated the entire Llanganuco Delta, including the village of Ranrahirca, where some 1800 people were killed. Similar to Yungay, large boulders barreled across the valley, pulverizing everything in their path. Up to 66 ft (20 m) of debris was left blanketing the Ranrahirca area (Figure 8.9F).

The debris flow crossed and dammed the Río Santa, which caused it to back up and form a lake 1.2 mi (2 km) long. Debris traveled as high as 272 ft (83 m) up the opposite valley wall and destroyed part of Matacoto village, before falling back down into the river. At this point, the flow was 22 mi (35 km) from its source. Less than 30 minutes later, this landslide dam overflowed. The mud and debris then swept downstream for about 62 mi (100 km) at speeds up to 22 mph (35 kmph), causing damage on the riverbanks. The remainder of the debris flood reached the Pacific Ocean, 112 mi (180 km) from the source.

The area covered in debris was well in excess of 3700 acres (1500 ha) and the flow was as wide as 2.7 mi (4.3 km) in places. The thickness of the mud and debris reached 66 ft (20 m) in the Ranrahirca area, but it was at least 6.5 ft (2 m) everywhere else. This meant that whole towns were chest deep in mud and rock. The flow destroyed the towns of Yungay and Ranrahirca and 10 nearby villages. It is difficult to determine the exact number of victims because the number of corpses buried so deeply in the mud and debris cannot be determined. In addition, victims were found floating in the Pacific Ocean and could have been washed away. The most cited death toll is 18 000, but estimates range between 15 000 and 25 000.

FIGURE 8.9C Photograph of a boulder that was carried in the avalanche. Note the mudfield around it from rock flour. *Source*: Courtesy of the US Geological Survey.

FIGURE 8.9D Photograph of an impact crater 48 ft (15 m) long and 16 ft (5 m) deep from a boulder hitting the slope during the avalanche. *Source*: Courtesy of the US Geological Survey.

FIGURE 8.9E Photograph showing the debris and mudflow in featureless light gray covering a former village. *Source*: Courtesy of the US Geological Survey.

FIGURE 8.9F Photograph showing the edge of the debris flow and the mountain slope covered in gray mud. *Source*: Courtesy of the US Geological Survey.

References

Mori, J. and Somerville, P. (2006). Seismology and Strong Ground Motions in the 2004 Niigata Ken Chuetsu, Japan, Earthquake. Earthquake Spectra - EARTHQ SPECTRA. 22. 10.1193/1.2172914.

Ou, Q., Kulikova, G., Yu, J., Elliott, A., Parsons, B., & Walker, R. (2020). Magnitude of the 1920 Haiyuan earthquake reestimated using seismological and geomorphological methods. *Journal of Geophysical Research: Solid Earth*, 125, e2019JB019244. https://doi.org/10.1029/2019JB019244.

Wakamatsu, K. and Matsuoka, M. (2006). Development of the 7.5-arc-second Engineering Geomorphologic Classification Database and its Application to Seismic Microzoning; Bull. Earthq. Res. Inst. Univ. Tokyo, 81, 317–324.

CHAPTER 9

Killer Tsunamis

CHAPTER OUTLINE

9.1 Terror from the Sea 158

9.2 Generation of a Tsunami 158

9.3 Quantifying Tsunami Intensity 160

Earth's Fury: The Science of Natural Disasters, First Edition. Alexander Gates.
© 2022 John Wiley & Sons Ltd. Published 2022 by John Wiley & Sons Ltd.
Companion website: www.wiley.com/go/gates/earthsfury

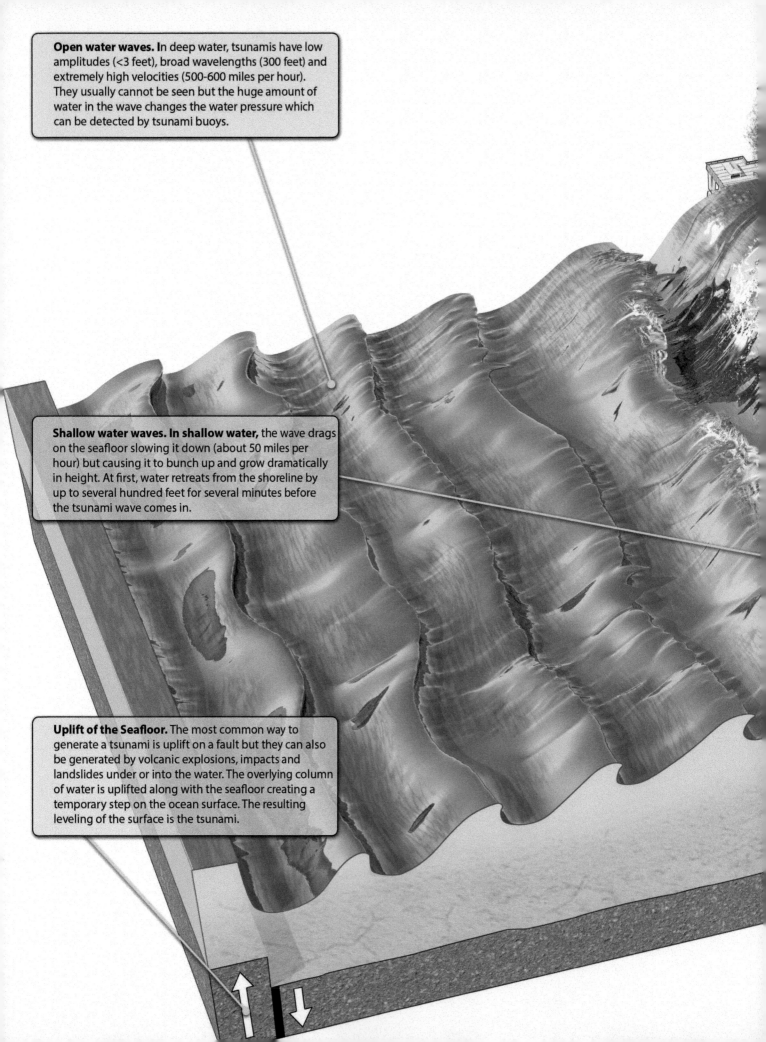

Open water waves. In deep water, tsunamis have low amplitudes (<3 feet), broad wavelengths (300 feet) and extremely high velocities (500-600 miles per hour). They usually cannot be seen but the huge amount of water in the wave changes the water pressure which can be detected by tsunami buoys.

Shallow water waves. In shallow water, the wave drags on the seafloor slowing it down (about 50 miles per hour) but causing it to bunch up and grow dramatically in height. At first, water retreats from the shoreline by up to several hundred feet for several minutes before the tsunami wave comes in.

Uplift of the Seafloor. The most common way to generate a tsunami is uplift on a fault but they can also be generated by volcanic explosions, impacts and landslides under or into the water. The overlying column of water is uplifted along with the seafloor creating a temporary step on the ocean surface. The resulting leveling of the surface is the tsunami.

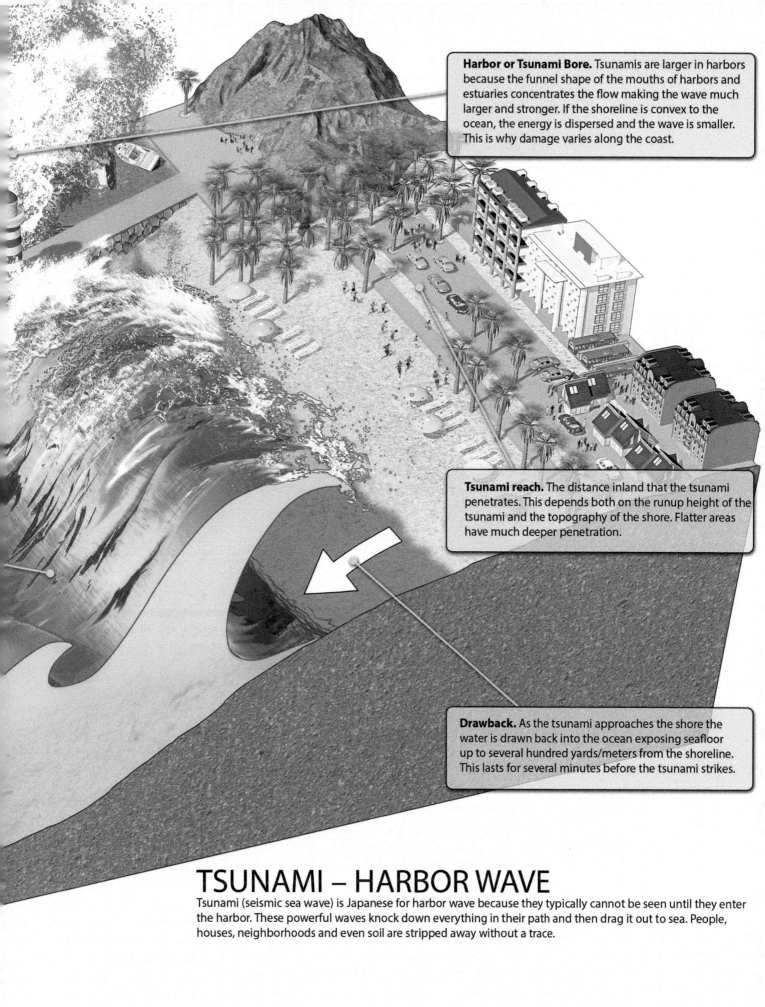

Words You Should Know:

Amplitude – The height of the tsunami wave above mean sea level at a particular time.

Drawback – Tsunamis commonly arrive trough first, so water is drawn away from the shoreline 10–100 ft or more before the arrival of the destructive tsunami wave.

Flow depth: The height of a tsunami wave above the ground surface at any point.

Inundation – The area of land that is covered by water by the run-up of the tsunami.

Megatsunami – A Hollywood-type description of a very large tsunami.

Meteotsunami – A tsunami formed by meteorological events such as a hurricane. A controversial term not covered in this chapter.

Microtsunami – A tsunami so small that it can only be detected instrumentally.

Run-up height – The maximum vertical height above sea level that a tsunami reaches As it comes ashore, also known as inundation height.

Seiche – An earthquake-generated wave that sloshes back and forth across a lake or other body of water.

Teletsunami – A tsunami that crosses an entire ocean basin.

Tsunami bore – An anomalously high and quickly moving tsunami wave as a result of being funneled into a river mouth or estuary.

Tsunami forerunner – Any oscillations of water level foretelling the arrival of the main tsunami waves.

Tsunamigenic – Any event, natural or man-made, that is capable of generating a tsunami.

Tsunami reach – Tsunami reach or inundation line is the maximum inland penetration of the wave measured from the mean sea level line.

Wavelength – The distance between the same point on consecutive waves (crest or trough).

Wave train – There is usually one main tsunami wave, but it typically has a number of associated waves with it forming a sequence or train.

9.1 Terror from the Sea

Tsunamis used to be called "tidal waves," but the term was dropped because they are not related to tides. Tsunamis were also called "seismic sea waves," but that term is not in favor because it only refers to tsunamis generated by earthquakes. *Tsunami* is a Japanese word that means "harbor wave." It is called a harbor wave because it can be a clear, calm day when a monster wave emerges in the harbor from apparently nowhere and wreaks havoc on the coastal communities.

9.2 Generation of a Tsunami

Tsunamis are extremely deadly waves generated in the ocean. They can be generated by earthquakes, underwater landslides, surface to water landslides, volcanoes, extraterrestrial impacts, possibly weather, and large bomb blasts. By far the most common of these is earthquakes. The basic mechanics of most tsunamis is that a rapid change in the height of the seafloor uplifts or drops the column of water above it. Water cannot form a step on its surface, so it quickly re-equilibrates in a wave. Tsunamis generated by land to sea avalanches are more analogous to splashing water because they act on the water surface instead of depth. Explosions from volcanoes or bombs force the water away in all directions at high speed through rapid expansion. Extraterrestrial impacts have aspects of all of these mechanisms.

Once the tsunami is generated, it travels through deep water at phenomenal velocities. They can travel at speeds of 600 mph (965 kmph), which is how they can cross ocean basins so quickly. In deep water, they have very low amplitudes, generally much less than 3.3 ft (1 m). However, they have very broad wavelengths of 300 ft (91 m) or more. This broad wavelength means that even though they are low, they displace an enormous amount of water compared to typical waves no matter how high they are.

Tsunamis behave much differently in shallow water. Wave motion penetrates the surface of the oceans to a depth of one half the wavelength. This means that the motion of a tsunami wave with a wavelength of 300 ft (91 m) will penetrate to a depth of 150 ft (46 m). The line between movement and no movement is called the wave base. As the tsunami approaches the shore, the seafloor is ever shallower. At some point, the wave base of the tsunami touches the seafloor and the friction causes the wave to slow down. The friction causes the front of the wave to travel slower than the back and the wave compresses to a shorter wavelength and a much higher amplitude. This is especially true where the coastline funnels the wave energy into a tighter area like a harbor or estuary creating a tsunami bore. This is where tsunamis get their name.

As the tsunami wave approaches the shore, it leads with a trough. This pulls the seawater away from the shore up to several hundred feet depending upon the size of the wave. This process is called drawback and it unnaturally exposes the seafloor which can draw attention of onlookers who run to inspect this oddity. Drawback can precede the tsunami by as much as three to four minutes. When the tsunami wave appears, it is at peak height and still traveling at 50–60 mph (80–96 km). The onlookers cannot outrun the wave and are typically its first casualties.

The height that the tsunami reaches above sea level as it comes ashore is the run-up (Figure 9.1). The height of the water at any given point is the flow depth. The land covered by the tsunami flood is the inundation area. The farthest point that the tsunami penetrates inland is the tsunami reach or the inundation limit or line. Once all that water is transported on land, it quickly flows back out to sea in the process of flowback. Although flowback appears harmless relative to the tsunami, it can drag debris and people far offshore stronger than any rip current.

If a tsunami is too small to see, it is called a microtsunami and can only be detected by sea level monitoring instruments. If the tsunami only impacts the area right around the source, it is considered a local tsunami. If it is strong enough to impact other areas in the ocean basin, it is a remote tsunami. As a subset of remote tsunamis, those that cross an entire ocean basin and impact communities on the opposite shores are teletsunamis.

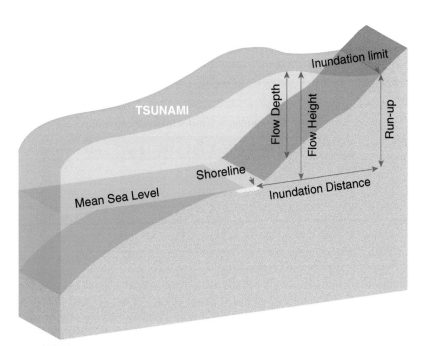

FIGURE 9.1 Block diagram showing the parameters of a tsunami that has struck the shore.

9.3 Quantifying Tsunami Intensity

Just like most natural hazards, there have been several intensity scales proposed for tsunamis. None of them has really become well accepted like the Saffir–Simpson Scale for hurricanes, the Fujita Scale for tornadoes, or the magnitude (Richter) and intensity (Mercalli) scales for earthquakes. These are consistently quoted in the news and have become common knowledge for much of the general public. It appears that tsunamis are closer to having an accepted scale in recent years.

The oldest scale was developed in 1923 by Sieberg, but it covered all anomalous ocean waves rather than just tsunamis. It was later improved by Ambraseys in 1962 to form the Modified Sieberg Scale as follows:

TABLE 9.1 Modified Sieberg Scale for Tsunamis.

Category	Description
1.	**Very light**: Wave is only perceptible on tide gauges.
2.	**Light**: Wave noticed by shore communities familiar with the sea. Generally noticed on flat shorefaces.
3.	**Rather strong**: Generally noticed. Flooding on shallow sloped coasts. Small boats carried away. Slight damage to small structures along the coast. Reversal of the river flow in estuaries for some distance upstream.
4.	**Strong**: Local flooding along the shoreline to a significant depth. Light scouring of soils. Damage to embankments and dikes. Damage to all small structures along the coast. Minor damage to large structures along the coast. Large sailing vessels and small ships carried inland or out to sea. Coastlines littered with debris.
5.	**Very strong**: General flooding along the shoreline to significant depth. Damage to breakwater walls and large structures along the coast. Total destruction of small structures. Severe scouring of cultivated land. Coastlines littered with floating debris. Besides large ships, all vessels carried inland or out to sea. Large wave bores in estuaries. Harbors damaged. People drowned.
6.	**Disastrous**: Partial or complete destruction of all man-made structures well inland from the coast. Significant flooding of all coastlines. Severe damage to large ships. Trees uprooted or broken. Many casualties.

The Sieberg Scale was considered too descriptive and unspecific. The same wave might have different scales depending upon the coast that it impacted. In an effort to create a more uniform system, the Soloviev–Imamura System was developed in 1970. This tsunami intensity scale is based on a formula that considers only relative height. The formula yields:

TABLE 9.2 Soloviev-Imamura Intensity System.

Intensity	Average height of wave
2	9.2 ft (2.8 m)
3	18 ft (5.5 m)
4	36 ft (11 m)
5	74 ft (22.5 m)

Unlike the Siebert Scale, the intensities are unlimited because the successive height can just be entered into a formula. However, this scale was also not popular because the heights of waves could just be reported rather than an intensity which was the common practice. Instead, in 2001, two geologists (Papadopoulos and Imamura) suggested that the scale should reflect the Modified Mercalli Scale used to characterize the intensity of earthquakes based on the damage they produced rather than a simple height scale which is similar to a magnitude measurement. They proposed a scale that would use Roman numerals, be based on damage, and have 12 increments similar to the Modified Mercalli Scale. The scale is as follows:

I. Not felt:
Not felt even under the most favorable circumstances and no resulting damage.

II. Scarcely felt:
Felt by few people on small vessels. Not observed on the coast and no damage.

III. Weak:
Felt by most people on small vessels. Observed by a few people along the coast, but no damage.

IV. Largely observed:
Felt by all people on small vessels and by a few people on large vessels. Observed by most people along the coast. A few small vessels moved onshore. No damage.

V. Strong:
Felt by all people on large vessels and observed by all people along the coast. Several people become frightened and seek higher ground. Many small vessels transported onshore with several damaged. Possible sand layers deposited under favorable conditions. Limited flooding of cultivated land and outdoor facilities of shoreline structures.

VI. Slightly damaging:
Many people frightened and evacuate to higher ground. Small vessels transported violently onshore and damaged. Damage and flooding in a few wooden structures.

VII. Damaging:
Most people frightened and evacuate to higher ground. Many small vessels damaged. Several large vessels rock violently. Sand layers and pebbles cover parts of the surface. Many wooden structures are damaged, and several are demolished or washed away. Damage and flooding in a few masonry buildings.

VIII. Heavily damaging:
All people escape to higher ground and several are washed away. Most small vessels are damaged, many are washed away. Several large vessels are moved ashore or damaged from collisions. Beaches are eroded and littered with debris. Extensive flooding along the coast and minor flooding inland. Most wooden structures are washed away or demolished. Damage of a few masonry buildings.

IX. Destructive:
Many people are washed away. Most small vessels are destroyed or washed away. Many large vessels are beached, and a few are destroyed. Beaches are extensively eroded and covered in debris. Subsidence of the ground surface locally. Many masonry buildings are damaged.

X. Very destructive:
There is general panic of the population. Most people are washed away along shorelines. Most large vessels are beached, and many are destroyed or badly damaged. Rocks up to small boulders are moved ashore and inland. Cars are overturned. Extensive ground subsidence. Damage of many masonry buildings. Artificial embankments collapse, and port water breaks are damaged.

XI. Devastating:
All lifelines are interrupted. There are extensive fires. Backwash of tsunami floods carries cars and other debris out to sea. Large boulders from the seafloor are swept ashore and inland. Heavy damage to even the strongest of masonry buildings.

XII. Completely devastating:
Essentially all masonry buildings are demolished. Extensive death and flooding.

After major tsunamis in 2004 and 2011, there was renewed interest in developing a broadly accepted tsunami intensity scale. This reevaluation took many of the ideas of the 2001 scale and the result was the Integrated Tsunami Intensity Scale (ITIS-2012) which was published in 2013. The ITIS-2012 is a 12-level scale defined with Roman numerals. However, it is based on the quantitative assessment of objective criteria that are readily accessible. There are six groups of observational features in assessment. The determination of intensity is based on: (i) the amounts and sizes of the phenomena, (ii) the direct impact on humans, (iii) the impact on objects that are movable, (iv) the impacts on coastal infrastructure, (v) the impact on the environment, and (vi) the impact on constructed structures.

Although ITIS-2012 has been applied to several research studies, it has not been as widely adopted as was hoped. The number of variables and descriptors make it cumbersome to apply in many cases. In addition, another branch of research in intensity scales developed. This evolved from an effort to revise and improve the Modified Mercalli Scale. This group devised the Environmental Seismic Intensity (ESI 2007) Scale which was also based on 12 levels of intensity. This group devised the Tsunami Environmental Effects (TEE-16) Scale which also has 12 levels designated with Roman numerals and with similarly specific quantitative and observational features for each level.

CASE STUDY 9.1 1960 Valdivia (Great Chilean) Earthquake and Tsunami

The Largest on Record

The strongest earthquake on record is actually best known for the tsunami it produced. The Great Chilean or Valdivia earthquake occurred on the South American Subduction Zone off the coast of Chile at 3:11 p.m. on 22 May 1960 (Figure 9.2). The magnitude of the main shock was 9.5–9.6 and the surface waves produced a maximum shaking intensity of XII (maximum) on the Modified Mercalli Scale. The duration of the shock was more than three minutes. The rupture focus was 21 mi (33 km) deep, and the epicenter was just offshore of Valdivia in the Pacific Ocean. Because the subduction surface has the same geometry as a thrust fault, it is referred to as a megathrust. The rupture zone is estimated to be from 311 mi (500 km) to

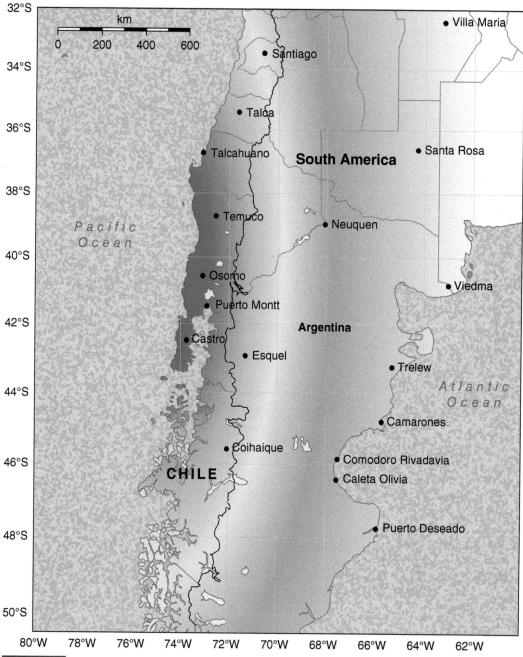

FIGURE 9.2 Map of Southern South America with Chile and the shaking intensity from the 1960 Valdivia earthquake. *Source*: M 9.5 - 1960 Great Chilean Earthquake (Valdivia Earthquake), shakeMap, US Geological Survey.

621 mi (1000 km) long and experienced as much as 110 ft (30 m) of thrust movement.

The rapid uplift of the seafloor generated a very large tsunami that crossed the entire Pacific Ocean (Figure 9.2A). In contrast to most tsunamis that only impact local areas, a tsunami that crosses an ocean basin and can still be recognized on tide gauges is a teletsunami. Tsunamis travel at high velocities in open water and grow into large waves as they approach a coast. Commonly, tsunamis are only recognizable near their source. It takes a powerful tsunami to cross an entire ocean.

The maximum run-up of the tsunami was 85 ft (26 m) on the Isla Mocha, Chile (Figure 9.2B). Run-up is the difference between the elevation of the tsunami height and sea level. The tsunami waves had heights of 33 ft (10 m) on the Hawaiian coast, and it killed 61 people, injured 43, and caused extensive damage, especially at Hilo (Figure 9.2C). Waves as high as 18 ft (5.6 m) struck Honshu, Japan 22 hours after the main shock and it destroyed some 1600 houses and leaving 185 people dead or missing (Figure 9.2D). The tsunami struck the Philippines 24 hours after the earthquake killing an additional 32 people. Tsunami waves were 16–33 ft (5–10 m) high in New Zealand, and even Russia recorded waves on tide gauges in the Atlantic Ocean in England, Bermuda, and South Africa and in the Indian Ocean in Mauritius and Western Australia, making it the first global tsunami.

The number of deaths in Chile from the earthquake and tsunami is between 5700 and 7000. The official Chilean estimates are that more than 2 million people were left homeless and damage was US$550 million (in 1960) (Figure 9.2E). The tsunami caused at least $23.5 million in damage in Hawaii. Waves of 3–6.5 ft (1–2 m) even caused two deaths, four injuries, and damage of $1 million on the United States West Coast. Total damage for the event is $400–800 million or $3.46–6.91 billion in 2019 dollars.

Just four years after the 1960 Great Chilean earthquake, the 1964 Good Friday earthquake also caused a massive tsunami. These two disasters convinced officials to develop the international Pacific Tsunami Warning System (PTWS). The Pacific Ocean Rim contains the most seismically and volcanically active zones on Earth making it the most prone to large tsunamis. Residents of the Pacific Ocean coastal communities live with constant threat of devastation. Over the next several years, the technology to sense tsunamis advanced greatly and installed in the Pacific Ocean in addition to coordination centers and warning and evacuation systems, collectively called the PTWS.

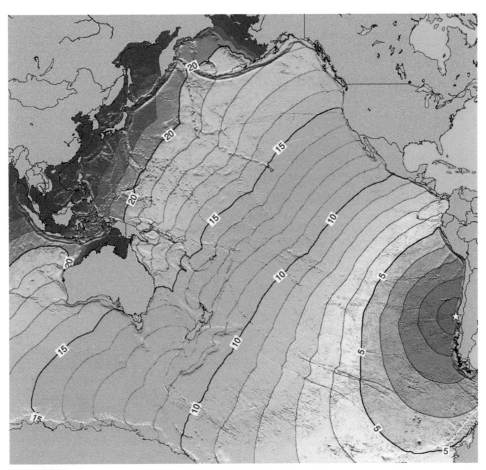

FIGURE 9.2A Map showing the dispersion of the 1960 Valdivia tsunami across the Pacific Ocean. The lines are travel time contours and the numbers show hours from generation. *Source*: Map courtesy of NOAA.

FIGURE 9.2B Before (top) and after (bottom) photos of the 1960 Valdivia earthquake and tsunami in a small harbor in Chile. *Source*: Courtesy of NOAA.

PTWS utilizes Deep-Ocean Assessment and Reporting of Tsunamis (DART) program technology. DART is an international early warning system for tsunamis while they are still far from coastal areas. The equipment consists of three main parts. Seafloor seismic sensors detect a tsunami as it passes over. Sensors record changes in the weight of water above which the tsunami is detected. The information is beamed to a buoy on the ocean surface above it. The buoy then beams the information to the Geostationary Operational Environmental Satellite (GEOS) which broadcasts it to land-based receivers around the Pacific Ocean. The United States has the Pacific Tsunami Warning Center in Hawaii, and the Alaska Tsunami Warning Center and the West Coast Tsunami Warning Center in California. Coupled with the global seismic network, DART warns coastal communities about impending tsunamis to allow evacuations well in advance of when they strike. Automatic earthquake alerts are issued to the centers for magnitudes greater than 6.5 and warning bulletins are issued for earthquakes greater than 7.5 (7.0 for Alaska). Evacuation routes are well marked and known by residents in PTWS-designated coastal communities.

FIGURE 9.2C Photo of the 1960 Valdivia tsunami damage on Hilo, Hawaii. *Source*: Courtesy of the US Geological Survey.

FIGURE 9.2D Photo of the 1960 Valdivia tsunami striking a community in Japan. *Source*: Courtesy of the US Geological Survey.

FIGURE 9.2E Photo of the 1960 Valdivia tsunami damage on the coast of Chile. *Source*: Courtesy of the US Geological Survey.

CASE STUDY 9.2 2004 Indian Ocean Tsunami

In Disasters, Timing Is Everything

Many people become complacent about natural disasters especially in developed countries. That complacency was deadly on 26 December 2004. The Christmas holiday week led vacationers to fill the resort towns around the Indian Ocean. During the morning following Christmas parties, people left their hotel rooms to relax on the beaches when a huge natural disaster struck. At 7:58 a.m., a huge submarine earthquake of magnitude 9.2 occurred 100 mi (167 km) into the Indian Ocean off the west coast of Northern Sumatra, Indonesia near Banda Aceh (Figure 9.3). Not since the 1964 Good Friday earthquake had an earthquake been this strong, and it was the third most powerful earthquake ever recorded. The focus was at 19-mi (30 km) depth.

The fault that generated the earthquake is a megathrust in the subduction zone that forms the Sunda Trench where the India Ocean and Burma (Asia) Plates collide. A 720-mi (1200-km) length of this fault slipped up to 50 ft (15 m) during this event with most of the movement concentrated along a 240-mi (400-km) segment. The seafloor was lifted and in turn so was the ocean water above it. It was the redistribution of the ocean surface that formed a huge teletsunami that devastated the entire Indian Ocean Basin.

US Geological Survey geophysicists realized the severity of this earthquake as the arrivals appeared on their seismographs. They immediately suspected that a major tsunami was possible. The problem was that unlike the Pacific Ocean which had a real-time tsunami monitoring system, the Indian Ocean did not. Further, there was no warning system around the ocean basin, so alerts could not even be sent out. The agency would have to call every coastal town. Even if they could make that many phone calls in a very short time, at 7:30–8:00 a.m. on the morning after Christmas (Boxing Day), no one would be available to receive them. As a result, the overbooked coastal resorts were struck with no warning.

The largest waves of the tsunami struck the Indonesian coast less than one hour later (Figure 9.3A). They were 50–90 ft (15–27 m) in height and penetrated inland 0.6 mi (0.9 km) to more than 1 mi (1.7 km) depending upon topography. The waves progressed around the Indian Ocean for the rest of the morning slowly losing height and velocity but still causing death and destruction (Figure 9.3B). Resorts in Thailand, especially Phuket, were demolished as the 40-ft (12 m) waves came ashore (Figure 9.3C). As the wave approached, it drew the water away from the coast (drawback) exposing the seabed for several minutes before the wave struck. Curious tourists ran to see the exposed seabed only to be crushed as the wave struck. In about two hours, the tsunami struck India and Sri Lanka at 30 ft (9 m) high and continued inland 0.6 mi (0.9 km). An eight-year-old girl in India had done a report on tsunamis the previous semester in her school. When she saw the water retreat, she screamed that everyone should leave the beach and saved about 100 people. The waves devastated the ocean islands they encountered and caused death and destruction on the east coast of Africa about eight hours later, thousands of kilometers away.

The death toll estimates were revised several times, but settled at 227 900. A staggering 1.7 million people were left homeless. The worst damage occurred in Indonesia where there were 130 000 confirmed dead and 37 000 missing and presumed dead. The uncertainty in the numbers is expected with a tsunami. In land-based earthquakes, the collapsed buildings are right over the

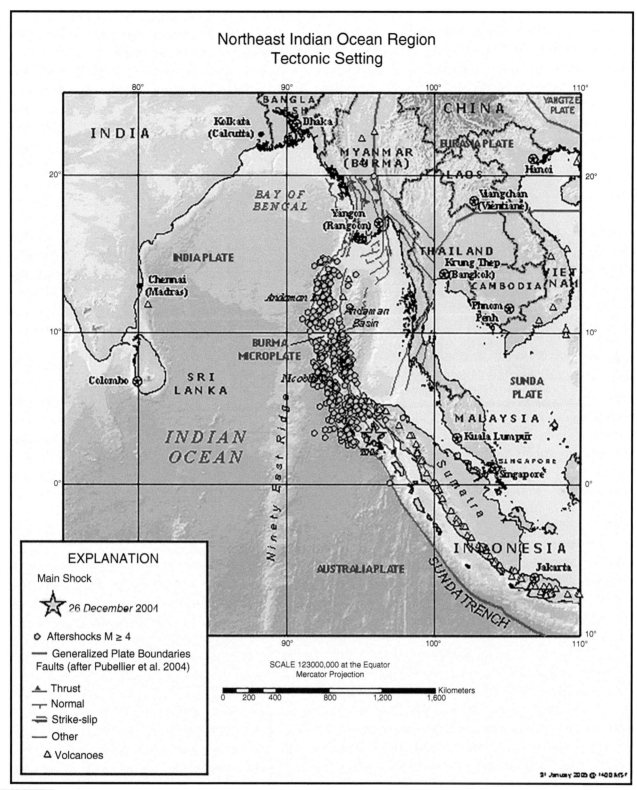

FIGURE 9.3 Map of Southeast Asia showing the location of the epicenter of the earthquake that generated the Indian Ocean tsunami and its aftershocks. *Source*: North Indian Ocean Region Tectonic Setting, United States Geological Survey.

FIGURE 9.3A Photo of the drawback and bathers investigating the seafloor at Phuket, Thailand with the 40-ft (12-m)-high roiling white wall of water of the 2004 Indian Ocean tsunami racing toward them. *Source*: Getty Images.

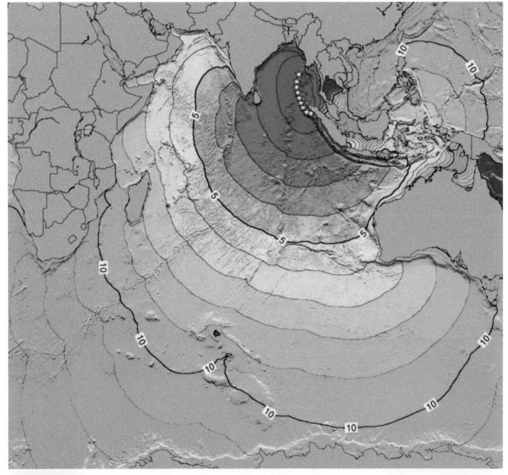

FIGURE 9.3B Map showing the dispersion of the 2004 Indian Ocean tsunami. The lines are travel time contours and the numbers show hours from generation. *Source*: Image courtesy of NOAA.

FIGURE 9.3C Photo of a flooded village in Thailand from the 2004 Indian Ocean tsunami. *Source*: AP Photo/Gemunu Amarasinghe.

FIGURE 9.3D Satellite image of an agricultural area in Indonesia struck by the 2004 Indian Ocean tsunami, (1) before and (2) after. *Source*: Courtesy of NASA.

FIGURE 9.3D Continued

FIGURE 9.3E Panoramic view of an area of Indonesia impacted by the 2004 Indian Ocean tsunami. *Source*: Courtesy of the US Geological Survey.

foundation and the bodies are still present. In a tsunami, the bodies, houses, and neighborhoods are swept out to sea (Figure 9.3D).

The media coverage for this event was more complete than for any other natural disaster to that point. Television and Internet were flooded with tsunami pictures and videos for weeks (Figure 9.3E). They were amateur videos by tourists that commonly ended with a mad scramble as the wave approached. There were dead and missing from so many countries that it was a worldwide incident. There were reports of rescued people who had been clinging to trees for days and miles out in the ocean, and presumed dead people being located by their relatives and friends.

CASE STUDY 9.3 2011 Tōhoku Earthquake (Great East Japan Earthquake)

Earthquake-Generated Nuclear Accident

In addition to the direct disaster of the tsunami, collateral disasters can be spawned by the waves that exacerbate the situation. The Tōhoku earthquake not only caused a devastating tsunami but also the second worst commercial nuclear disaster in history. A major earthquake struck just offshore of Northeast Honshu, Japan on Friday, 11 March 2011 at 2:46 p.m. The magnitude was 9.0 making it the strongest earthquake recorded in Japan and the fourth strongest recorded earthquake ever. The epicenter was 80 mi (129 km) east of Sendai, Japan and 231 mi (373 km) northeast of Tokyo (Figure 9.4). The focus was at a moderate depth of 19.9 mi (32 km).

The earthquake was generated from a megathrust of the subduction of the Pacific Plate beneath the Japanese Arc/Asian Plate. The rupture on the fault plane was 186 mi (300 km) long and 93 mi (150 km) wide with movement of 98–131 ft (30–40 m) at its maximum point. This produced rapid uplift of the seafloor of 8–26 ft (2.4–8 m) locally. The uplift produced a deadly tsunami that struck Northeastern Japan (Figure 9.4A). The earthquake caused subsidence of as much as 2.0 ft (0.6 m) along 250 mi (400 km) of coastline.

Despite the power of the earthquake, the most destructive aspect was the tsunami which had run-up heights to 127 ft (38.9 m) and inland penetration of up to 6 mi (10 km) (Figure 9.4B). After the earthquake struck, submarine sensors detected the tsunami and an alert for the northeast coast of Japan was broadcasted. However, the tsunami arrived in just 10–30 minutes and flooded the Sendai Airport in less than an hour. This warning was too short for many residents to evacuate and safeguards to be enacted. Even though many of the harbors and inlets had tsunami protection walls, the waves overtopped them (Figure 9.4C). The flooding impacted an area of 181.5 mi^2 (470 km^2) of

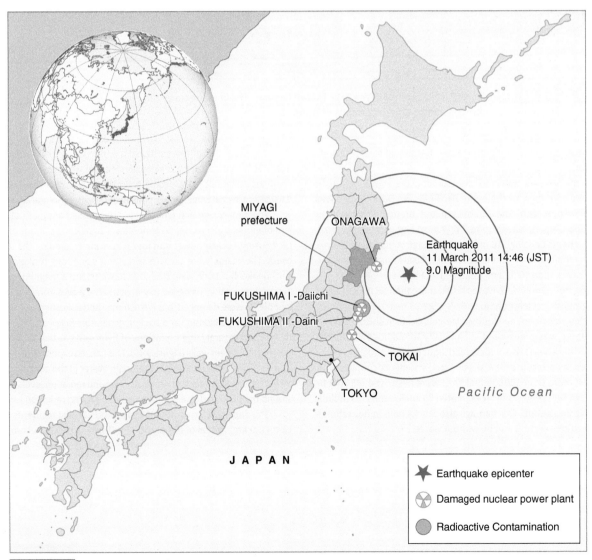

FIGURE 9.4 Map of Japan showing the location of the epicenter of the 2011 Tōhoku earthquake and the Fukushima nuclear power plants damaged by the tsunami. Inset shows location of the area on the earth. *Source*: Adapted from The Geological Society, Tohoku Earthquake, Japan: © Maximilian Dörrbecker.

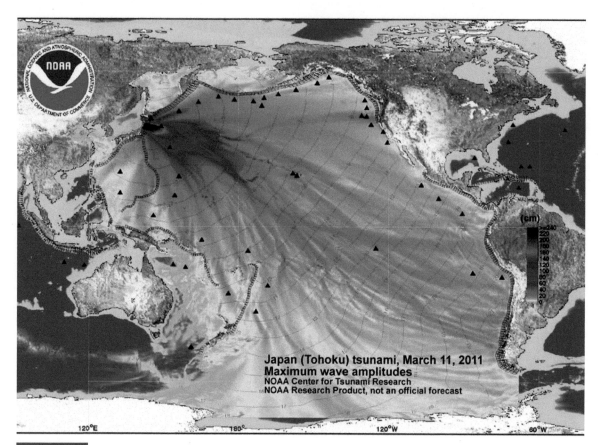

FIGURE 9.4A Map showing the dispersion of the 2011 Tōhoku tsunami. The lines are travel time contours and the numbers show hours from generation. *Source*: Courtesy of NOAA.

coast and drowned many residents of areas that should have been safe (Figure 9.4D).

When the sensors detected the tsunami, the PTWS Center in Hawaii issued warnings for the entire Pacific Ocean Basin. There was damage in harbors and coastal communities around the basin. In California and Oregon, waves up to 8 ft (2.4 m) high damaged docks, harbors, and boats causing $10 million in damage. A man in California was killed by them. Even in Chile, 11 000 mi (17 000 km) away from the epicenter, waves up to 10 ft (3 m) damaged 200 houses.

Shaking from the surface waves lasted more than six minutes and produced ground acceleration up to 2.99 g (29.33 m s^{-2}). The intensity of the shaking was IX on the Modified Mercalli Scale in shore communities, but there was widespread VIII intensity throughout Northeastern Japan. There was also widespread liquefaction and landslides on hillslopes. As a result of the shaking and its effects coupled with the tsunami, more than 190 000 buildings were destroyed (45 700) or damaged (144 300) generating an estimated 25 million tons (23 million mt) of rubble and debris. This damage also left 4.4 million households without electricity and 1.5 million without water.

The casualties of the earthquake and tsunami were 15 901 people dead, 6157 injured, and 2529 people missing. This means that the final death toll was about 18 500 people. Estimates on the cost of this disaster are $360 billion, which would make it the most expensive natural disaster ever.

The earthquake and tsunami disaster caused another major problem of global concern. Japan generates a significant amount of its electricity using nuclear power and four plants are located on the northeast coast. The plants were built to withstand significant shaking from an earthquake, but none were designed to withstand a magnitude-9 event. In no way were they designed to withstand the waves from a tsunami. All four plants were damaged, but Fukushima I and Fukushima II plants were impacted by the tsunami and backup generators failed as a result. Fukushima I had several large explosions of hydrogen gas that built up in the outer containment walls (Figure 9.4E). This caused exposure of the reactor core and leaking of radioactive materials. Water in the containment vessel had radiation levels 1000 times background radiation levels and areas around the plant had levels eight times background radiation levels.

The Japanese government evacuated 200 000 residents within a 12-mi (20 km) radius of the plant. The US government recommended

FIGURE 9.4B Photo of the 2011 Tōhoku tsunami and cars overtopping the tsunami protection walls in Sendai, Japan. *Source*: Handout/ REUTERS.

FIGURE 9.4C Photo of the 2011 Tōhoku tsunami coming ashore in Sendai, Japan. *Source*: Kyodo News/Getty Images.

that citizens should remain at least 50 mi (80 km) away from the plant. Fortunately, the weather changed, and winds blew most of the airborne radioactive release out to sea. However, radioactive iodine, cesium, and strontium were detected in soil and even in tap water around the plant. Consumption of food from the area was banned and various levels of panic ensued. In California, store shelves were stripped of potassium pills in fear that radiation would reach the United States. Although this was an overreaction, the Fukushima I leak is second only to Chernobyl as the worst civilian nuclear disaster in history.

174 CHAPTER 9 Killer Tsunamis

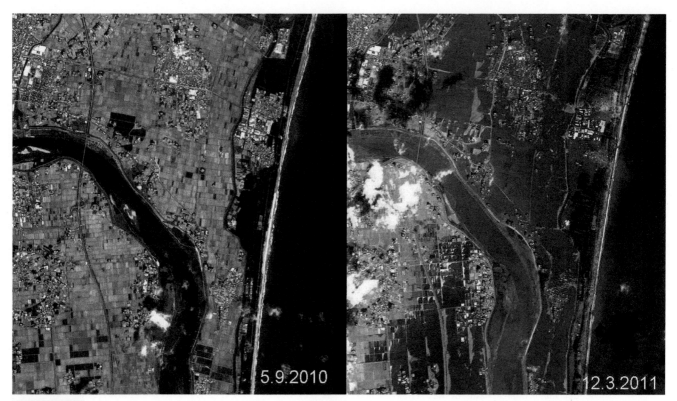

FIGURE 9.4D Satellite image of northeast coast of Japan struck by the 2011 tsunami, left (before) and right (after). *Source*: Courtesy of NASA.

FIGURE 9.4E Aerial photo of smoke emanating from the fires at the Fukushima nuclear power plant damaged by the 2011 Tōhoku tsunami. *Source*: Photo 12/Ann Ronan Picture Library/Alamy Stock Photo.

CASE STUDY 9.4 Cumbre Vieja Volcano, Canary Islands

Potential Hollywood-Scale Disaster

After the 2004 Indian Ocean tsunami and the 2011 Tōhoku tsunami, alarm about coastal safety against tsunamis increased dramatically. Although there had been no recent occurrences on the Atlantic coastal areas, they were included in the alarm. There were recorded tsunamis in the Atlantic Ocean in 1927 in New England and Eastern Canada, in Puerto Rico in 1918, and even a teletsunami from the 1755 Lisbon earthquake that reached the Caribbean. However, none of these events was even remotely as deadly as the recent events.

The Cumbre Vieja volcano is relatively unremarkable and part of the Canary Islands which are the result of a mantle plume (Figure 9.5). This kind of volcano only produces dangerous eruptions on very rare instances. Further, the Canary Islands are off the African west coast and quite a distance from the United States East Coast. All of these reasons make it an unlikely threat. However, the two recent teletsunamis wreaked havoc all the way across large ocean basins, making anything seem possible.

During the 1949 eruption of Cumbre Vieja, the western flank of the volcano began a slow collapse, in a developing landslide (Figure 9.5A). During the 1971 eruption, the mass slid farther to an even more precarious situation. A research team from the United Kingdom speculated that the flank could collapse during a subsequent eruption producing a catastrophic avalanche into the Atlantic Ocean. This avalanche could produce a situation like the 1958 event in Lituya Bay, Alaska, in which it generated a huge wave that sloshed 1700 ft (518 m) up the facing slope. The research group speculates that a teletsunami 300 ft (91 m) high could travel across the Atlantic Ocean and devastate the US East Coast (Figure 9.5B). It could also wreak havoc in the Caribbean Sea islands and even west coast of Africa.

If such a disaster occurs, the death and destruction would make the 2004 Indian Ocean and 2011 Tōhoku, Japan disasters pale in comparison. On the other hand, the Canary Islands have been in existence and erupting for many millions of years, and yet there is no evidence on the Atlantic coast or Caribbean area ever having been struck by a large tsunami much less a major teletsunami.

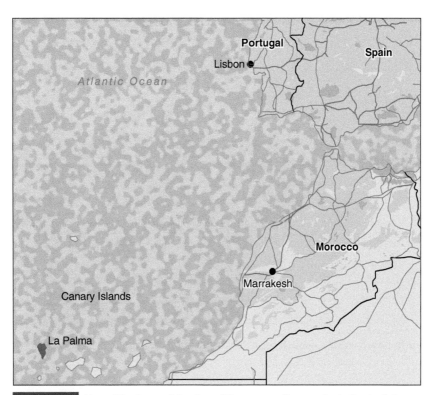

FIGURE 9.5 Map of the Canary Islands and the surrounding area including La Palma which contains Cumbre de Vieja.

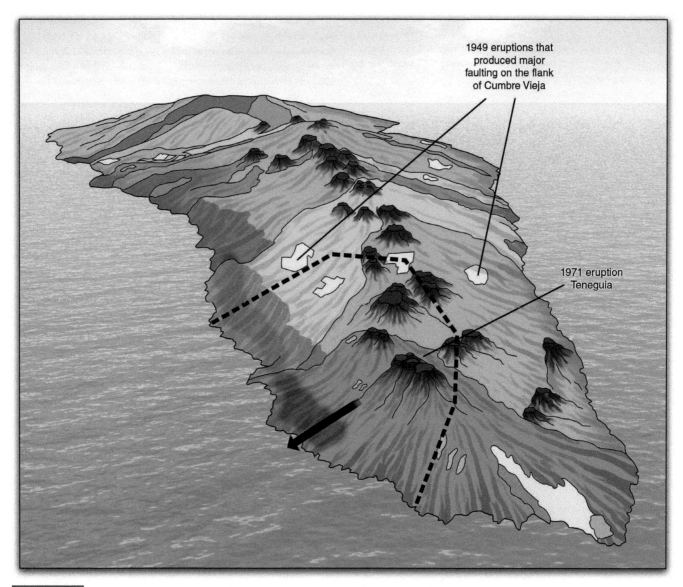

FIGURE 9.5A Diagram showing the section of Cumbre de Vieja that is developing the landslide. *Source*: Adapted from Slideplayer, La Palma.

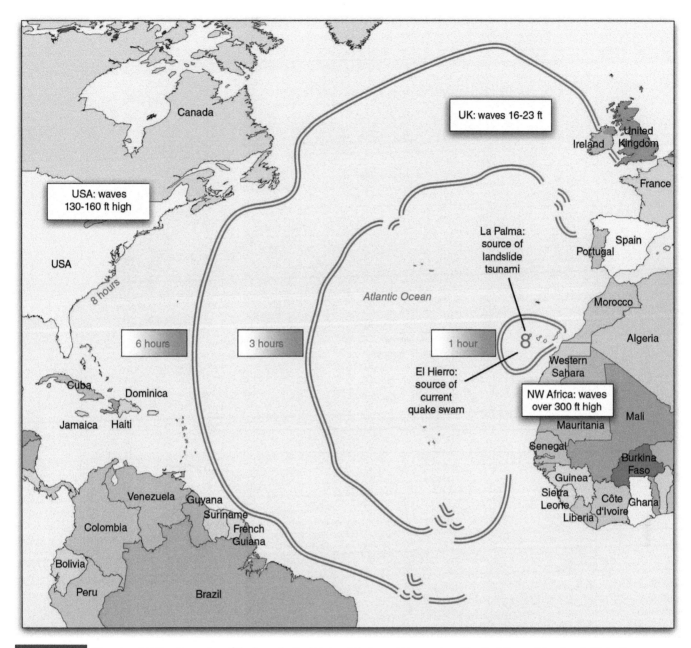

FIGURE 9.5B Map showing the dispersion of the hypothetical tsunami that would be produced by the failure of Cumbre de Vieja. *Source*: Adapted from Holly Deyo, 2011, standeye.com/Dare ToPrepare.com.

CASE STUDY 9.5 1958 Lituya Bay, Alaska Megatsunami

The Biggest Wave Ever

There are numerous processes that create large ocean waves of various sizes and destructive capability. There are numerous examples of them, but, by far and away, the largest wave in recorded history was the 1958 Lituya Bay megatsunami. This wave was produced by very unusual processes involving earthquakes and mass movement. It also provides the background for many documentaries and movies on doomsday scenarios.

On 9 July 1958 at 10:16 p.m., a major earthquake struck the Lituya Bay area, 120 mi (200 km) west of Juneau, Alaska (Figure 9.6). The magnitude was a strong 7.8 and the maximum intensity from surface wave shaking was XI, or extreme, on the Modified Mercalli Scale. The focus was at 22 mi (35 km) depth on the Fairweather Fault and the epicenter was on the northeastern edge of the Lituya Bay (Figure 9.6A). The area is so desolate that this event normally would not have been noticed by more than a few people. However, the shaking loosened a large mass of rock from a hillslope along the northeastern wall of the bay (Figure 9.6B). The rock mass was 40 million cubic yards (30.6 million m^3) and weighed 90 million tons (82 million metric tons). It plunged approximately 3000 ft (914 m) down and into the Gilbert Inlet of Lituya Bay (Figure 9.6A) as a rockfall and rockslide that hit the water at more

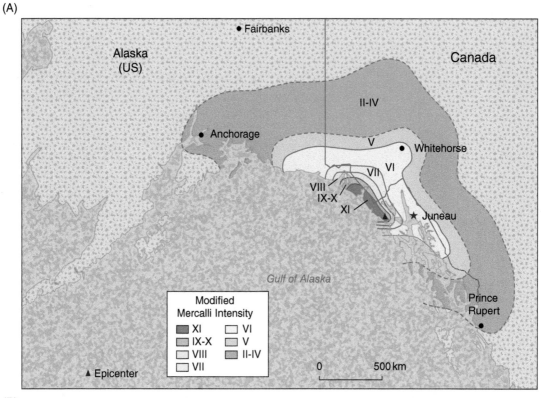

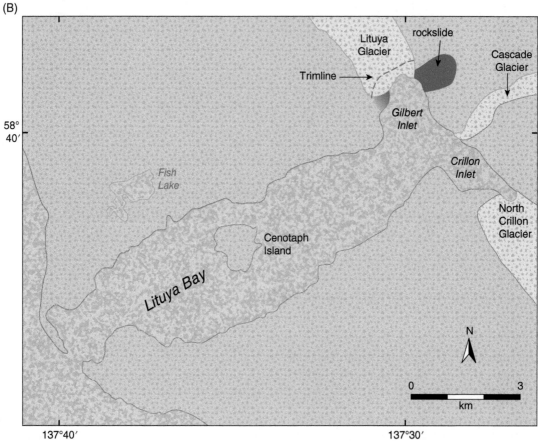

FIGURE 9.6 Maps showing (A) the location of the epicenter and intensity of the 1958 earthquake in Alaska and (B) the Lituya Bay area.

FIGURE 9.6A Photo of Lituya Bay showing the left side shoreline stripped of vegetation by the 1958 wave. *Source*: Courtesy of the US Geological Survey.

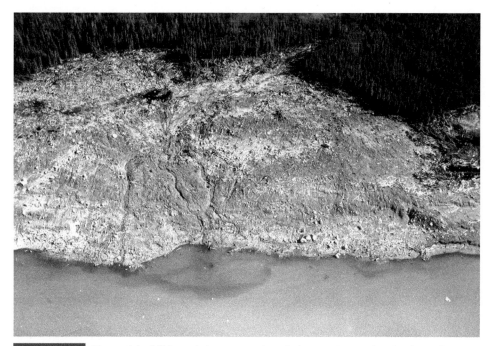

FIGURE 9.6B Photo of the hill face of Lituya Bay stripped of vegetation to a height of 1700 ft (518 m) by the 1958 wave. *Source*: Courtesy of the US Geological Survey.

than 100 mph (161 kmph). The sound of the impact was heard 50 mi (80 km) in all directions, and it apparently caused around 400 m (1312 ft) of ice to break off the Lituya Glacier. The glacier released a massive amount of sediment that also plunged into the bay forming a double slide. This combined mass caused the bay water to slosh forward in a megatsunami or megaseiche that was the largest ever measured.

The 100-ft (30-m)-high wave drove across the bay with such power that it continued up the facing slope to a maximum elevation

FIGURE 9.6C Photo of the Lituya Bay shoreline stripped of vegetation by the 1958 wave. *Source*: Courtesy of the US Geological Survey.

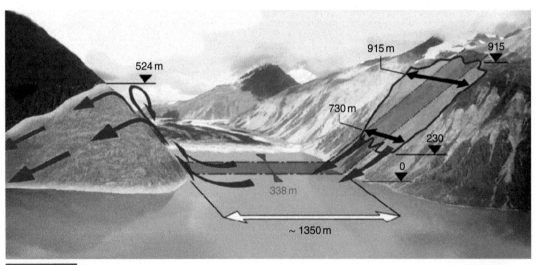

FIGURE 9.6D Diagram showing the collapse of the hill face of Lituya Bay sliding into the water and generating the wave. *Source*: Pararas-Carayannis, G. (1999).

of 1720 ft (524 m) (Figure 9.6B). It is therefore not actually a 1720-ft (524-m) wave. The water stripped out trees and soil right down to the bedrock along the entire slope. The trees and other vegetation were washed back into the bay and filled it with debris and ice (Figure 9.6C). Several fishing vessels were in the bay at the time, some of which survived and others did not, leading to a death toll of five. There is even a story of a boat riding the wave and being deposited high on the hillslope. The wave sloshed back and forth several times in a seiche (Figure 9.6D).

A megatsunami is a Hollywood-type description of a very large tsunami. Television scare-tactic documentaries use the huge Lituya Bay wave as evidence for the megatsunami concept. However, there is no evidence that a wave anywhere near this size has ever existed in a large ocean basin. They seem to be restricted to relatively small water bodies. An asteroid impact in a large ocean might produce a megatsunami wave, but such large impacts are extremely unlikely. Very large marine volcanic eruptions could also generate huge waves, but not as big as Lituya Bay.

CASE STUDY 9.6 1998 Sissano Tsunami

Underwater Landslide

Tsunamis can be produced by a number of sources from asteroid impacts to volcanic explosions. Most tsunamis are produced by sudden uplift or depression of the seafloor as the result of submarine surface faulting during an earthquake. Much less common are tsunamis produced by mass movements. There was a major wave produced by a surface avalanche into the Lituya Bay, Alaska in 1958. However, proving that underwater landslides could also produce tsunamis was not so easy. The 1998 Sissano or Aitape earthquake and tsunami was probably the most definitive example at the time. Since then, many other landslide-generated tsunamis have been identified and several historic tsunamis have been reinterpreted as landslide-generated.

On 17 July 1998, at 5:49 p.m., a magnitude 7.0–7.1 earthquake struck the Aitape coast on the island of Papua New Guinea. The epicenter was 16 mi (25 km) off the coast near Aitape and the depth of the focus was 19 mi (30 km) (Figure 9.7). The surface waves produced a maximum shaking intensity of VII on the Modified Mercalli Scale which is considered severe. The shaking lasted about 19 seconds. Following this main shock, there were four moderate to strong aftershocks that struck in rapid succession. The focal mechanism of the main shock was either a low-angle megathrust or a compensating steep reverse fault. The source of the stress is the oblique subduction between the Pacific and Australian Plates at the New Guinea Trench at convergence rate of 4.3 in. (11.1 cm) per year.

The earthquake, however, was not the catastrophic part of this event. Between 10 and 25 minutes after the main shock, three large tsunami waves devastated the villages of Sissano, Warapu, Arop, and Malomo on the Papua New Guinea north coast (Figure 9.7A). The three waves are estimated to have been an average of 13 ft (4 m) high. The second of the three waves, however, averaged between 33 and 49 ft (10–15 m) high and caused most damage. The peak wave heights and greatest destruction were along a stretch of coast 8.7 mi (14 km) long around the Sissano Lagoon with as much as 1640 ft (500 m) inland penetration of the waves (Figure 9.7B). In this area, all buildings were destroyed and 20–40% of the people perished (Figure 9.7C). Less-intense destruction extended southeast 14 mi (23 km) and northwest about 5 mi (8 km). The tsunamis were felt up to 155 mi (250 km) to the west-northwest.

Villagers reported hearing a loud boom and observing bubbling of the sea. The ocean receded from the shoreline by about 164 ft (50 m), exposing the seabed for four to five minutes. Then a wave became visible as a distant dark line on a sea surface. The wave began to break about 656–984 ft (200–300 m) from the shore. Almost all of the fleeing villagers were overtaken and battered in the debris-filled water. Some were able to cling to floating debris, but most were swept out to sea (Figure 9.7D). Because many bodies were swept away, the death toll is a matter of speculation. At least 1600 people perished, and the maximum is estimated at 2700 with the most accepted death toll of 2200. There were at least 1000 seriously injured and 10 000 left homeless.

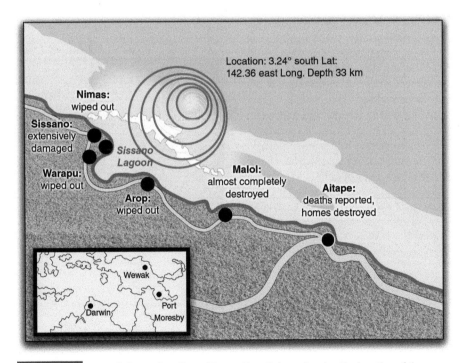

FIGURE 9.7 Map of Aitape shoreline of Papua New Guinea showing the location of the epicenter of the 1998 earthquake. Inset shows location of the area in the South Pacific.
Source: Modified after DrGeorgePC.com Tsunami page.

FIGURE 9.7A Photo showing onshore damage caused by the 1998 Sissano tsunami. *Source*: Image courtesy of NOAA.

However, there were problems with the tsunamis having been generated by the earthquake. The first issue was that the wave heights were greater than expected from a magnitude-7 earthquake. These heights were more consistent with an earthquake greater than magnitude 7.5. The second irregularity was that even though the epicenter was close to shore, there were 20 minutes between the earthquake and the tsunami. This is 10 minutes later than it should have arrived. The distribution of the tsunami heights was also anomalous with the highest waves far from the epicenter. There was also 20–27.5 in. (50–70 cm) of coastal subsidence in the area of most destruction. Researchers concluded that the inconsistencies might be explained by a submarine landslide generating the tsunami rather than the earthquake. The earthquake triggered the submarine landslide.

Scientists uncovered submarine landslide deposits and extensional fissures in the seafloor supporting the theory. In response, a multichannel seismic reflection profile was run in the area and imaged the submarine landslide scar. The landslide is estimated to have had a maximum thickness of 2493 ft (760 m), a length of 2.8 mi (4.5 km), and a width of 1.55–1.86 mi (2.5–3.0 km). The cross-sectional area of the slide was 0.89 mi^2 (2.3 km^2) and the volume was 0.9–1.1 mi^3 (3.8–4.6 km^3). This formed a 6.1-mi (10-km)-wide submarine amphitheater. The distance the slide moved along the failure plane is estimated at 3215 ft (980 m). Modeling of the event suggests that the movement took about 45–50 seconds.

This new discovery caused researchers to investigate tsunami hazards in many areas that were previously not considered to be at risk of earthquake-generated landslide tsunamis. Therefore, the 1998 tsunami is considered an important event because it demonstrated that even small and deep-water submarine mass failures (SMFs) can generate devastating tsunamis that can strike without warning.

FIGURE 9.7B Photo showing an area impacted by the 1998 Sissano tsunami. *Source*: Image courtesy of NOAA.

FIGURE 9.7C Photo showing a house swept out to sea by flowback of the 1998 Sissano tsunami. *Source*: Image courtesy of NOAA.

FIGURE 9.7D Photo showing an area stripped of vegetation by the 1998 Sissano tsunami. *Source*: Image courtesy of NOAA.

Reference

Pararas-Carayannis, G. (1999). Analysis of mechanism of tsunami generation in Lituya Bay. *Science of Tsunami Hazards* 17(3): 193–206.

CHAPTER 10

Predicting Earthquakes and Reducing Hazards

CHAPTER OUTLINE

10.1 Where: Easy, When: Not So Easy 187

10.2 Cockroaches and Other Precursors 188

10.3 Earthquake Hazard Reduction 194

10.4 Big Data: The New Hope 200

Earth's Fury: The Science of Natural Disasters, First Edition. Alexander Gates.
© 2022 John Wiley & Sons Ltd. Published 2022 by John Wiley & Sons Ltd.
Companion website: www.wiley.com/go/gates/earthsfury

Words You Should Know:

Base isolation – Devices and building practices that reduce the transmission of shaking of the earth from buildings or the effect of shaking on buildings.

Precursor – Feature or change in the environment that indicates the possible coming of an earthquake.

Recurrence interval – The average time interval between earthquakes of similar size in a particular area.

Structural reinforcement – Building techniques that reduce the impact of shaking on houses, buildings, and other infrastructure.

Tsunami gates – Gates across river mouths, valleys, and low-lying areas that move into place when tsunamis are sensed.

10.1 Where: Easy, When: Not So Easy

Being able to predict earthquakes has been a goal of society for centuries. Many have claimed the ability to predict them, but there has been no consistency even with solid identification of precursors. New technology is showing great possibilities of being effective, but it has yet to be proven. Instead, society has chosen to protect against them as much as possible. This usually depends on how much money is available to build protective infrastructure.

Considering the dangerous and unpredictable nature of earthquakes, being able to predict them is an important goal. Earthquakes are the most unpredictable and the most destructive of all natural disasters. There are three main aspects of prediction: location including the epicenter and depth of focus, power of the event including the magnitude and intensity, and, most importantly, the timing of the main shock. Methods to determine the locations of future earthquakes are relatively sound. Plate tectonic theory constrains the most active areas on Earth. Communities and governments keep records on earthquakes in several document types including newspapers, death records in municipal and religious institutions, building permits, and diaries and journals. Historical records with enough information to estimate location and severity of the majority of earthquakes extend up to several thousands of years in some areas.

Ever since seismographs became reliable, several government offices and research and educational institutions compile information on every sizable earthquake. The US Geological Survey maintains a database on all earthquakes that have magnitudes greater than 2.0 including location, depth, time, and magnitude where possible. These records are used to identify any and all active faults. These databases form the basis to improve prediction.

10.1.1 Location

All active plate tectonic margins produce regular earthquakes. More than 99% of all earthquakes and nearly 100% of powerful earthquakes occur at plate boundaries (Figure 10.1). In these areas, active faults are well-mapped as are the associated earthquake hazards. Intraplate earthquakes occur in all areas other than plate margins on faults that can lie inactive for hundreds of years or more. For this reason, only a few of them are identified and well-mapped and the vast majority remains unknown. Fortunately, the earthquakes they produce are overwhelmingly minor causing no to minimal damage and only rare casualties. For this reason, the areas they strike are ill-prepared to deal with them. Large earthquakes in these areas are catastrophic.

10.1.2 Strength

The power of earthquakes can be measured or estimated using historical records. This information determines the potential strength of possible future earthquakes. In areas with regular seismic activity, data from modern equipment are used to estimate the power of the earthquakes that might occur. Methods have been developed to use historical records to estimate the location, intensity, and magnitude of ancient earthquakes. These methods utilize damage reports, observations of repairs to the damage, and the geographic size of the affected area. This information is compared with modern earthquakes of known magnitude and intensity in an area to determine the power of the historic earthquake. These predictions can be relatively accurate in active margins and notoriously poor for intraplate settings.

10.1.3 Timing

Projecting when an earthquake will occur is the most difficult and most important aspect of prediction. This is also addressed using a combination of modern and historical records. By observing the frequency of earthquakes in an area over time, a recurrence interval for earthquakes is established. Recurrence interval is the average time span in years between earthquakes of a certain

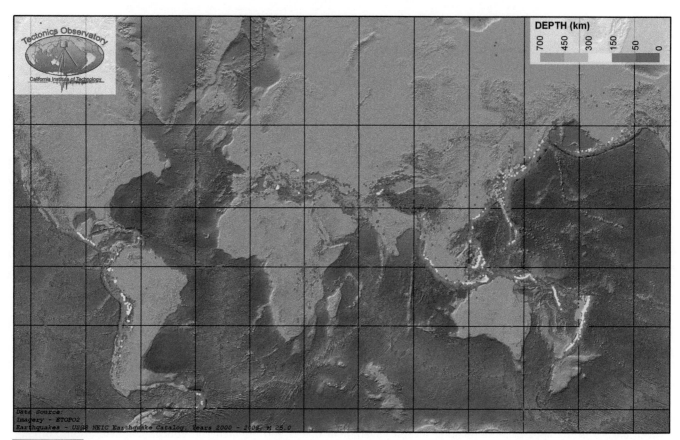

FIGURE 10.1 Map of the earth showing the locations of epicenters of earthquakes from 1960 to 2000. *Source*: Image courtesy of the National Science Foundation.

magnitude. There are different recurrence intervals for each magnitude and they are specific to the area. For example, the recurrence interval for a certain city might be 50 years for a magnitude-7.0 earthquake, but 500 years for a magnitude-8.0 earthquake. Recurrence intervals are among the better tools for prediction of earthquake timing because the buildup of tectonic stress on a fault is constant and the strength of the rocks does not vary. If the historical records extend a long time and are complete, a reasonably precise recurrence interval may be determined. The problem is that uncertainty of 20 or 25 years is considered scientifically accurate, but it may not be acceptable to most communities and residents.

In active margins like California, Japan, New Zealand, the Philippines, and Indonesia, the abundance of earthquakes makes it possible to determine relatively accurate recurrence intervals. On the other hand, earthquakes are so infrequent and poorly understood in intraplate areas that recurrence intervals can have errors in the hundreds of percent.

The error increases dramatically for recurrence intervals of specific-size earthquakes. It is a certainty that California will experience an earthquake over the next few months, but when the next magnitude 7.0 or higher earthquake will occur, is far from certain. It can probably be predicted to occur within the next 25–50 years, which is excellent in geological terms, but not useful for society or individuals. Emergency management planning can only accommodate uncertainty of a few days to one week at most, and residents might tolerate a month or more. In response, earthquake precursors are studied to try to determine a tighter time scale for impending earthquakes.

10.2 Cockroaches and Other Precursors

There are several accepted precursors to predict an earthquake, but they can rarely be identified (Figure 10.2). Activity of small earthquakes tends to increase before a large earthquake. The problem is that small earthquakes do not always mean that a large one is imminent. Small earthquakes can persist for months or years and the large earthquake may not occur. Even if the large earthquake does occur, its size is unpredictable. Other precursors include increased radon gas emissions, changes in water levels in wells and springs, changes in oil levels and oil and natural gas flow rates, changes in groundwater chemistry, changes in seismic velocity of the rocks, rapid local uplift or subsidence, and strange behavior of insects (such as cockroaches), birds, and animals at various lead times. The changes in radon, water, gas, and oil levels and flow are thought to be the result of dilation of small fractures as the stress builds up. This may be the cause of the strange animal behavior, but there is also a theory that the stressing of crystals in the rocks produces piezoelectric (minor electric impulses) effects that can be sensed by animals.

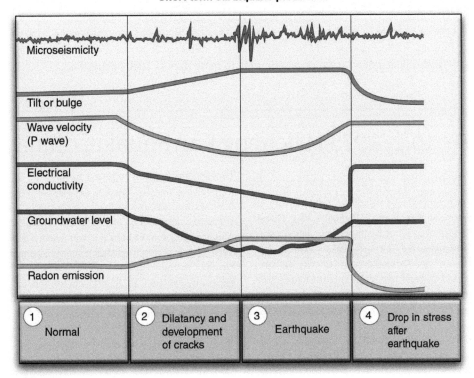

FIGURE 10.2 Traditional earthquake precursors and their changes with time leading up to an earthquake.

In addition to the fact that they may not occur, the problem with most of these other precursors is monitoring. The cost involved in detailed monitoring of every active fault for radon, water levels, water chemistry, and seismic velocity is prohibitive. There may also be other causes for changes in many of the precursors, such as weather, usage, and climate change. The warning times for the precursors are also not known. There is no exact science to monitor strange behavior of animals and insects and determine what constitutes strange behavior.

In Haicheng, China, there were multiple excellent precursors that were acted on with large-scale evacuations which saved tens of thousands of lives. On 4 February 1975, a magnitude-7.3 earthquake struck the heavily populated but evacuated Haicheng area and only 1300 people were killed and 17 000 injured (Figure 10.3). Without the evacuations, it is estimated that more than 150 000

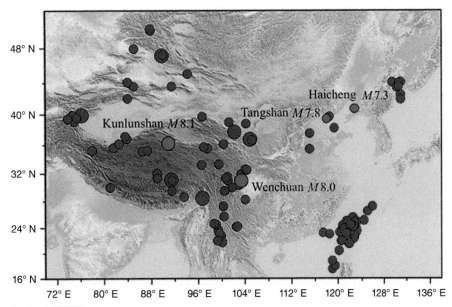

FIGURE 10.3 Map of China showing the locations of instrumental major earthquakes including the 1975 Haicheng and 1976 Tangshan earthquakes. *Source*: Fuqiong Huang, et. al., (2017), with permission from ELSEVIER.

people would have perished. This success awed Western seismologists into believing that the Chinese developed reliable methods to predict earthquakes. This awe was short-lived because on 28 July 1976, with no warning, double main shock earthquakes with high magnitudes struck the even more heavily populated area of Tangshan, China. This event caused as many as 655 000 deaths and more than 800 000 injuries. Perhaps because of the secrecy of the Cold War or perhaps because the prediction methods had proven clearly unreliable, the Chinese government never admitted to more than 255 000 having died.

CASE STUDY 10.1   1975 Haicheng Earthquake, China

A Well-Predicted Earthquake

A powerful earthquake of magnitude 7.3 struck the industrial city of Haicheng, China at 7:36 p.m. on 4 February 1975 (Figure 10.3). It was a damaging and deadly earthquake, but its importance is the catastrophic loss of life that was prevented. The earthquake was the first to be accurately predicted using scientific methods and the city was successfully evacuated. The swift and effective actions of the Chinese government saved hundreds of thousands of lives. If this was not done, it would have been included in the list of most catastrophic earthquakes of all time.

The epicenter was located in Chagou Village, Haicheng, Liaoning Province, China, just northeast of Beijing and the area of highest damage was 293 mi^2 (760 km^2) around it. Shaking from surface waves caused intensity of IX on the Modified Mercalli Scale over an area of 32 mi (51 km) by 14 mi (22 km) and maximum intensity reached XII in the epicentral area. Minor damage was reported in Seoul, South Korea and it was felt in Primorsky Kray, USSR and Kyushu, Japan. The earthquake was caused by movement on a previously unknown, left-lateral strike-slip blind fault with a northwest orientation.

The Haicheng earthquake exhibited a remarkable series of precursors that led to the main shock. The precursors were in four stages, long-, middle-, and short-term and imminent beginning about a year in advance. The initial precursor was an increase in seismicity noted by seismologists and reported to officials. Other precursors began in following months including geodetic or land surface deformation, changes in groundwater levels, color, and chemistry of well water, pressure changes in oil and gas wells, and peculiar animal behavior. This unusual animal behavior was widely recognized by December 1974. Residents reported dazed snakes and rats that appeared to be "frozen" to the roads. The unusual animal behavior became widespread by February 1975 and included restlessness and agitation of cows, horses, and other large animals. Chickens refused to enter their coops and rats acted as if they were drunk.

Physical and chemical anomalies began in the surface features of the region. Several areas of unusually warm water temperature melted sections of an otherwise frozen reservoir during the very cold winter of 1974–1975. Abnormally high fluid pressure was recorded in groundwater including waterspouts as high as 8 ft (2.3 m) two hours before the earthquake and the pressure in an oil well increased from 118 bars on 8 October to 139 bars on 11 October 1974. Sulfur gas emissions and "earthquake light" appeared shortly before the main shock.

However, the foreshock sequence was the most important precursor. There was a surge in minor earthquake activity beginning on 3 February 1975. The area seismic observatory recorded 33 small earthquakes of steadily increasing magnitude in the main shock epicenter area. Levels in water wells started dropping. A well went dry that previously had a water level 3.3 ft (1 m) below the surface. The water level in

FIGURE 10.3A   Photo of damage suffered in the 1975 Haicheng earthquake. *Source:* Adams 1976/with permission of John Wiley & Sons.

FIGURE 10.3B Photo of destruction in the 1975 Haicheng earthquake. *Source*: Wang et al. 2006/with permission of Seismological Society of America.

one community well dropped by 1.6 ft (0.5 m) and it dropped by 2.3 ft (70 cm) in another well. The water turned muddy and bitter-tasting in all of these wells. After that, the foreshocks abruptly stopped.

Scientists and officials reacted to the precursors. On 13 January 1975, the Seismological Bureau predicted an earthquake of magnitude 5.5–6 in the Yingkou–Dairen–Tantung region of Liaoning Province during the first six months of 1975. When the foreshock sequence began, they concluded that a major earthquake was imminent and initiated plans for evacuation. When the foreshocks stopped, they concluded that stress was building up for the main shock and a massive evacuation of the population was ordered on the afternoon and evening of 4 February. This was not easy because the population of Haicheng was more than one million and there were more than three million people in the area. Warnings were the responsibilities of local governments and they were not uniformly applied making the evacuation patchy.

Most of the population had been evacuated when the main shock struck. The evacuation saved these people because the surface waves caused intense shaking, extensive surface failure, and widespread liquefaction which devastated buildings and infrastructure (Figure 10.3A). It damaged or destroyed more than 54 680 665 ft^2 (5 080 000 m^2) of urban housing, 867 000 rooms in rural houses, 1038 mi (1670 km) of transport pipelines, over 2000 bridges, and more than 700 hydraulic facilities. Sand boils from liquefaction covered more than 69 mi^2 (180 km^2) of farmland. The total economic loss was 0.8 billion Chinese Yuan.

These staggering structural losses should have been associated with tens of thousands of deaths and yet there were only 1328 deaths and 16 980 injuries. Including casualties from fires started by the earthquake and exposure from the loss of shelter, the official total death toll is 2041 and 24 538 injuries. It has been estimated that the deaths and injuries would have exceeded 150 000 without the evacuation.

The town of Dashiqiao provides an example of the evacuation and its effectiveness. Local authorities prohibited shopping and public entertainment, relocated hotel guests and hospital patients, and dispersed people with loudspeakers before the main shock. Even though the town experienced a shaking intensity of IX which caused a staggering 66% of the buildings to collapse (Figure 10.3B), only 21 people died, from a population of 72 000 people. Similarly, the Yingluo Commune of Haicheng was located in the most severely damaged region, but the residents were evacuated outdoors using loudspeakers. As a result, even though 95% of the 28 027 rooms in the commune were totally destroyed, only 44 people out of a population of 35 786 died.

CASE STUDY 10.2 2009 L'Aquila Earthquake, Italy

Danger of Discounting Earthquake Predictions

A strong earthquake of magnitude 6.3 struck the Abruzzo region of Italy at 3:32 a.m. on 6 April 2009. The epicenter was located in the small town of L'Aquila about 55 mi (85 km) northeast of Rome (Figure 10.4). The focus was at a relatively shallow depth of 5.5 mi (8.8 km). There were reports of earthquake light. The area has a record of strong earthquakes in 1315, 1349, 1452, 1501, 1646, 1703, and 1706. The city is built on the soft sediments of an old lake bed, which amplify seismic waves causing increased damage. The population of the town is about 80 000.

The earthquake was produced by movement on a northwest-oriented normal fault that crosses the Apennine Mountains. The plate tectonics of the region include the subduction of a microplate beneath Italy, the collision of Africa with Southern Europe, and the opening of the Tyrrhenian Basin to the west. The cause of the earthquake, however, was the east–west extension that occurs along the Apennine range.

Surface waves caused the collapse of numerous buildings including several historic structures (Figure 10.4A). Nearly 11 000 buildings were damaged or destroyed. The collapsed buildings caused the death of 308 people with 1500 injuries and 65 000 homeless (Figure 10.4B). Nearly 40 000 people were forced to live in tents for extended periods, whereas others lodged with relatives or in hotels (Figure 10.4C).

The government response to the disaster was rapid. Nearly 11 700 rescue workers provided emergency rescue and relief to the survivors.

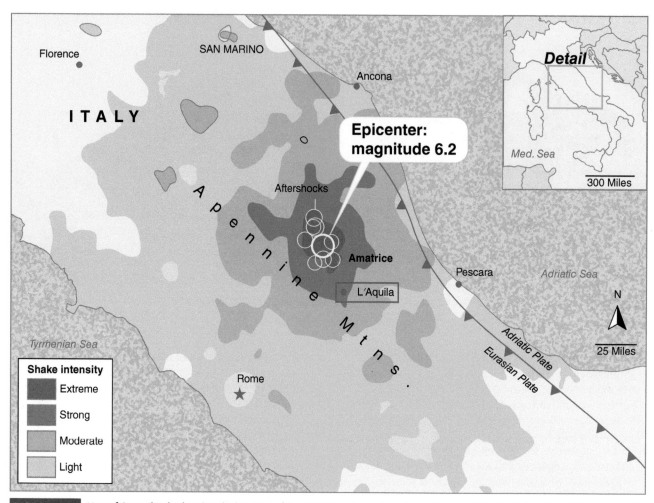

FIGURE 10.4 Map of Central Italy showing the location of the 2009 L'Aquila earthquake and areas of shaking intensity with inset map of the Italian Peninsula. *Source*: Adapted from Tim Meko and Ashley Wu/The Washington Post; information from the US Geological Survey.

FIGURE 10.4A Aerial photograph of earthquake destruction in L'Aquila. *Source*: AP Photo/Gregorio Borgia.

Many countries offered aid to Italy, but only some offered by the United States was accepted. Italy primarily handled the disaster on its own.

The L'Aquila earthquake occurred during a period of increased seismic activity in the area. There were foreshocks for several months prior to the earthquake including a magnitude-4 event the preceding day. The main shock was followed by several strong aftershocks including magnitudes of 5.3 on 7 April and 5.1 on 9 April.

About a month before the main shock, volcanologist Giampaolo Giuliani measured increased radon emissions in the area and predicted a large earthquake in the region. Increased radon is a long-accepted precursor to earthquakes, but difficult to monitor. His warnings, however, were disregarded by city officials and he was charged with causing a public panic and ordered not to spread rumors. A meeting was held on 31 March 2009 by the Civil Protection Agency including

FIGURE 10.4B Photograph of a street damaged by the 2009 L'Aquila earthquake. *Source*: puckillustrations/Adobe Stock Photos.

FIGURE 10.4C Photo of a street destroyed by the 2009 L'Aquila earthquake. *Source*: ZUMA/ZUMA Press, Inc./Alamy Stock Photo.

seismologists to discuss the potential for a large earthquake. The board concluded that a large earthquake was unlikely but not impossible.

When the earthquake struck, public opinion reversed abruptly. The six seismologists from the Italian National Institute of Geophysics and Volcanology and a senior official from the Civil Protection Agency who attended the meeting on 31 March were accused of failing to warn residents about the impending earthquake. They were charged with giving "inexact, incomplete, and contradictory information" about the potential of an earthquake. Charges of manslaughter and unintentionally causing injury carrying prison sentences of up to 15 years were brought against the seven defendants. The city of L'Aquila filed a separate civil lawsuit against the seven for €50 million (about US$67 million) in compensation. It was the first instance that scientists were held responsible for not predicting an earthquake.

There was a strong reaction to the trial from the international scientific community. In June of 2011, the American Association for the Advancement of Science (AAAS) wrote a letter to the president of Italy expressing concern about the trial. Later that year, 5200 international scientists signed a petition supporting the six seismologists. The Seismological Society of America wrote a letter of protest to the president of Italy stating: "Years of research worldwide have shown that there is currently no scientifically accepted method for short-term earthquake prediction that can reliably be used by civil protection authorities for rapid and effective emergency actions." Questions were raised by investigators of the disaster about the adherence to regional building codes. There was excessive damage to newer buildings, bringing the quality of building inspection into question. Eventually, the scientists were acquitted but not before many months of great worry.

10.3 Earthquake Hazard Reduction

Although there has been much research, reliable methods to predict earthquakes have not yet been developed. The only way to avoid the impact of powerful earthquakes has been to design and build structures to withstand them, and even with improved prediction techniques, this will not change. This is the approach of the United States and other advanced countries where strict building codes are established in seismically active areas. This approach has been successful in reducing an immense number of casualties in earthquakes. The 1989 Loma Prieta earthquake near San Francisco, California had a 6.9 magnitude, but only had 63 fatalities, 42 of which resulted from the collapse of one elevated highway. By contrast, the 2010 Port-au-Prince earthquake in Haiti had a similar magnitude, depth, and fault type, yet resulted in a death toll of up to 316 000. This astounding difference is because California has very strictly enforced building codes that are reviewed and revised after every major earthquake. By contrast, earthquake hazard reduction in Haiti is poor and adherence to building codes is lax. Effective earthquake hazard reduction is the difference between a catastrophe and a managed event. There are several methods of earthquake hazard reduction such as protection against mass wasting, protection against tsunamis, protection against shaking, and protection against liquefaction.

10.3.1 Protection Against Mass Wasting

Mass wasting caused by the shaking is among the most destructive aspects of an earthquake in rugged terrains. In the 1970 Nevados Huascarán earthquake, part of a glacier broke loose from 22 000 ft (6706 m) elevation, free fell for 3000 ft (914 m), and then roared down the slope as an avalanche at speeds of up to 250 mph (402 km) for 22 mi (35.4 km). It caused the deaths of more than 18 000 people. In China, there are loess deposits from the deposition of windblown dust from the Gobi Desert. Many Chinese resided in burrows (yaodongs) dug into these deposits. The shaking from earthquakes can cause them to collapse burying the occupants and sending landslides into the valleys below burying the above ground residents as well. The 1556 Shaanxi earthquake of China may have been the worst natural disaster in history with more than 800 000 victims almost exclusively as the result of this kind of collapse.

There are a number of landscaping and construction methods that can reduce damage from slope failure.

10.3.1.1 Slope Reduction
Even low slopes can fail with the intense shaking from an earthquake, but steeper slopes contain more material and instability to create larger mass movements. Steeper slopes pose more of a hazard because failure is more likely to be in the form of rockfalls or debris avalanches which are at much higher velocity than earth flows or slumps which tend to be slow. Slope reduction therefore can involve the reduction of the height of hills, the reduction of slope angle, and covering of steep slope faces with soil. All modifications require heavy machinery and the removal of a significant amount of rock and soil.

10.3.1.2 Vegetation
Depending upon the rock, sediment, and soil making up the slope and its angle, the planting of vegetation can reduce or even prevent slope failure. The mat of root systems of the plants can stabilize the soils from erosion and failure, in some cases. Deeper root systems of trees and certain bushes can reduce or prevent slump and earthflow on shallow to moderate

slopes during an earthquake even with accompanying liquefaction. Vegetation, however, cannot prevent rockslides, rockfalls, or avalanches.

10.3.1.3 Retaining Walls and Fences

Stabilization of low bedrock cliffs against rockfalls and rockslides can be accomplished using retaining walls, fencing, rock bolts, or concrete coating in some cases. This is typically only done in locations to prevent severe damage and loss of life involving moving vehicles such as along highways and railroads. There are several ways to stabilize bedrock cliff faces depending upon height and need. The rock face can be cemented to seal up cracks and reduce breakage during shaking; use of rock bolts to stabilize the wall or covering it with steel fencing can be done to restrict broken rock fragments to falling straight down. On railroad lines, these fences are typically fitted with sensors that alert train crews about the rockfalls. In high exposures of fractured rock or in areas subject to regular earthquakes, solid vertical walls can be constructed in front of them and the space between is back filled with gravel. This too prevents rockfalls.

Steep soil-covered slopes can also be stabilized with walls. However, if they are at the base of hills, the force of groundwater can be immense. If the walls are made of solid materials like concrete, or railroad ties, drains, or weeps must be installed through the walls and the hill should be landscaped to control the drainage. Otherwise, the wall will be toppled over very quickly. Buried stabilizers called "dead men" also help to keep the wall from shifting.

10.3.2 Protection Against Tsunamis

Tsunamis are among the deadliest of earthquake hazards. They form very low broad waves in deep water that travel at speeds of 600 mph (966 km). As they approach the coast, the base of the wave drags on the seafloor which compresses it into a much narrower but taller wave traveling at 50 mph (80 km). The momentum of the wave makes it especially deadly because it can destroy any human structure and it penetrates deeply onshore. Most people are killed by being battered by the wave and the debris it carries. If they survive the wave, they may still be carried far out to sea and drowned as the water drains back. The deadliest tsunami in recent history struck the Indian Ocean in 2004 and had a death toll of more than 220 000 people because the area was completely unprepared. Even if an area is well prepared, such as Japan, they can still be deadly. The 2011 Great East Japan earthquake (Tōhoku earthquake) generated a tsunami that killed more than 25 000 people and caused the second greatest civilian nuclear disaster in history.

10.3.2.1 Early Warning Systems

After the two strongest recorded earthquakes struck the Ring of Fire in Chile in 1960 (M = 9.6) and Alaska in 1964 (M = 9.2) and generated destructive tsunamis, a tsunami early warning system was established for the Pacific Ocean. The system includes the placement of detection buoys around the ocean, a satellite transmission system, a siren warning system, and evacuation plans for the surrounding communities. The buoy is attached to a pressure sensor on the seafloor because even though the wave is low, it is so broad that it changes the water pressure beneath it. Once a tsunami is detected, the buoy transmits the information to the satellite which then activates the warning system throughout the Pacific where residents follow evacuation routes. Tsunami evacuation is typically vertical to get above the waves. After the 2004 tsunami disaster, the Indian Ocean was also equipped with an early warning system. The Atlantic Ocean does not have a sophisticated early warning system because the margins are not active and tsunamis are rare.

10.3.2.2 Tsunami Walls and Gates

Japan has had so many devastating tsunamis that the government constructed tsunami walls and gates in key harbors and waterfronts. Walls are primarily on the shoreline or at the mouth of harbors. Shoreline walls are made of concrete and some combination of a sloped and vertical barrier that are about 14.8 ft (4.5 m) high. Harbor walls extend from the shore into the harbor mouth as breakwaters that restrict access of waves into the harbor. The walls typically have a 300 ft (91 m) wide and 45 ft (13.7 m) high base of rocks and a 20 ft (6.1 m) high and 25 ft (7.6 m) wide armored cap made of concrete or steel. The cap may also be composed of concrete tetrapods which restrict rather than block water passage. The trouble with the walls is they are disruptive to local ecology and the sediment system and can require costly refurbishment projects. They are also not that effective against large waves. The 2011 Great East Japan earthquake tsunami topped the tsunami walls.

Tsunami gates are steel doors that attach to the tsunami walls and swing or slide shut to protect harbors and waterfronts. On land, they are about 5–6 ft (1.5–1.8 m) high, so of limited effectiveness, but those in the water across harbors may be 25 ft (7.6 m) or more. There is limited usage of these gates because the earthquake may disrupt the power supply to the gates and they are difficult to operate manually.

FIGURE 10.5 Building damaged in the 1985 Mexico City earthquake with the middle floors crushed. *Source*: Image courtesy of USGS.

10.3.2.3 Protection Against Shaking

The most damaging effect of earthquakes to built structures is the shaking caused by surface waves. Structures are designed to withstand the vertical force of gravity. Forces in other directions and at other strengths are not anticipated for most buildings and therefore not considered unless absolutely necessary. Surface waves are enhanced in sediment and fill relative to bedrock and shake laterally if they are Love waves or vertically if they are Rayleigh waves. Rayleigh waves caused middle floors of office buildings to collapse as is well illustrated in the 1985 Mexico City earthquake (Figure 10.5), and Love waves toppled a major elevated highway sideways in the 1995 Kobe earthquake, Japan. Establishing and enforcing building codes to resist these surface waves in an area determine whether a strong earthquake will be damaging or catastrophic.

10.3.2.4 Structural Reinforcement

In economically disadvantaged areas, buildings are typically constructed of block or brick walls and poured concrete. These

FIGURE 10.6 Steel-rebar-reinforced concrete column deformed during the 1971 San Fernando earthquake. *Source*: Image courtesy of USGS.

are brittle and crumble if flexed. The passage of surface waves can cause these buildings to collapse leading to many casualties. To strengthen the concrete, steel reinforcement bar or "rebar" is imbedded (Figure 10.6). The rebar not only strengthens the concrete but is much more flexible. Rebar can be designed to resist flexing in any direction.

10.3.2.5 Structural Design

In addition to lack of reinforcement, the other main reason buildings fail is that they are designed primarily to resist the force of gravity. If construction is primarily vertical and horizontal as in simple wood and especially brick houses, it will provide minimal resistance to the flexing and lateral movement caused by surface waves and result in collapse. Once supporting walls collapse, every floor of a building drops and crushes the one beneath it resulting in horrific casualties. Cross bracing or cross members consist of crossing supports in an "X" pattern that are installed within the typical orthogonal beams and joists of a wood house (Figure 10.7). They greatly strengthen the structure to non-vertical and non-horizontal stresses.

Similarly, corner braces are diagonal braces installed at the corners of house frames to strengthen the structure against shearing motions. In wood, brick, and concrete structures, another method is the installation of shear walls to resist lateral and shear forces (Figure 10.8). Shear walls can be as simple as sheets of plywood for a wood house that are firmly attached to the joists and beams to maintain the angles between. This prevents the walls from twisting or tipping over so that they will not collapse as easily. These construction practices are common.

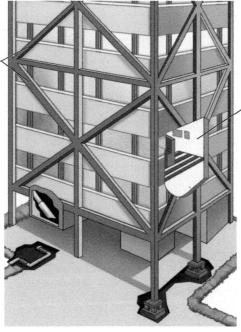

FIGURE 10.7 Diagram of a building with cross bracing on the faces.

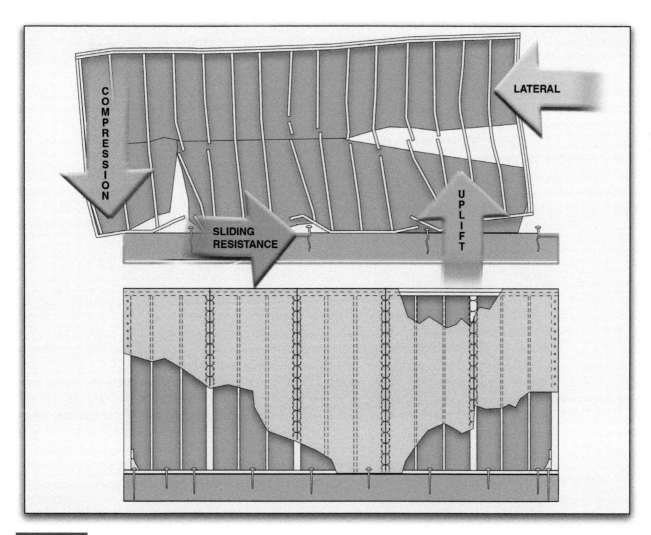

FIGURE 10.8 Diagram showing deformation of walls without shear walls (top) and the installation of shear walls (bottom) to prevent it.

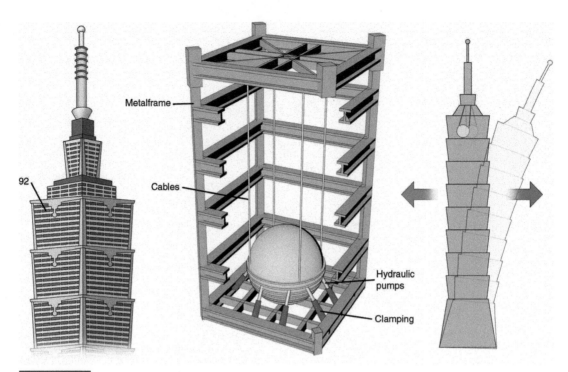

FIGURE 10.9 Diagram showing the position of the tuned mass damper in the Taipei 101 building (left), an enlargement showing the details of the tuned mass damper (center), and the action of the tuned mass damper during flexing of the building (right).

Another problem that buildings can experience is that the shaking frequency of the surface waves can match the resonant frequency of the building. If this happens, the building vibrates extensively like a tuning fork and can suffer great damage or be destroyed. Resonant frequency is a function of height of the building and its mass distribution.

One method to reduce swaying and resonant frequency issues is by installing a tuned mass damper. To do this, an enormous weight is suspended in the mid to upper floors and momentum causes it to act in opposition to the shaking, thereby balancing the stresses. The best example of a tuned mass damper is in the Taipei 101 skyscraper in Taiwan (Figure 10.9). The building features a 728-ton (660-mt) steel ball suspended between the 87th and 91st floors. It not only stabilizes the tower against seismic waves but also typhoon winds as high as 135 mph (216 kmph).

Another method to reduce the impact of shaking on upper floors is building elevation control. In this design, the upper part of tall buildings is constructed in a pyramid shape. This both strengthens the building and reduces the mass at higher elevation which reduces swaying of the upper parts of the building. The best example of this construction is the iconic Transamerica building in San Francisco, California (Figure 10.10). The shape is functional in the earthquake-prone city but also in areas of hurricanes and tropical storms for the same reason.

10.3.2.6 Base Isolation

If a building can be even partially isolated from the shaking of surface waves, there is a better chance that it will survive. Base isolation decouples the base of the building from the ground below, so it absorbs surface wave shaking and returns the building to its original position after the earthquake has passed (Figure 10.11). Even though base isolation usually involves complex engineering, simple forms of it exist in ancient structures. By perching a building on multiple stacked, flat slabs of rock or other material, slippage between the slabs can absorb some of the shaking. The famous Inca citadel of Machu Picchu in Peru was built around 1450 AD but contains primitive base isolation with flat and polished lower slabs that are able to slide against each other during shaking. This type of base isolation is called a friction-pendulum system and it is still in wide use with combinations of materials to enhance the slippage and bowl-shaped basal plates to return the building to its original position after the earthquake ends (Figure 10.12). These are called slider-bowl units and are enclosed within a cylinder of metal and synthetic composites to restrict lateral movement. Sliders are typically composed of a metal arm with a ceramic head and the sliding surface is commonly metal with a composite covering to reduce friction.

There are other base isolation techniques that are also effective. Surface wave shaking may be absorbed by a damper between the building and the foundation that is stiff vertically but weak horizontally. One type is a lead rubber bearing (LRB) composed of alternating layers of flat steel plates and rubber with a cylindrical lead core (Figure 10.13). The bearing acts like a shock absorber in a car with the rubber limiting movement and returning the building to its original position like a spring. The steel plates allow only lateral movement, so the building will not flex or lean. Another base isolation technique uses roller bearings in bowl-shaped plates mounted between the building and the foundation. Other base isolators are actual springs with

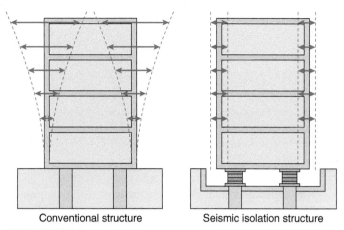

FIGURE 10.11 Diagram of building vibrations from seismic waves in conventional construction (left) and in construction with base isolation (right).

FIGURE 10.10 Photograph of the Transamerica building in San Francisco, California. *Source*: Courtesy of the US Library of Congress.

dampers to keep them from wobbling mounted between the building and foundation. They include angle limiters that keep the building upright.

10.3.3 Protection Against Liquefaction

Liquefaction occurs if surface waves shake unconsolidated sediments or soil in areas where the water table is shallow. The shaking drives groundwater to the surface where the soil and sediment are turned to mud. The lack of support of the building foundations by the mud causes them to fail and tilt or fall over. This can occur in any coastal, lacustrine, or fluvial region or in fill dirt. The most famous case of foundation failure from liquefaction was in 1964 in Niigata, Japan where earthquake-proof buildings tilted and fell over on their side completely intact.

FIGURE 10.12 Diagram of slider-bowl base isolation unit.

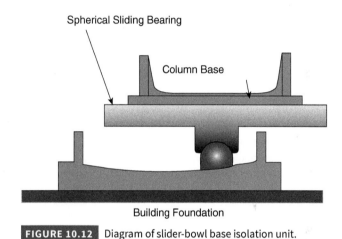

FIGURE 10.13 Diagram of lead rubber bearing base isolation unit.

10.3.3.1 Pilings

Many house and building foundations sit on a slab directly on soil or they are one-story deep basements. Both of these can easily shift if exposed to liquefaction. By attaching the foundations of buildings to deep pilings, this problem can be largely averted. Pilings are long beams or columns that are driven deeply into the ground so that they can act as keels to the buildings. Only the near surface becomes unstable in liquefaction. The soil and sediment for the hundred or more feet of depth into which the pilings are driven remain stable. Pilings enable tall buildings to be built in cities that have no bedrock, such as New Orleans, Louisiana.

10.4 Big Data: The New Hope

With the rapid advancement of technology, the amount of data that can be collected and analyzed in a timely manner has skyrocketed. This has resulted in the development of new technologies that may make earthquake prediction possible in the near future. Satellites can monitor many parameters of the near-surface on a much larger scale than surface devices. They cannot monitor radon emissions or changes in seismic velocity like the traditional precursors, but they can monitor other important parameters that may be even better predictors.

High-precision global positioning systems (GPS) are now sensitive enough to detect even minor changes in the surface elevation. Using satellites, Interferometric Synthetic Aperture Radar (InSAR) monitoring of the earth surface is sensitive enough to detect even very slow elevation changes of 0.04 in. (1 mm) per year. Subsidence and uplift commonly foretell an earthquake and can be mapped by satellites on a continuous basis (Figure 10.14). Satellites can also map infrared radiation (IR) and can detect even minor changes in heat flow from an area. There are documented changes in surface temperature of 7.2 °F (4 °C) just prior to the major 2001 Gujarat earthquake in India. Another potential precursor of earthquakes is low-frequency magnetic signals (0.01–0.02 Hz). For example, the magnetic signals increased up to 20 times the normal levels for two weeks before the 1989 Loma Prieta earthquake in California and the elevated levels continued well after the earthquake.

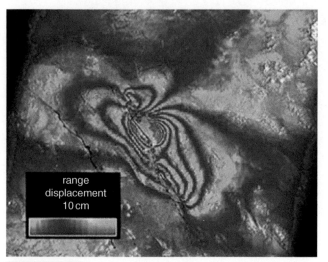

FIGURE 10.14 Color-enhanced surface deformation map during and around the time of the 1999 Hector Mine earthquake, California. *Source*: Image courtesy of NASA.

In 2003, a 20-year plan was announced to deploy the Global Earthquake Satellite System (GESS) which is a network of satellites to monitor active fault zones worldwide. Satellites were launched beginning that year to monitor several parameters but primarily vertical motions using InSAR (Figure 10.15). The next step is to analyze all of the information collected over the years on features leading up to the earthquakes and develop a system to predict them. This requires big data analysis. The company Terra Seismic, for example, claims that they can predict 90% of earthquakes within two to five months of their occurrence using big data analytics. They have shown that the information was available to predict many of the major earthquakes over the past 10 years. It appears that true earthquake prediction may be at hand.

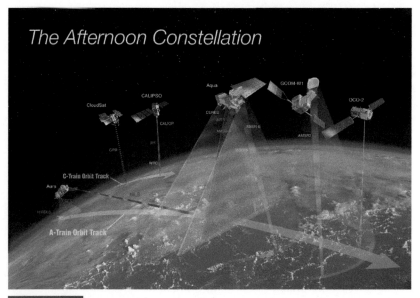

FIGURE 10.15 Diagram of the earthquake-precursor-sensing satellite system that will be used to predict earthquakes in the future. *Source*: Image courtesy of NASA.

References

Adams, R. D. (1976). The Haicheng, China, earthquake of 4 February 1975; the first successfully predicted major earthquake. First published: July/September 1976. https://doi.org/10.1002/eqe.4290040502.

Fuqiong, Huang, Mei Li, Yuchuan Ma, Yanyan Han, Lei Tian, Wei Yan, and Xiaofan Li (2017). Studies on earthquake precursors in China: A review for recent 50 years, *Geodesy and Geodynamics*. Elsevier. https://doi.org/10.1016/j.geog.2016.12.002.

Wang, K., Chen, Qi-fu., Sun, S., Wang, A. (2006). Predicting the 1975 Haicheng Earthquake. Published 1 June 2006. Doi:10.1785/0120050191.

CHAPTER 11

Avalanches and Landslides

CHAPTER OUTLINE

11.1 Gravity: What Goes Up Must Come Down 203

11.2 Rockfalls and Slides 204

11.3 Avalanches and High-Speed Movement 209

11.4 Low-Speed Movement 213

11.5 Karst Topography 219

Earth's Fury: The Science of Natural Disasters, First Edition. Alexander Gates.
© 2022 John Wiley & Sons Ltd. Published 2022 by John Wiley & Sons Ltd.
Companion website: www.wiley.com/go/gates/earthsfury

Words You Should Know:

Angle of repose – The maximum angle that a slope can attain before it fails.
Avalanche – A mass movement that is high speed and composed of dry material.
Debris flow – A moderate- to high-speed mass movement.
Earthflow – A slow-moving mass movement of soil and debris.
Karst – Topography that reflects chemical erosion of limestone terranes, marked by caves and sinkholes.
Landslides – A general term encompassing medium- and high-speed mass movements such as earthflow.
Mudflow – A high-speed downslope movement of fluidized mud.
Rockfall – Mass movement in which rocks fall free from an exposure to the ground.
Rockslide – Mass movement in which rock breaks away from an exposure and slides down a slope.
Sinkhole – A circular crater formed by the collapse of a cave roof.

11.1 Gravity: What Goes Up Must Come Down

Gravity is relentless in reducing all hills into flat plains. All slopes, no matter how shallow, are either unstable, metastable, or failing. Unstable slopes are on the verge of failure. Any vibration, rainstorm, or other alteration can cause downslope movement of material. This is defined as mass movement. Even slopes that appear stable are only temporarily stable or metastable. Eventually or under the right conditions, mass movements occur. The way in which material fails distinguishes mass movements.

The angle of repose is an important concept in mass movements. It is the maximum angle of a slope of unconsolidated material that can exist under a particular set of conditions before it fails. If you pour a bucket of dry sand on the floor, it forms a cone-shaped pile with the same slope angle no matter where it is measured. If you add more sand, the pile does not just get higher. Instead, most of it slides down the slopes (slope failure) and the pile grows wider at the same ratio as it gets higher. The angle of the larger pile slopes is the same as the smaller pile (Figure 11.1). Different materials (gravel, clay, boulders, etc.) form different angles of repose and the more friction between particles, the steeper the slopes.

The physical condition of the materials also controls the angle of repose. The angle of repose of moist sand is very steep. The moisture exerts surface tension and resulting capillary action among the grains holds them together. This is how sandcastles can be built on the beach. If you put enough water in the sand, it turns into a slurry and has a very low angle of repose. If a slope is frozen, it can form an even higher angle of repose than a wet one. In addition to temperature, wind can impact the slope angle as it can blow the slopes down as can earthquakes and other vibrations.

FIGURE 11.1 Photographs of various sized piles of gravel all showing the same angle, the angle of repose. *Source*: Gefufna/123RF Stock Photo.

FIGURE 11.2 Methods to stabilize slopes against failure. *Source*: Slope Failure Repair Options, March 29, 2012. Sinai Construction Engineering.

Vegetation can also affect the angle of repose for hillslopes. Trees have deep root systems that can hold soil and sediments together greatly increasing the angle of repose. The intergrown root systems of grass can hold the surface of slopes together but not as well as trees. Humans can impact the angle of repose through a variety of methods such as cement, retaining walls, gravel, netting, drainage ditches, and culverts among others (Figure 11.2). Typically, humans aim to increase the angle of repose to better stabilize them with varying degrees of success.

11.2 Rockfalls and Slides

An outcrop is where bedrock is exposed at the earth surface. If the outcrop is horizontal and nearly flat, it is called a pavement outcrop. Unless the outcrop is being actively undercut by caves or mining, the rock will be mechanically stable for a long period of time. However, if the outcrop forms a cliff or hillslope, it will eventually fail. If the outcrop is in a temperate climate, water that seeps into cracks in the rock will expand upon freezing, opening the crack even farther. This is called frost wedging and it is a major contributor to weakening of rock. In temperate and warmer climates, the infiltrating water is typically acidic and ion-bearing. It reacts with the crack walls and produces new larger minerals that can also force cracks open. Around rivers and in windy dry areas, the bedrock can be undercut leaving precarious overhangs. Construction projects can similarly produce cliffs and overhangs. All it takes is a minor earthquake or heavy rainstorm to bring these rocks down and cause a disaster. There are two main types of bedrock failure, a rockslide and a rockfall.

A rockslide is a rock mass that slides down a slope at high speed (Figure 11.3A). Commonly, the slopes are bedding planes or fractures. In sedimentary rocks, it is more common that bedding planes form the slopes. In this case, a rigid layer of rock, like limestone or sandstone, breaks and slides down an underlying soft layer like shale. In crystalline rocks, the slopes are formed

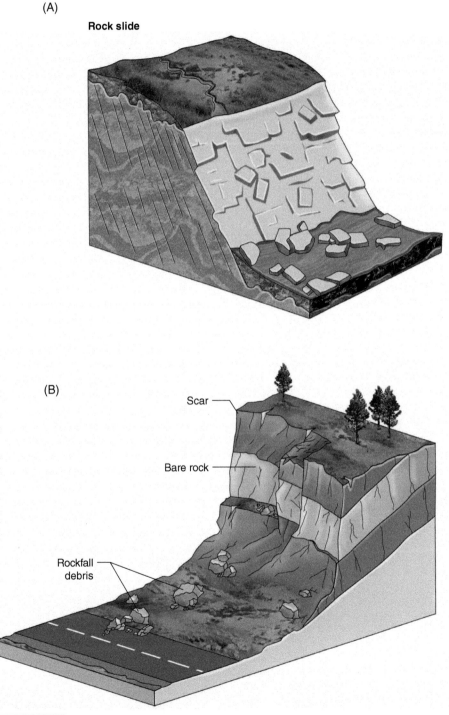

FIGURE 11.3 Block diagrams of types of mass movements. (A) Block diagram of a rockslide. (B) Block diagram of a rockfall with talus at the base of the slope. (C) Photo of a talus cone at the base of a cliff. *Source*: Courtesy of the US Geological Survey.

(C)

FIGURE 11.3 (Continued)

by joints and faults. The rock masses can break loose along steep fractures and joints, but they slide on shallower joints and faults. Rockslides are especially common during earthquakes and heavy precipitation events. Surface waves can shake loose even relatively stable rock and they slide into the valleys below producing disastrous results. Rapid infiltration of rainwater along breaks in the rock changes the friction condition, and causes rock masses to slide.

In contrast to rockslides which slide along a slope throughout their movement, rockfalls are masses of rock that break loose and fall to the slope below (Figure 11.3B). Most of this travel is a free fall, but masses may bounce as well and transition into avalanches. Rockfalls typically break from the bedrock along steep joints and fractures. Undercutting of slopes by rivers, wind, or human activities almost certainly produces rockfalls. Toppling is a type of rockfall in which the rock mass is rotated away from the exposure as it falls. Rockfalls are the leading cause of death from earthquakes in mountainous regions.

When the rockslide or rockfall occurs, a rock mass breaks free and slides or falls to the earth below as a coherent mass or in pieces. When the mass strikes the ground, it typically breaks into a pile of rubble. This pile of rubble at the foot of a hill or cliff is called talus. With repeated earthquakes or other events, this talus accumulates into a talus slope which is a cone-shaped deposit (Figure 11.3C). Talus slopes are common in areas of repeated earthquake activity.

CASE STUDY 11.1 1963 Vaiont Dam Disaster

Rockslide and Flood Wave

One of the tallest dams in the world was built across the Vaiont River in a narrow limestone canyon, in the Dolomite region of the Italian Alps (Figure 11.4). The dam was built between 1957 and 1960 with construction completed in October 1959, and authorization to fill the reservoir in February 1960. The dam was built to a height of 860 ft (262 m). The volume of the dam construction was 12 713 280 ft³ (360 000 m³) and it was built to contain 5 958 113 995 ft³ (168 715 000 m³) of water. This engineering wonder would be the site of one of the worst dam disasters primarily because the geology was not properly respected.

Even early in the project, there were indications that there might be problems in the future. During construction of the dam, on 22 March 1959, a landslide into the reservoir at the nearby Pontesei Dam generated a 66-ft (20-m)-high wave that caused a death. Later, on 4 November 1960, with the water in the reservoir at 620 ft (190 m) of the 860 ft (262 m) full level, a landslide of about 1 million cubic yards (800 000 m³) slid into the reservoir. As a result, filling of the reservoir was stopped and the water level was lowered by 160 ft (50 m), and construction was initiated to keep the reservoir usable even with landslides. In April and May of 1962, there were five minor earthquakes with the basin while water level was at 705 ft (215 m) depth. In July 1962, model-based

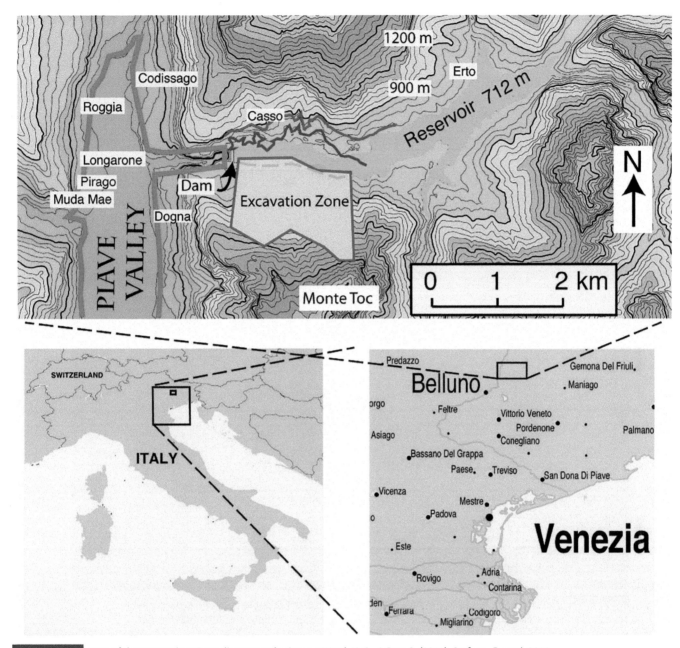

FIGURE 11.4 Map of the 1963 Vaiont Dam disaster, Italy. *Source*: Ward, N. S., & Day, S. (2011). Stefano, B. et al. 2011.

experiments indicated that landslides could generate waves that could top the dam if the water level was within 66 ft (20 m) of the dam crest. To address this, the water level was lowered to 82 ft (25 m) below the crest. During the summer of 1963, the basin was at this level because surface mass movements were common. On 15 September 1963, a large section of the mountain slid by 8.7 in. (22 cm). In response, on 26 September, the decision was made to slowly empty the water level to 790 ft (240 m). However, the slow collapse continued including 3.3 ft (1 m) of movement in one day.

The instability did not escape the attention of the local residents. In the first week of October, a local mayor issued a warning, urging the villagers to evacuate. A few residents evacuated, but there was little concern. Another mayor ordered the evacuation of the basin slopes and even posted notices of damaging waves from an expected landslide. At about 9 p.m. on 9 October 1963, traffic was restricted on roads below the dam, and telephone calls were made warning of possible small amounts of water over the dam.

However, at 10:39 p.m. on 9 October 1963, a massive rockslide dropped at up to 68 mph (110 kmph) into the reservoir which was 782 ft (238 m) deep and had a volume of 109 000 acre-ft (0.13 km^3). The rock mass was 1.2 mi (2 km) long by 1.0 mi (1.6 km) wide, which was twice the volume of water impounded behind the dam. Its volume was estimated at 340 million cubic yards (260 million m^3). In 45 seconds, the rockslide filled the reservoir with debris to 590 ft (180 m) above the water level for a distance of 1.2 mi (2 km) upriver. The impact of the slide generated a wave that drove 850 ft (260 m) up the opposite side of the reservoir stripping trees, rocks, and soil from the facing slope. The mass also displaced the water in the reservoir over the top of the still intact dam by 509 ft (155 m).

As the water overtopped the dam, it produced an impact crater 200 ft (60 m) deep and 260 ft (80 m) wide at the foot of the dam. It also generated an enormous 820-ft (250-m)-high wave of at least 1.77 billion ft^3 (50 million m^3) of water. The flood wave rushed down the canyon at 10:43 p.m., generated wind strong enough to break windows and drove strong earth tremors. The flood wave destroyed a number of downstream towns including Longarone, Pirago, Rivalta, Villanova, and Fae (Figure 11.4A). It arrived in Longarone at 10:43 p.m., about four minutes after the rockslide, and completely leveled the town. Of the population of 1328 in Longarone, there were 1269 fatalities, meaning there was a fatality rate of 94%. By 10:55 p.m., the flood wave had passed, and the valley was silent (Figure 11.4B).

It is estimated that 3000 people lost their lives in this disaster, but with the nature of the damage, the number is a best estimate. The rumor is that when the dam chief engineer learned of the disaster, he committed suicide.

FIGURE 11.4A Photo of the destruction immediately downstream of the Vaiont Dam (chasm in background). *Source*: Ispettorato Vigili del Fuoco; https://commons.wikimedia.org/w/index.php?curid=29389748#/media/File:Vajont1963_5.jpg.

FIGURE 11.4B Before (left) and after (right) photos of a town downstream of the Vaiont Dam. *Source*: Interesting Engineering.

11.3 Avalanches and High-Speed Movement

Mass movements can be divided into two types, dry and with a water content of greater than 20% (Table 11.1). Each of these is then subdivided by speed of the movement. The fast-moving flows are sturzstroms, debris avalanches, and debris flows.

If a large rock mass slides down a steep but rough slope, it may shatter the mass into fragments all moving at high speeds greater than 330 ft (100 m) per second. This flow is a sturzstrom and it is the fastest form of mass wasting, equivalent to a super debris avalanche. Sturzstroms can contain water, but movement is more commonly facilitated by the debris particles colliding with each other internal to the mass. The internal collisions cause the mass to move like a nearly frictionless cloud. It has been found that in some cases, they can hit the slope so hard that they melt it and harden into a glassy rock called *frictionite* because the heat is generated by friction. The impacts of the rocks hitting the slope can produce shatter cones and even shocked quartz, which is more typical of meteorite impacts. The damage from a sturzstrom is devastating.

Debris avalanches are very fast-moving bodies on steep mountain slopes (Figure 11.5). They can be primarily snow (snow avalanche) or composed of a mixture of soil, rocks, trees, and ice comprising debris. Debris avalanches travel at velocities of 33 and 333 ft (10–100 m) per second. They occur on mountain slopes and earthquakes and volcanoes are the most common cause. They can begin with a free fall where the falling debris can attain very high velocities and result in long travel distances. The flow of debris can produce winds of hurricane force that knock over trees. Debris avalanches are devastating to all life in their path. The debris avalanche scars are shaped like long narrow funnels. They leave trough-like chutes that look smooth from a distance but are pock marked from bouncing boulders.

Mudflows are a type of mass wasting with high fluid contents and high flow velocities. They are fast-moving rivers of mud that can inundate whole towns causing death and destruction. Mudflows occur in many areas as the result of heavy rain. However, mudflows can also occur in volcanic eruptions. A lahar is a boiling volcanic mudflow made of hot ash and melted alpine glacial ice or heavy precipitation at the time of eruption. The hot, fine ash makes it susceptible to forming a boiling mudflow with heavy rains for as much as a decade after the eruption.

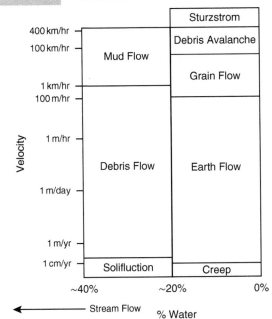

TABLE 11.1 The various types of mass movements based on water content and speed of movement.

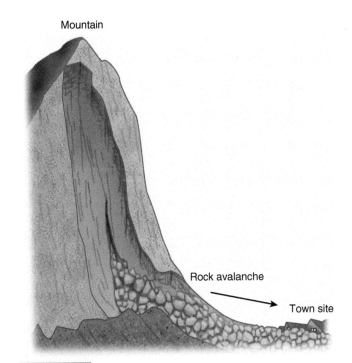

FIGURE 11.5 Diagram of a debris avalanche.

CASE STUDY 11.2 1962 Huascarán Avalanche, Peru

Avalanche to Debris Flow

A series of cataclysmic events devastated a coastal area of Peru beginning with a melting glacier. On 10 January 1962 at 6:13 p.m., between 3 and 6 million tons (3–6.1 billion kg) of ice from Glacier 511 on Nevados Huascarán broke loose at an elevation of 21 834 ft (6655 m). As is usually the case in this region, the breakage made a loud noise that was heard in the surrounding towns (Figure 11.6). Normally, there is a 20–30-minute delay between the sound of the ice breaking and the ensuing avalanche. This delay gives residents time to move to higher ground and relative safety. This avalanche, however, traveled 9.5 mi (15.3 km) in less than seven minutes, an initial speed of over 82 mph (133 kmph).

This mass barreled down the steep slopes of the Andes Mountains. It entrained a large amount of rock and soil and then fell into the gorge of Callejón de Huaylas attaining speeds of up to 111 mph (179 kmph) which is nearly a sturzstrom. The wind generated by the fast-moving flow reached hurricane force along the edges, sweeping trees, animals, and debris into the air. The flow entrained larger objects including houses from the town of Yanamachico where it killed all 800 residents. It slowed to 60 mph (96.6 kmph) classifying it as a debris avalanche where it demolished the town of Ranrahirca with 2700 residents. Then a debris flow continued down the Rio Santa at decreased velocity and it did not completely stop until the port of Chimbote, Peru, some 60 mi (96.6 km) away where bodies resulting from the disaster were found.

The volume of the initial mass was about 106 million ft^3 (3 million m^3) with 35 million ft^3 of rock (1 million m^3) and 71 million ft^3 (2 million m^3) of ice. However, as the 40 ft (12.2 m) high and 0.6 mi (1 km) wide mass barreled down the slopes with rocks and ice crashing together into warmer areas, the ice melted. The impacting rocks pulverized their edges into rock flour. The flow scooped up soil, rock, and debris which mixed with the water and rock flour to create a mass of mud and debris. The towns of Ranrahirca and Huarascucho were buried under 40 ft (12.2 m) of this mixture of ice, mud, trees, boulders, and other debris. As a result of this process, the initial 106 million ft^3 (3 million m^3) wound up at a final volume of 459 million ft^3 (13 million m^3).

The mass movement destroyed nine towns and seven smaller villages and left waist-deep mud throughout the impacted area. The death toll from this disaster cannot be accurately determined because many bodies were buried in the mud or swept out to sea. The best estimate is that between 3500 and 4000 people lost their lives. In addition, an estimated 10 000 farm animals were also killed. A mere eight years later, an even more powerful mass movement would strike the same area (see 1970 Nevados Huascarán).

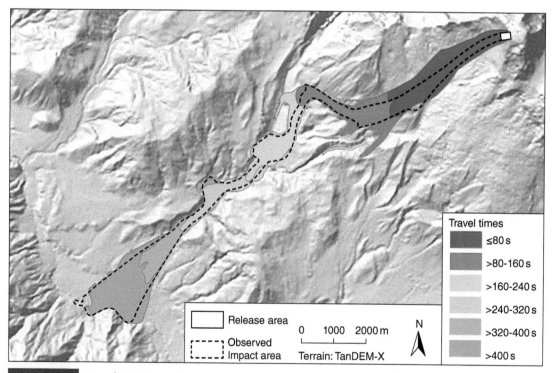

FIGURE 11.6 Map of 1962 Huascarán avalanche, Peru to debris flow. *Source*: Mergili, M., Frank, B., Fischer, J.-T., Huggel, C., & Pudasaini, S. P. (2018).

CASE STUDY 11.3 1916 White Friday Avalanches

Glacier Warfare

The word "avalanche" usually brings snow avalanches to mind. For this book, the natural disasters described primarily involve significant loss of life and property damage. Snow avalanches are commonly deadly and can destroy property, but they are localized and involve very limited numbers of casualties. One of the few exceptions to this generality is the White Friday disaster. The only problem is that it was not completely a natural disaster, but primarily a human-induced natural disaster.

World War I began in 1914 and European countries squared off to battle each other. The main conflict involved Austria, Hungary, and Germany against Russia, France, and Britain. The 1882 Triple Alliance among Italy, Germany, and Austro-Hungary allowed Italy to remain neutral at least at the beginning of the war. The fighting escalated into 1915 and the Russia–France–Britain alliance pressured the Italians to join them. They lured Italy with awarding them the Italian-speaking Tyrol region of Southwest Austria at the war's end. It would take two months of negotiations to convince Italy to declare war on Austro-Hungary later in 1915.

In the spring of 1916, the Austrians launched a major offensive sweeping south into the mountains of the Italian Alps (Figure 11.7). The objective was to attack and take the Italian plain, march into Venice, and encircle much of the Italian Army. However, the Italian Army reacted quickly and stopped the Austrians in the Alps where the border between them remained for the next two years. The conditions for the soldiers were absolutely miserable with snow, ice, and frigid temperatures. Each side attempted to gain the high ground over the other by climbing ever higher with their heavy artillery in tow (Figure 11.7A). Each army climbed as high as they could and dug in, blasting tunnels into the glaciers and mountainsides and building barracks to protect soldiers from the cold. An estimated 600 000 Italians and 400 000 Austrians would die on the Alpine front during World War I and a large number were from the cold which was dubbed "white death."

Compounding the soldier's misery, the heaviest snowfall of the twentieth century for the Alps was in the winter of 1916–1917. Snow began early and intense; by the beginning of December, the high-altitude snow already ranged from 26 to 40 ft (8–12 m) deep. A sudden thaw in the Alps and quick refreezing created conditions that were ripe for avalanches. This was certainly the case at Mount Marmolada, Italy.

An Austro-Hungarian military barracks was built in August 1916 on the Gran Poz summit of Mount Marmolada approximately 11 000 ft (3 553 m) above sea level. The wooden barracks were situated to be protected from Italian attack. They were built along cliffs to protect them from enemy fire as well as keeping them out of high-angle mortar range. The encampment also provided defense of Mount Marmolada which was contested at the time.

After the heavy snowfall in early December 1916, these barracks wound up directly under a steep mountainside of thick and unstable snow. The Austro-Hungarian commander of the barracks saw that his company was in imminent danger. He therefore wrote a request to his superior on 5 December 1916 to vacate the base at the Gran Poz summit to avoid a disaster. The appeal was denied and during the following eight days, continued heavy snowfall cut telephone lines and left the base cutoff from the command, stranded and low on supplies.

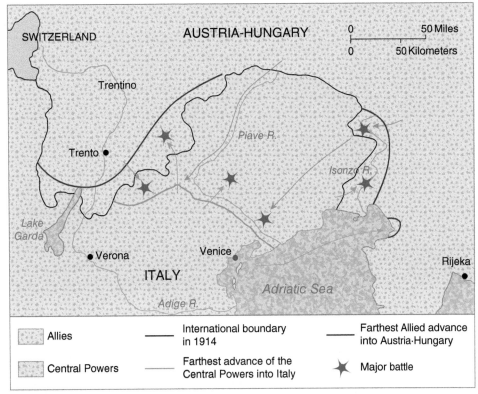

FIGURE 11.7 Map of the battle fronts in the Dolomite Mountains of the Italian Alps in World War I.
Source: Adapted from The *World Book Encyclopedia*, © The World Book.

212 CHAPTER 11 Avalanches and Landslides

FIGURE 11.7A Photograph showing the arduous task of lifting of artillery into the mountains for battle. *Source*: SOTK2011 / Alamy Stock Photo.

13 December is marked Saint Lucia Day, a religious holiday for Italian Catholics. It was on this day, Wednesday, 13 December 1916, at 5:30 a.m., that more than 200 000 tons (181 000 mt) or approximately 35 314 666 million ft³ (1 million m³) of snow and ice let go and barreled down the mountainside and onto the Gran Poz barracks (Figure 11.7B). The wooden buildings were packed with sleeping soldiers at the time. They collapsed under the huge weight of the avalanche, and the occupants were crushed to death. Of the 321 troops present, only a few were pulled to safety, while 270 soldiers were crushed or buried alive. Only 40 bodies were ever recovered. This was a purely natural disaster.

This was only the first of numerous deadly avalanches to occur that day and over the next several weeks. Once the troops learned about the instability of the steep slopes and the potential for using the avalanches as weapons, they began firing mortars into the hills above enemy emplacements rather than directly at them (Figure 11.7C). Defenses were built to protect from enemy fire, not from avalanches from above. Many of the avalanches were therefore not caused naturally, but instead from enemy fire. In particular, on the night of 13 December, an avalanche struck an Italian division in their mountain barracks where several hundred soldiers were also said to have perished. However, it is unclear if this was natural or human-induced.

The number of casualties from the White Friday avalanches cannot be accurately determined. The best estimate based on historical documents indicates that a minimum of 2000 Austrian and Italian soldiers perished on this one day as well as a number of civilians. Even though the killer avalanches in the Dolomite Mountains of the Italian Alps occurred on a Wednesday, the disastrous day was dubbed "White Friday." Naturally induced avalanches and White Friday warfare continued throughout December 1916 leading to the deaths of nearly 10 000 soldiers on both sides. It is the greatest number of deaths caused by snow and ice from avalanches in history.

FIGURE 11.7B Photograph of soldiers inspecting a scar left by an avalanche that recently passed through. *Source*: Austrian National Library.

FIGURE 11.7C Artist's rendition of soldiers being killed in an avalanche during battle. *Source*: Universal History Archive/UIG via Getty images.

11.4 Low-Speed Movement

Low-speed mass movements can also be generated by earthquakes, though they are commonly generated by other means as well. They are more commonly composed of soils and can be coherent. Precipitation hastens low-speed failure, though they can be dry as well (Figure 11.8).

Debris flows or slides are masses of unconsolidated surficial material that, like rockslides, slide down a sloped bedrock surface like a bedding plane or a fault plane. Debris flows are most common where thin layers of soils mantle slopes and become water saturated by precipitation or thaw. Earthquakes can generate debris slides and flows and they are common on the slopes of volcanoes where layers of ash fail.

An earthflow is a viscous flow of saturated, fine-grained materials that moves downslope at speeds ranging from barely perceptible to about 3.3 in. (0.1 m) per second (Figure 11.8A). Clay, fine sand and silt, and fine pyroclastic material are susceptible to earthflow. Earthflow velocity and distance traveled are controlled by water content with faster and longer travel distances with higher water contents. Earthflows can move for years. They normally begin when water content increases, which can be from volcanic eruptions or liquefaction during an earthquake. Shaking from surface waves may initiate flow in a saturated soil. The flows bulge in the center as they travel, which channels more fluid to that area as the edges dry out. Earthflows stop moving if water content decreases.

Slumping is the downhill sliding of a rock mass or unconsolidated soil or debris along a concave-upward failure plane usually involving backward rotation of material during movement (Figure 11.8B). Slump may happen for several reasons such as heavy rain, melting snow and ice, removal of vegetation by disease, fire, or human activity by shaking, or liquefaction associated with seismic activity. Slumping commonly produces a scarp and fissures along the top margin of the slumped material. Most fissures produced in earthquakes are from slumping.

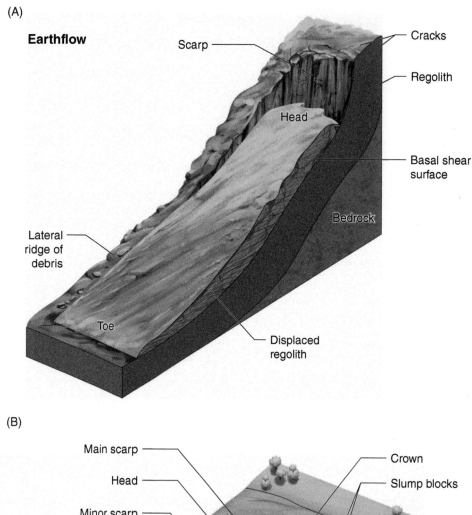

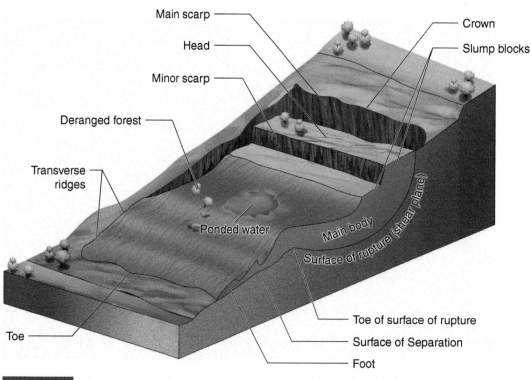

FIGURE 11.8 Block diagrams of slow-velocity mass movements. (A) Earthflow. (B) Slump.

The slow flowing of water-saturated soil and debris downhill is solifluction. It is wet creep and flow rates are imperceptibly slow at 0.5–1 in. (1–2 cm) per week at fastest. Solifluction is common in cold climates where the top few centimeters of frozen soil can be thawed and saturated with meltwater. The soil has the consistency of cement, but moves faster toward its center, forming lobes on the hill slopes.

CASE STUDY 11.4 1999 Vargas Mud and Debris Flow

Largest Rain-Produced Flow

Debris flows from volcanoes can be very large compared to those from rainstorms which tend to be much smaller. Nonetheless, in certain cases, debris flows induced by storms can be disastrous. This was the case with the 1999 mudflow and debris flow in the state of Vargas, Venezuela (Figure 11.9). This disaster was caused by a once-in-1000-year flood that killed tens of thousands of people, destroyed thousands of homes, and led to the collapse of the infrastructure in that region.

The Vargas coast of Venezuela is very rugged with steep slopes. The Sierra de Ávila Mountains are just 3.7–6.2 mi (6–10 km) from the coastline and yet reach an impressive elevation of 8.858 ft (2700 m). The rivers and streams drain north from the mountains through steep-walled canyons that feed into alluvial fans before emptying into the Caribbean Sea (Figure 11.9A). These alluvial fans are built of sediments deposited by debris flows and floods sourced by the sediment and debris in the canyons. Once free of the canyons, the sediments form the only extensive flat areas along the otherwise mountainous coastline of Northcentral Venezuela. For that reason, they have been extensively developed and urbanized and now have a high population density. The estimated population of this area was 300 000 in 1999.

In 1999, Northern Venezuela had a particularly wet month of December. On 2 and 3 December, a strong storm produced 7.9 in. (20 cm) of rain along the north coast. This heavy rain soaked and softened the soils and sediments on the mountain slopes. Just two weeks later, during 14–16 December, an amazing 35.9 in. (91.1 cm) of rain fell in a short 52-hour period. This amount is about the average total rainfall for one year in the region. Between 6 and 7 a.m. on 16 December alone, 2.8 in. (5.8 cm) of rain fell in one hour. The rate of precipitation on both 15 December and 16 December exceeded the 1000-year flood event for the area. These torrential rains produced extensive flash floods throughout the area.

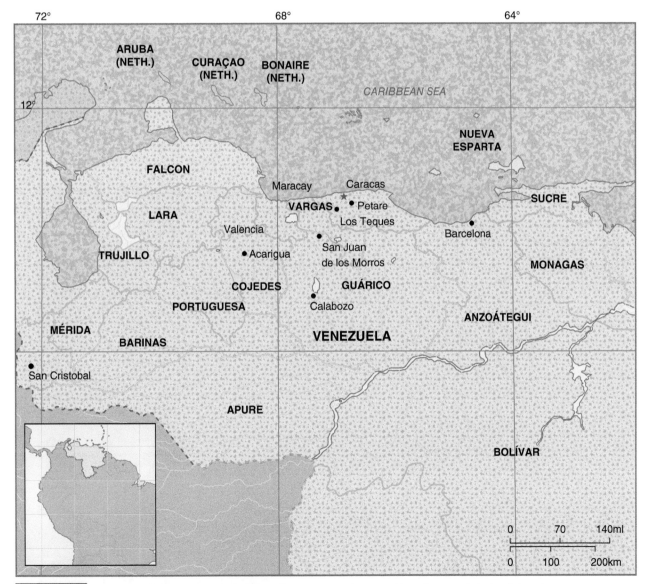

FIGURE 11.9   Map of Venezuela showing the location of Vargas. Inset shows position of Venezuela in South America.

FIGURE 11.9A Photo showing the city of Vargas on an alluvial fan as viewed through the valley that funneled the mudflow. *Source*: Wikimedia/https://upload.wikimedia.org/wikipedia/commons/f/f4/Monta%C3%B1asdeslave.jpg.

Observers reported that flooding began just after 8 p.m. on 15 December. Runoff entered stream channels and rushed to the sea eroding the deposited sediments on the way. Especially hard hit was the Caraballeda alluvial fan. The residents began fleeing the rivers as they overtopped their banks. After the initial stream flooding, the torrential rains triggered shallow landslides all across the area that stripped soil and rock off the mountain slopes. Continued rainfall liquefied the landslides into debris flows of rock, mud, and vegetation (Figure 11.9B). The first accounts of the debris flows were at 8:30 p.m. Eyewitnesses reported crashing rocks on the Quebrada San Julián.

FIGURE 11.9B Photo showing the mudflow engulfing part of Vargas. *Source*: Courtesy of the US Geological Survey.

FIGURE 11.9C Photo showing the mudflow deposit covering part of the city of Vargas. *Source*: Courtesy of the US Geological Survey.

Later, between 2 and 3 a.m. on 16 December, residents reported rumbling noises and vibrations as more debris flows appeared. The next major series of debris flows was observed between 5 and 7 a.m. and the last series was reported between 8 and 9 a.m. Flooding was also reported in a number of channels between 7 and 9 a.m. and it lasted until late in the afternoon that day. These later floods contained less sediment and incised new channels into the fresh debris flow deposits.

The landslides initiated on slopes that ranged in steepness from 30° to greater than 60° with the average slope failure at 42°. The width of slides varied greatly averaging 65 ft (20 m) with a maximum width of about 130 ft (40 m). These steep slopes caused the flows to move very rapidly with average flow velocities ranging from 11 to 48 ft/s (4–14.5 m/s). These fast flows carried very large boulders in addition to the soil and gravel (Figure 11.9C). The average size of boulders deposited upstream was between 2.5 and 3.3 ft (0.75–1 m), with the largest boulders downstream at sizes up to >13 ft (4 m) (Figure 11.9D).

The fast-moving flash floods and debris flows destroyed thousands of houses, bridges, and other structures. They cut new channels, several meters deep, into all alluvial fans on the coastline, and blanketed them with thick sediment deposits. In particular, the Caraballeda fan was heavily urbanized, including high-rise buildings and multistory houses. It lies at the mouth of the Quebrada San Julián which produced the largest boulders and thickest deposits. Once the debris flows emerged from the ravine, they overflowed the channels and rapidly spread debris throughout the community. These flows demolished two-story houses and multistory apartment buildings (Figure 11.9E). The neighborhood of Los Corales was buried under 9.8 ft (3 m) of mud and many of homes were swept into the ocean. In other areas, entire towns including Cerro Grande and Carmen de Uria completely disappeared. Farther down the fan, the debris flows were

FIGURE 11.9D Photo showing an example of the boulders carried by the mudflow.
Source: Courtesy of the US Geological Survey.

FIGURE 11.9E Photo showing the collapse of a building undermined by the mudflow in Vargas.
Source: Courtesy of the US Geological Survey.

contained to streets where they thinned to about 3.25 ft (1 m). About one third of the Caraballeda fan was covered by debris flows.

The deposit thicknesses were as much as 13–16.3 ft (4–5 m) to a maximum of 17.5 ft (5.3 m) near the center of the fan to less than 1.7 ft (0.5 m) at the edges. Where the flows entered the Caribbean Sea at Caraballeda, they extended the shoreline by 133–200 ft (40–60 m) along a 37-mi (60 km) stretch of the coastline in Vargas. The total volume of the Vargas mudflow and debris deposit is estimated to be 67 million ft^3 (1.9 million m^3). To put this into perspective, an Alluvial Fan Sedimentation Scale is used to rank such events (Table 11.2).

The intensity uses a magnitude scale, M, which is the log of depositional volume. Using this scale, the volume of debris flow deposits on the Caraballeda fan yields M = 6.3 which is among the largest ever rainfall-induced debris flows.

The death toll has been difficult to determine and has ranged from 10 000 to 50 000. The exact number was difficult to determine

TABLE 11.2 Alluvial Fan Sedimentation Scale (Keaton et al. 1988; Keaton and Matthewson 1988).

Intensity	Damage	Description
0	None	No damage
1	Negligible	Damage to landscape and access; no damage to structures; minor scour and/or sediment deposition
2	Slight	Sediment generally <3.3 ft (1 m) thick deposited against buildings without structural damage; sediment flooded around parked vehicles
3	Moderate	Sediment generally >3.3 ft (1 m) thick deposited against buildings with easily repairable structural damage; basements partially filled with sediment; repairable damage to parked vehicles shoved by flow
4	Severe	Sediment >3.3 ft (1 m) deposited against buildings with repairable structural damage; basements completely filled with sediment; wood structures detached from foundations; nonrepairable damage to parked vehicles shoved by flow
5	Extreme	Sediment >3.3 ft (1 m) thick deposited against buildings with nonrepairable damage; structures collapsed by force of flow; wood structures shoved from foundations; parked vehicles so badly damaged that they have small salvage value

because the census data were unreliable at the time. In addition, only about 1000 bodies were recovered, because most were buried in the landslides or swept out to sea by the mudflows. The figure of 30 000 deaths is now generally accepted, which was 10% of the population of the area at the time.

The disaster is estimated to have inflicted damages totaling US$1.79–3.5 billion. This accounts for more than 8000 homes and 700 apartment buildings destroyed in Vargas. This destruction displaced up to 75 000 people. In all, more than 70% of the population of Vargas was impacted by the disaster.

Venezuela received tens of millions of dollars from international organizations to address the disaster. However, in spite of initial dispersals of funds, and announcements of reconstruction plans, not much was done to help the survivors. Then President Chávez was distracted by political squabbles and abandoned recovery efforts. The survivors of the Vargas disaster eventually left their refugee camps and attempted to rebuild their homes on their own. By 2006, the population had recovered, and the damaged infrastructure was slowly being rebuilt. However, more than a decade after the disaster, thousands still remained homeless.

11.5 Karst Topography

Limestone is a rock that dissolves in surface and groundwater that is even weakly acidic. In temperate and tropical areas, rotting vegetation at the surface makes water acidic. It can seep into the ground and dissolve limestone bedrock. In these areas, acidic groundwater dissolves caves and caverns into the bedrock where streams flow through them rather than on the surface (Figure 11.10). The underground streams dissolve and abrade the bedrock which enlarges the caves. Rock dissolves at the surface and the cave is enlarged underground until the roof of the cave thins to the point where it cannot support its own weight. The roof collapses and forms a sinkhole at the surface.

A sinkhole is a crater-like surface depression that forms as rock or soil beneath it is removed. Sinkholes form after a heavy rainstorm or an earthquake or they can form by themselves. They can be devastating if they develop quickly, but most form quite slowly over months to years. Sinkholes can form in city streets or on driveways as supporting soil beneath them is eroded away. True sinkholes, however, lead to caves in limestone. The surface features of karst include few or no streams, multiple circular depressions, and circular ponds and lakes from water-filled sinkholes.

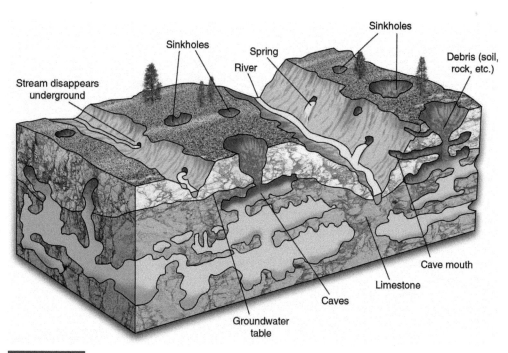

FIGURE 11.10 Block diagram of karst topography and features.

FIGURE 11.10A Photo of a sinkhole in a residential area. *Source*: AP/Images Anonymous.

11.5.1 Dangers from Sinkholes

Sinkholes in open fields are not dangerous, but if they form under man-made structures, they can be destructive and hazardous. Slow-forming sinkholes can crack house foundations and cause roads to sink and crumble, which eventually destroys them requiring costly repairs. If they form as catastrophic collapses of cave roofs, they can be deadly. Sinkholes have formed in residential communities, swallowing homes and causing injury or deaths of inhabitants (Figure 11.10A).

Sinkholes can form at almost any time in areas that contain large limestone units including the state of Florida and in large areas of Kentucky, West Virginia, Virginia, Ohio, Illinois, New York, and Pennsylvania among others. Areas with extensive mining operations can also contain sinkholes as a result of the collapse of mine roofs similar to cave roofs with equally dangerous results.

References

Keaton, J.R., Anderson, L.R., and Mathewson, C.C. (1988). Assessing debris flow hazards on alluvial fans in Davis County, Utah. In: *Twenty-fourth Annual Symposium on Engineering Geology and Soils Engineering: Publications and Printing* (ed. R.J. Fragaszy), 98–108. Pullman, Washington: Washington State University.

Keaton, J.R. and Mathewson, C.C. (1988). Stratigraphy of alluvial fan flood deposits. In: *Hydraulic Engineering, Proceedings of the 1988 National Conference, Colorado Springs, Colorado* (eds. S.R. Abt and J. Gessler), 149–154. New York: American Society of Civil Engineers.

Mergili, M., Frank, B., Fischer, J.-T., Huggel, C., and Pudasaini, S. P. (2018). Computational experiments on the 1962 and 1970 landslide events at Huascarán (Peru) with r.avaflow: Lessons learned for predictive mass flow simulations. *Geomorphology*, 322, 15–28. https://doi.org/10.1016/j.geomorph.2018.08.032. Licensed under CC BY-4.0.0005279566.

Ward, N. S. and Day, S. (2011). The 1963 Landslide and Flood at Vaiont Reservoir Italy. A tsunami ball simulation [JB]. *Italian Journal of Geosciences*, 1, 16–26. https://doi.org/10.3301/IJG.2010.21.

Stefano, B., Mauro, C., and Gianluca, G. (2011). Geological evolution of a complex basaltic stratovolcano: Mount Etna, Italy. *Italian Journal of Geosciences*, 130, 306–317. 10.3301/IJG.2011.13.

CHAPTER 12

Weather and Storms

CHAPTER OUTLINE

12.1 Why Do We Have Weather? 222

12.2 Climate Systems of The Earth 223

12.3 Air Masses and the Fronts that Separate Them 226

12.4 Generation of Severe Weather 228

Earth's Fury: The Science of Natural Disasters, First Edition. Alexander Gates.
© 2022 John Wiley & Sons Ltd. Published 2022 by John Wiley & Sons Ltd.
Companion website: www.wiley.com/go/gates/earthsfury

Words You Should Know:

Air masses – A large package of air in the atmosphere that takes on the physical characteristics of the area where it develops.

Climate – The long-term weather patterns of an area.

Coriolis Effect – Deflection of objects and sense of rotation generated by the rotating of the earth and the resulting relative speed by latitude.

Derecho – A fast-moving, long, bow-shaped line of storms marked by extreme winds.

Fronts – The boundaries between two air masses.

Hail – Frozen precipitation caused by vigorous circulation in a supercell where condensed droplets can be driven to high enough altitudes where they freeze.

Humidity – The percentage of moisture in air relative to the maximum amount possible at a given temperature and pressure.

Lightning – The rapid release of static electricity in a single discharge or bolt.

Precipitation – The condensation of water vapor into droplets and the falling of the water to the earth in any state.

Thunder – The sound of the explosion caused by a bolt of lightning.

12.1 Why Do We Have Weather?

Weather and storms cause some of the worst natural disasters on the earth. The majority of these disaster-producing storms are described in separate chapters. However, it is the same processes that control regular weather as those that control the devastating storms. In this chapter, the processes controlling weather are described as well as weather disasters other than hurricanes/cyclones, tornadoes, and desertification.

The basic controls on weather are simple. First, warmer air holds more moisture. The humidity level is the measure of the moisture contained in the air. If the warm air is cooled enough, the moisture condenses into droplets of water. This is nicely illustrated by a cold drink on a humid summer day. The outside of the glass appears to sweat as the moisture in the air condenses on the cold glass. The fogging up of a mirror in a bathroom with the shower on or the fogging of eyeglasses when walking from the cold outdoors into a warm room happens for the same reason.

Second, the lower atmosphere decreases in temperature as a function of altitude (Figure 12.1). That is why many mountains are snowcapped, whereas the lower altitude of the mountain is not. It is colder when a person travels from the lowlands to the

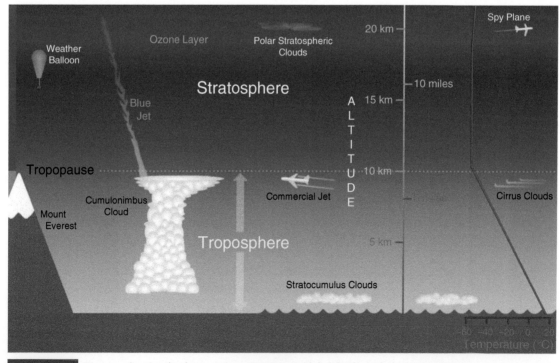

FIGURE 12.1 Diagram showing the decrease in temperature with height through the troposphere.

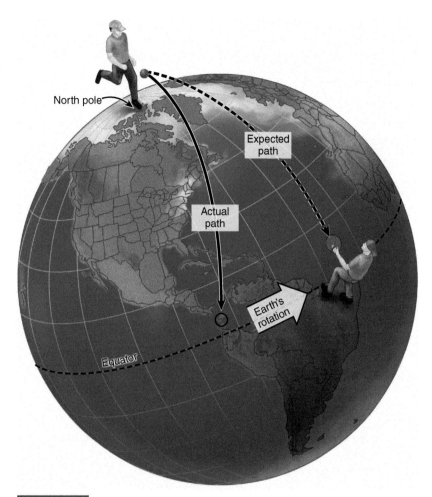

FIGURE 12.2 Diagram showing the straight path of a thrown ball with a nonrotating earth and the curved path of a thrown ball with a rotating earth illustrating the Coriolis Effect.

mountains. Based on the first basic control, the air also holds less moisture at higher altitude than it does in the lowlands as reflected by the humidity. The temperature is also generally distributed geographically on the earth. The temperature is warmer the closer to the equator and colder toward the poles. The humidity follows this distribution as well.

The air pressure is also distributed by altitude. The air pressure is the weight of the air above you being held to the planet by gravity. It is all around us at about 14 lb per square inch, 30 in. of mercury, or 1 bar of pressure at sea level. People are used to the pressure and hence do not notice. If they travel from sea level to the mountains, they notice how "thin" the air is and have a hard time getting enough air. The movement of air adds to this pressure. Air sinking down toward the surface increases the pressure (high pressure), whereas rising air decreases the air pressure (low pressure). This upward movement of air in low pressure carries the warmer, moister air to high altitudes where it is colder. That causes the moisture in the air to condense, and precipitation occurs. The downward movement of air in high pressure carries the colder, drier air to lower altitudes where it is warmer and can hold more water, so clear skies prevail.

The final component is the Coriolis Effect (Figure 12.2). The earth rotates once each day no matter where on the planet. However, from a vantage in space, the speed you will move depends where you are on the planet. At the equator, you would travel all the way around the earth, that is, 24 000 mi (38 624 km), meaning that you would travel at 1000 mph (1609 kmph). At the pole, you would just spin around one full turn in 24 hours. That means points closer to the equator always move faster than those poleward. This difference deflects southward-moving airborne objects to the west and causes weather systems to rotate.

12.2 Climate Systems of The Earth

The controls described control not only local weather features but also the large-scale weather patterns of the earth. These large-scale weather patterns are called climate. The main driver of climate is the rising of air at the equator because it is heated and becomes less dense, and the cooling of air at the poles makes it denser and it sinks. This creates a conveyor belt effect or convection cell of the rising air at the equator moving poleward at high altitude and the sinking air at the poles moving toward the equator along the surface (Figure 12.3).

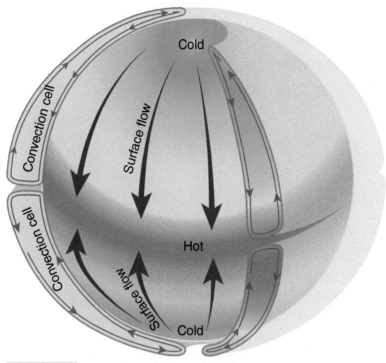

FIGURE 12.3 Diagram showing the simplified circulation of the atmosphere in a nonrotating earth with a single convection cell from equator to pole.

This simple convection system is complicated by the Coriolis Effect which twists the flow and breaks it into three distinct convection cells per hemisphere called Hadley, Ferrel, and Polar cells (Figure 12.4). The rising air at the equator translates to low pressure and condensation which signal precipitation. The tropical rain forests occur near the equator around the world for that reason (Figure 12.5). The Hadley cells drive cold, dry air toward the surface at 30° north and south latitudes producing dry, clear, and calm weather. It is for this reason that the deserts of the world are concentrated in these areas. For example, the Sahara and Arabian deserts occur in that cell. The confluence of the next poleward cells produces rising air at 60° north and south latitudes producing the temperate rain forests

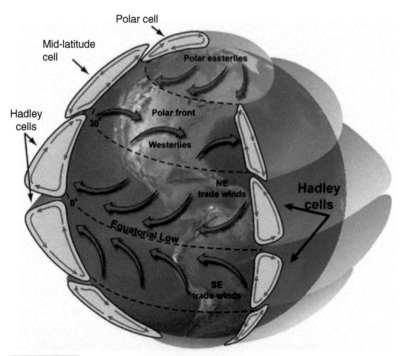

FIGURE 12.4 Diagram showing the actual convection cells of the rotating earth with the Hadley cells closest to the equator and the major wind patterns shown. *Source*: Image courtesy of the US National Weather Service.

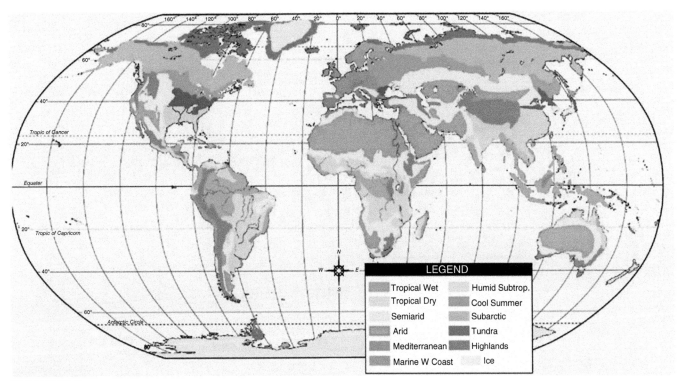

FIGURE 12.5 Map of the earth showing the major climatic belts. Note that tropical rain forests occur around the equator and the deserts occur at 30° north and south latitude. *Source*: Maps courtesy of www.theodora.com/maps used with permission.

like those in the northwest coast of North America. The air still falls at the poles producing a desert. It is not intuitive that the poles are deserts as they are covered by ice and snow, but because they never melt, these accumulations cover thousands of years.

The twisting of the cells by the Coriolis Effect produces wind patterns along the surface that are more east–west than north–south as would be expected in a simple circulation model (Figure 12.4). The sinking air in the high-pressure confluences slides along the surface as wind and lifts back upward in the low-pressure areas. These surface winds are the easterlies and westerlies depending upon location.

In addition to the major climates, local climates and microclimates also develop governed by surface features. Oceans tend to modulate the local temperatures and are especially impactful if they contain strong currents. The Gulf Stream is an ocean current that moves warm water from the Gulf of Mexico to Western Europe (Figure 12.6). It makes these areas warm to temperate when they should be much cooler. Oceans can also generate local disturbances that can change the precipitation from what should be dictated by the climate belts. The abundance of landmasses can reduce the amount of available water to be evaporated to the atmosphere from normally rainy areas to make them much drier than what the climate model dictates. For example, Siberia is dry because there so much landmass even though it is within the northern-temperate rainforest zone. Even local geography can change the climate. Perhaps the best example is in the Northwestern United States where moist air from the Pacific Ocean is driven up the west side of the Cascade Mountains (Figure 12.7). The driving of air up a mountain slope is the same as lifting it in low pressure, thereby producing heavy precipitation. This phenomenon is called orographic precipitation. In the Cascades, the west side of the mountains is a temperate rain forest from the orographic precipitation which dries out the air leaving the east side of the mountains, where the air is falling, as a desert.

An extreme climate type is glaciation which results from the accumulation of ice and snow on a long-term basis. The precipitation generally falls as snow and is compacted into glacial ice. There are two types of glaciation, continental and alpine, both of which involve the movement of ice bodies, and include continuous renewal and melting. Both types entrain rock, soil, and other debris, and form distinctive deposits. Alpine, however, is restricted to the mountains and is typically young, whereas continental is located on relatively flat land in polar regions. It only extends away from those areas during ice ages.

Alpine glaciers form in mountains as a function of climate, elevation, and latitude. If a mountain is in an area that receives snow and some part remains below freezing for enough of the year to support permanent ice, alpine glaciation occurs. Most high mountains have some glaciation. As soon as ice is thick enough, under the force of gravity, it begins to flow downhill very slowly in the zone of transport. As it moves, the ice moves to warmer temperatures at lower elevation and it melts. The point where melting back balances the downward movement of the ice is the termination of the glacier. The termination shifts downhill in winter. The edges and base of the glaciers entrain rock and soil called a moraine which collects at the base of the hill. There typically are rivers emanating from the ends of alpine glaciers.

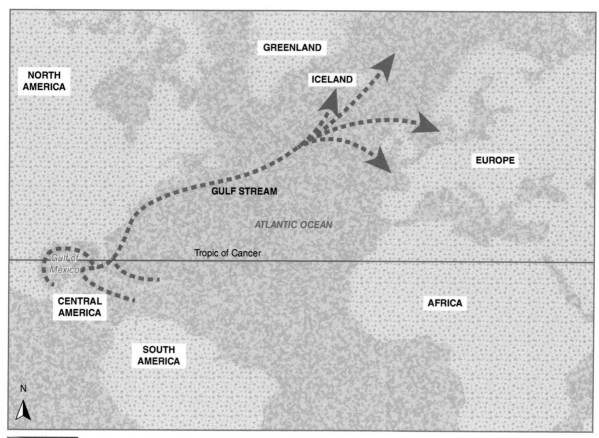

FIGURE 12.6 Map of the North Atlantic Ocean Basin showing the Gulf Stream ocean current.

FIGURE 12.7 Diagram showing the orographic effect of moist rising up the front side of a mountain causing precipitation and sinking down the back side of the mountain as dry air.

12.3 Air Masses and the Fronts that Separate Them

A package of air (air mass) that sits above an area for long enough takes on the area characteristics. For example, if the air mass develops over land in the north of the Northern Hemisphere, like Canada, it will be cold and dry. If it develops over the Gulf of Mexico, it will be warm and moist, and over the North Pacific, it will be cold and moist. These air masses can then move and interact with other air masses. The boundary where two of these air masses meet is called a front. There are several different types of fronts depending upon the air masses colliding. There are cold fronts, warm fronts, and occluded fronts.

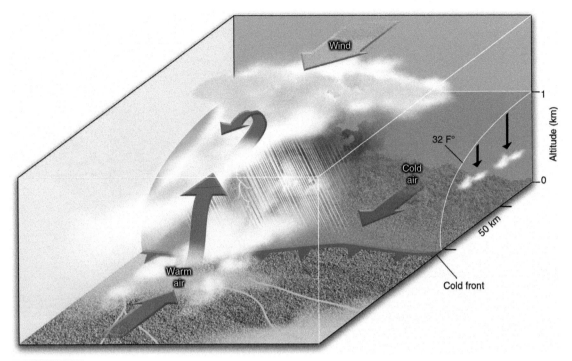

FIGURE 12.8 Block diagram of a cold front showing the front moving from back to front.

12.3.1 Cold Front

The passage of a cold front is where cold air replaces warm air in a particular area. Cold air masses primarily move from northwest to southeast across most of the United States or inland from the Pacific Ocean and less commonly inland from the Atlantic Ocean. Cold air masses from the north and northwest are drier than the warm air masses they contact. Cold air is denser than warm air, so it pushes beneath the warm air at the front (Figure 12.8). The front, however, is very steeply sloped because there is friction of the front on the ground as it moves. The steep boundary makes the weather at the cold front intense and short-lived.

Typical cold fronts on the edge of cold, dry air masses of the continental interior are marked by the appearance of thick cumulus clouds on a still and humid day. They are far more common and noticeable in the summer than in the winter. The thick, tall clouds build up quickly with abrupt heavy rain and strong wind (Figure 12.8). Thunderstorms are common and tornadoes are possible along these fronts in extreme situations. The heavy weather lasts a short period of time and the skies quickly clear once it passes. The humidity drops abruptly as does the temperature, though the wind may remain for a few days. With cold, wet air masses from oceans, there is less of a drop in humidity after passage of the front and fog may develop with chilly wet conditions after the heavier rain ends.

12.3.2 Warm Front

In the passage of a warm front, warm air replaces cold air at a given location. In the United States, if a warm air mass moves northward from the Gulf of Mexico, as is most common, it is moist. If it moves northward from the southwest, it is dry. Warm air is less dense than cold air, so it rises above the cold air at the ground surface (Figure 12.9). The warm air on the leading edge of the warm air mass forms a wedge above the trailing edge of the cold air. The passage of the front is therefore much more drawn out than a cold front.

The passage of a warm front with a humid air mass typically begins with a thin, high-level haze. Sun dogs (small partial rainbows on either side of the sun) can occur during the day and a halo may be seen around a full moon at night. After several hours, thin, high-stratus clouds appear and slowly build thicker and lower. Rain begins as drizzle and slowly intensifies lasting up to several days, clearing slowly.

12.3.3 Occluded Front

A cold front can overtake a leading warm front if it is moving quicker. If the two frontal types overlap, they can form an occluded front. In this situation, the steep-faced cold front overrides the shallow, wedge-shaped warm front (Figure 12.10). Cold air hugs the ground surface, forcing the wedge of warm air aloft. A result is that, at the surface, only two cold air masses are felt, and temperature remains about the same with the passage of the front.

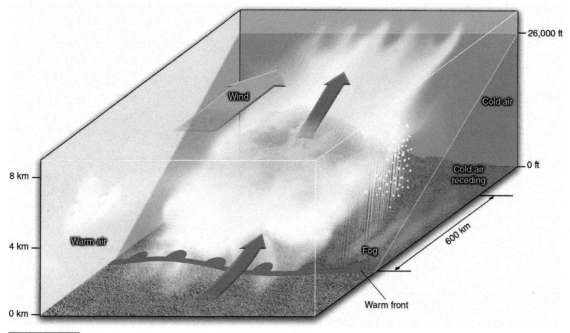

FIGURE 12.9 Block diagram of a warm front showing the front moving from front to back.

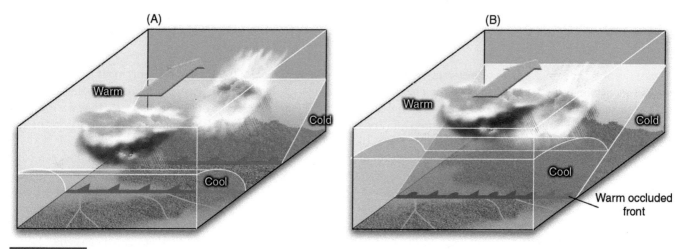

FIGURE 12.10 Block diagrams of (A) a cold front moving from front to back and overtaking a warm front to create (B) an occluded front.

These mixed fronts produce weak cloud and precipitation responses reflecting both cold and warm fronts. As the front approaches, the weather conditions resemble a warm front with the slow building of clouds and rain. The weather can then abruptly intensify when the steep cold front part of the system passes before the rain ends. Occluded fronts tend to move very slowly and inclement conditions can persist for several days.

12.4 Generation of Severe Weather

Severe weather can take on many forms including hurricanes, typhoons, cyclones, and monsoons. They can also produce flash floods. These topics are covered in other chapters. The weather listed is also not related to fronts. The severe weather described here is related to fronts, and largely cold fronts. Tornadoes are also primarily produced at cold fronts, but they are discussed in a separate chapter as well.

The generation of severe weather at cold fronts is primarily related to thunderstorms. The generation of strong thunderstorms requires a front separating air masses with strong differences in moisture and temperatures. Thunderstorms differ from normal rain in the vigorous vertical motions or circulation they include, which is a common product of strong cold fronts. The strong upward motion propels warm moist air to great heights where it condenses producing heavy rain and even ice. Strong downward motions produce the surface winds and even possible funnel clouds and tornadoes associated with thunderstorms. It is the relative strengths of the upward versus downward motions that determine the type and degree of danger that will be produced by the storm (Figure 12.11), though the dominance of the flow direction typically varies through the life of the storm. They can change from one type to another.

Strong upward motion produces heavy precipitation but less wind and lightning. The lightning and resulting thunder are produced by friction in the clouds and with the surface, so enhanced by both up and down motion. Lightning is an electrical discharge of the static electricity produced by the friction. The strong ascending and descending air also produces the most damaging effects such as tornadoes. If descending air dominates, there will be a lot of wind but very little precipitation.

The systems that produce the thunderstorms can also vary. There can be a single circulation cell in the system and these are aptly called single-cell storms. They are generally weaker with less damaging effects. If there are two or more circulation cells, they are called multicellular storms. They are typically much more powerful than single-cell storms and can produce tornadoes but typically only minor ones. If the circulation is especially vigorous and the system is large, they are called supercells. These supercell systems are responsible for the vast majority of the front-related damage and disasters.

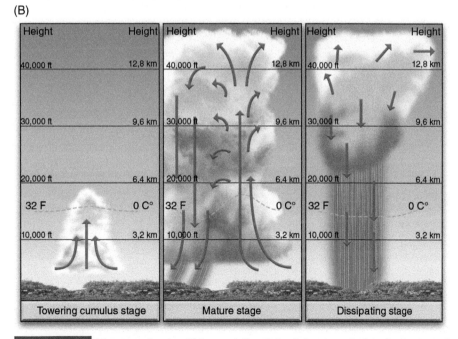

FIGURE 12.11 Diagrams showing (A) two varieties of circulation types in thunderstorms and (B) evolution of a thunderstorm through the three typical stages.

Wall Cloud. The base of the rotating mesocyclone extends below the cloud deck (cloud base) because of downdrafts. The funnel cloud/tornado is drawn out of the base of the wall cloud.

Hail. Frozen precipitation falls into humid air and water is condensed onto it. The strong updrafts sweep it back up to where the condensed water is also frozen increasing the size. This cycle is repeated until the hail becomes too heavy for the updraft and it falls to the ground.

Tornado. Strong downdrafts extend the base of the rotating mesocyclone and wall cloud towards the surface in a thin high velocity, rotating funnel cloud that becomes a tornado when it reaches the surface.

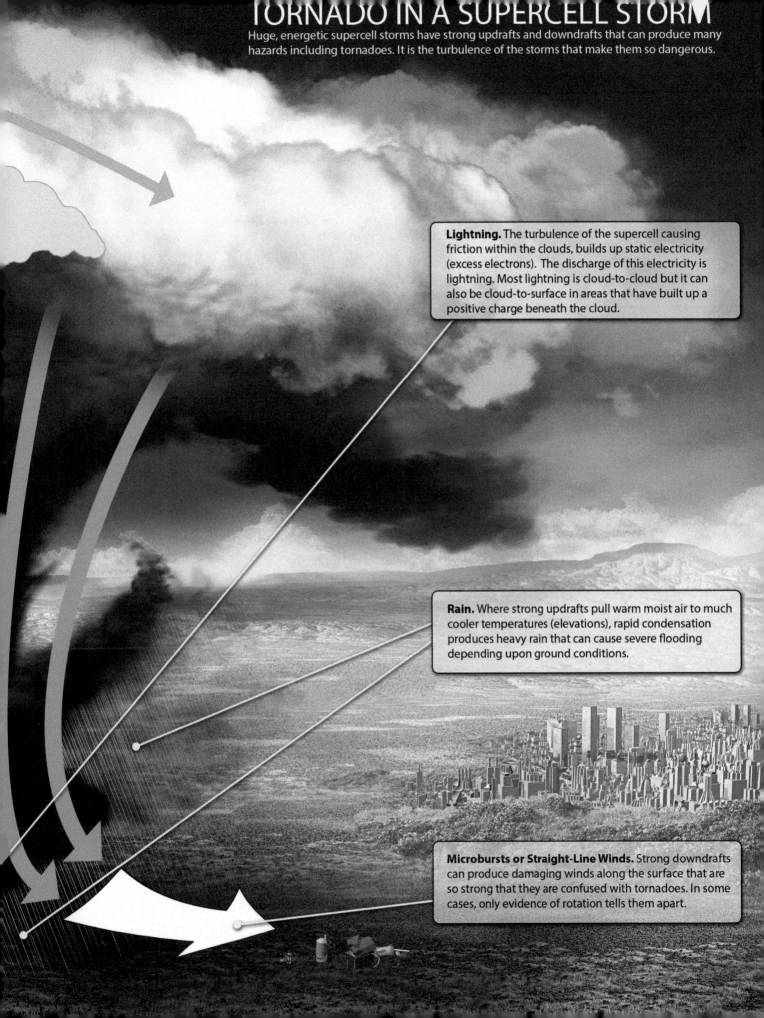

CASE STUDY 12.1 1985 Delta Flight 191 Crash

Downbursts and Wind Shear

The overwhelming focus of concern in supercell thunderstorms is tornadoes because they are so devastating. Hail can also be quite damaging and flash flooding can be deadly. The other aspects of a thunderstorm are usually ignored because they are inconsequential in comparison. However, this depends on the situation. Downbursts and other straight-line winds can be quite powerful, and although they mostly cause minimal damage to structures, under certain situations, they can be deadly.

Downbursts are especially deadly for air traffic. There were several deadly crashes as a result of downbursts or wind shear in thunderstorms. One of the deadliest was Delta Air Lines Flight 191 which crashed on 2 August 1985 upon final approach to Dallas/Fort Worth International Airport (Figure 12.12). A total of 135 people were killed in the disaster including 126 of the 152 passengers, 8 of 11 crew members, and one person on the ground.

A downburst is a column of rapidly descending air that strikes the ground and spreads out in all directions (Figure 12.12A). It can produce straight-line winds of over 150 mph (240 kmph), capable of producing extensive damage similar to tornadoes. The difference is that tornado damage shows convergent motion that is consistent with rotating winds rather than straight-line winds. Downbursts are from the downward circulation in the storm and are especially common in highly convective thunderstorms like those in supercells.

Downbursts are downdrafts of cool air that descend and accelerate as they approach the ground. There are two basic categories of downbursts and both kinds are associated convective thunderstorms. Dry downbursts occur in thunderstorms with very little rain. By contrast, wet downbursts occur in thunderstorms with large amounts of rain. The majority of downbursts are microbursts which are less than 2.5 mi (4 km) across. Downbursts that are larger than 2.5 mi (4 km) can be termed macrobursts. At all sizes, they tend to form in three stages progressing from downburst to outburst to cushion. The rapidly changing patterns of updrafts and downbursts make forecasting wet microbursts very difficult. They create vertical wind shears, which are hazardous to aviation, especially during landings.

FIGURE 12.12 View of the Dallas/Fort Worth Airport showing the track of the crash and the runway that it was scheduled to land on. *Source*: James Richard Covington.

FIGURE 12.12A Illustration of a downburst from a thunderstorm as an airplane approached an airport. *Source*: NASA artist's rendering of a microburst.

This was the case on 2 August 1985, for Delta Air Lines Flight 191 which flew from Fort Lauderdale, Florida to Los Angeles, California with a stopover at Dallas/Fort Worth International Airport in North Texas. The 7:00 a.m. daily weather report showed a southward-moving cold front that extended from the East Coast into the Southern Plains. There were cooler temperatures, widespread cloud cover, and light rain behind the front, but hot and humid air in front. In Dallas that day, temperatures climbed into the upper 90s °F (35–37 °C) and it was 101 °F (38 °C) at the airport. As a result of the cold front, cumulus clouds developed during the afternoon across North and Central Texas including some towering cumulonimbus along the frontal boundary near the Dallas/Fort Worth Airport (Figure 12.12B).

The Lockheed L-1011 TriStar of Flight 191 departed Fort Lauderdale at 2:10 p.m. Central Time. The flight weather forecast called for widely scattered rain showers and thunderstorms possible at Dallas. As the aircraft approached New Orleans, Louisiana, reports of intensifying Gulf of Mexico weather caused the flight crew to switch to a more northerly approach to Dallas. Weather radar imagery showed weak convection at around 5:00 p.m., but by 5:28 p.m. radar echoes began to intensify north of Dallas which were classified as category VIP 1 or "weak echo."

It was about this time that Flight 191 was approaching Dallas. At 5:43 p.m., the flight was cleared to descend to 10 000 ft (3000 m) and suggested to take a 250° heading toward the northern approach. The captain, however, wanted to avoid flying through a large storm cell and took a safer route. At 5:47 p.m., the flight descended to 9000 ft

FIGURE 12.12B Annotated satellite image of the weather at the time of the crash. *Source*: National Weather Service (NWS).

(2700 m), and at 5:51 p.m., it descended to 7000 ft (2100 m) and then slowed its speed to 180 knots (210 mph; 330 kmph). The crew prepared the aircraft for landing as it descended to 5000 ft (1500 m) and that is when it began to rain.

At this same time, weather radar was showing increasingly dangerous conditions. At 5:48 p.m., radar imagery showed a black "speck" in the southwestern storm cell, indicating that convection intensity increased to VIP 3 which is rated "strong." Intensifying thunderstorms were observed at the airport and satellite imagery showed areas of towering cumulus clouds. At 6:00 p.m., radar imagery showed that the storm had developed two adjacent cells with convective strengths of VIP 3 or "strong," just 2–4 mi (3.2–6.4 km) north of the runway on which Delta Air Lines Flight 191 was attempting to land. Microburst winds at this location were exceeding 80 mph (129 kmph) and a nearby observer recorded a wind gust of 53 mph (85 kmph) just minutes after the flight exited the rain curtain.

At 6:04 p.m., Flight 191 reduced its speed to 150 knots (170 mph; 274 kmph), and the captain radioed the tower saying, "Delta one ninety one heavy, out here in the rain, feels good." The tower reported winds of 5 knots (5.8 mph; 9.3 kmph) and gusts to 15 knots (17 mph; 27.4 kmph). The landing gear was lowered and the flaps set in preparation for landing, and the aircraft descended to 1000 ft (300 m). At 6:05 p.m., the cockpit voice recorder began capturing panicked voices and sounds. The aircraft began to descend at a steep 30° angle but with wild variations.

The aircraft was descending at more than 50 ft/s (15 m/s) when the ground proximity warning sounded. The crew did everything possible to right the aircraft, but at 6:06 p.m., it touched down on a plowed field 6336 ft (1931 m) north of the runway. The landing gear left depressions in the field for 240 ft (73 m) as the aircraft approached Texas State Highway 114. The nose gear touched down on the westbound lane and the plane skidded across the road at 200 mph (320 kmph) striking a car and killing the driver. The aircraft skidded along the ground, breaking into pieces. A fire broke out in the cabin and many passengers were doused in jet fuel and perished in flames, others were sucked out of the plane. The aircraft finally crashed into water tanks at the edge of the airport where the fuselage was engulfed in a fireball (Figure 12.12C).

The crash left 137 people dead and 28 injured, some severely. The National Transportation Safety Board (NTSB) performed a full investigation that concluded the crash resulted from the decision to fly through a thunderstorm, the lack of procedures to avoid microbursts, and the lack of warning about wind shear conditions. As a result of this and several other similar crashes, in 1988, the Federal Aviation Administration (FAA) required all commercial aircraft to be equipped with onboard wind shear detection systems by 1993. Many airports installed ground-based wind shear detection systems as well. As a result of this improved ability to detect downbursts and accompanying wind shear, aircraft accidents caused by wind shear have decreased dramatically since 1995.

FIGURE 12.12C Photo of the remaining tail section of the airplane. *Source*: AP Photo/Carlos Osorio.

CASE STUDY 12.2 1994 Dronka Lightning Strike

Resultant Lightning Disasters

Lightning strikes are deadly, but typically only to one or, at most, a few people for each direct strike. The worst known single deadliest direct strike was from a severe 23 December 1975 thunderstorm in part of Rhodesia that would become Zimbabwe, Africa. This area has very strong thunderstorms with frequent lightning strikes and a paucity of high-standing grounding features. This deadliest strike was into a hut crowded with 21 people trying to escape the storm. All were killed. However, if lightning strikes and damages a structure that houses a large number of people, called secondary or resultant damage, it can be far more deadly. This was the case in the 1994 Dronka lightning strike disaster, Egypt (Figure 12.13).

Lightning strikes are electric discharges between clouds or clouds and the ground. They are primarily restricted to cumulonimbus clouds in thunderstorms. Cloud-to-ground lightning terminates on the ground surface, though, rarely, they can propagate in the opposite direction in ground-to-cloud lightning. About 25% of all lightning strikes are cloud-to-ground. The remaining 75% of lightning is within clouds, intracloud, or cloud-to-cloud, never reaching the ground.

In cloud-to-ground lightning, a channel or "stepped leader" of negative charge propagates downward in zigzag path of about 165-ft (50 m) segments that terminate at forks in the path (Figure 12.13A). The stepped leader cannot be seen, but it shoots to the ground in milliseconds. Near the surface, this negatively charged stepped leader induces channels of positive charge to extend upward from taller objects depending upon their conductivity. When the negative stepped leader connects to the positive charge path, a powerful electrical current travels about 60 000 mi (96 560 km) per second back to the cloud in a channel of about 1 in. (2.5 cm) diameter. The channel of the lightning strike can be heated to as much as 50 000 °F (27 760 °C). A typical cloud-to-ground flash consists of 1 to up to 20 return strokes. Although a single-return stroke flash can last milliseconds, a multiple-return stroke occurred over Argentina on 4 March 2019 that lasted for 16.73 seconds.

Lightning strikes can travel great distances from the storm to the ground. A lightning strike called "bolt from the blue" begins inside a cloud, but emerges and travels horizontally some 6 mi (9.7 km) or more from the cloud before reaching the ground. It is called a bolt from the blue because it can strike an area on a clear day. Cloud-to-cloud discharges are miles long and can be upwards of tens of miles in length. The longest

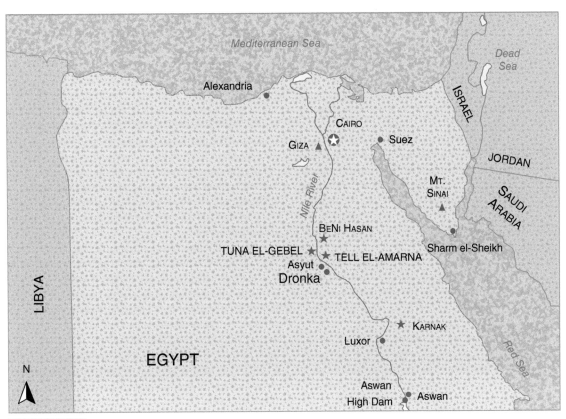

FIGURE 12.13 Map of Egypt showing the location of Dronka.

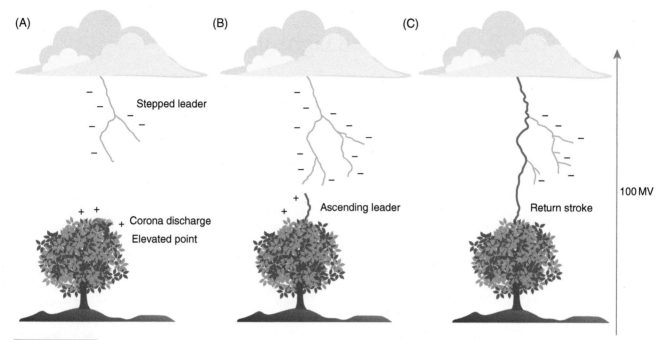

FIGURE 12.13A Illustration of the generation of a lightning bolt. (a) Formation of a stepped leader descending from a thunderstorm. (b) Formation of an ascending leader toward the cloud. (c) Generation of a lightning bolt between the earth and cloud. *Source*: Adapted from Kasparian et al. (2008).

lightning strike occurred over Southern Brazil on 31 October 2018. Satellite records show that it extended 440 mi (709 km). The surface location of the strike is referred to as the "termination."

If a person is within the discharge channel, it is called a direct strike. In this case, all of the electricity passes through the person, which results in explosions of flesh and bone, organ damage, and nervous system damage. It can be fatal or cause permanent injury and impairment. The mortality rate in a direct strike is between 10% and 30%, and as many as 80% of survivors suffer long-term injuries.

Another injury type occurs when a person is touching a conductor that is electrified by a strike. This is called a contact injury. A side splash injury is when branches of strike split or fork from the primary strike channel and injure a person. If as the earth surface charge races toward the strike channel, it may pass up a person's legs and through their body. This is a ground current or "step potential" injury. If the strike occurs too close to a person, they can be injured by the shock wave, known as blast injuries.

These injuries are inflicted by the strike, so they are direct. Secondary or resultant injuries occur when the strike causes some other structure or feature to inflict the damage and injury. These mainly include explosions, fires, and accidents. There are numerous cases of lightning striking aircraft and inflicting extensive damage.

There are approximately 240 000 lightning strike incidents annually. In the United States, the most common area for lightning strikes is in Florida. Annual death toll from lightning strikes varies, but is estimated to range from 2000 to 6000, globally. In the United States, approximately 51 fatalities from lightning strike occur annually.

The deadliest case of a secondary or resultant lightning disaster in recent history occurred in Dronka, Egypt on 2 November 1994. Dronka is a village of about 10 000 people about 200 mi (320 km) south of Cairo (Figure 12.13). The evening was marked by the development of very severe thunderstorms throughout that part of Egypt in well-developed supercells. These storms persisted for up to five hours locally and caused significant damage and flash flooding. Toward the end of the thunderstorms, a bolt of lightning struck a complex of eight fuel tanks owned by the Egyptian Army and located along a railway line. Three of the storage tanks, each holding about 5000 tons of aircraft or diesel fuel, were ignited in the strike. The storage tanks exploded and split open, spilling burning fuel throughout the area. The tanks and train depot subsequently collapsed as flash flood waters built up behind them and they swept the flaming fuel into the village of Dronka. It was later described as resembling napalm.

More than 200 houses were destroyed in Dronka including two five-story structures from which there were no survivors (Figure 12.13B). One of the fuel tanks appeared to threaten igniting the other five if the wind changed direction, as they were only about 213 ft (65 m) apart. In response, at least 20 000 terrified residents from Dronka and the surrounding towns fled the area to Assiut City, the provincial capital, some 5 mi (8 km) away. Even though black smoke covered the entire area, firefighters let the fire burn itself out for fear of the other tanks igniting. The number of people killed in this disaster is unclear. One main source claims a death toll of 469, but there are reports of more than 500 deaths.

FIGURE 12.13B Photo showing damage from the lightning disaster in Dronka. *Source*: Stringer/Reuters.

CASE STUDY 12.3 2017 Denver Hailstorm

Damage in Hail Alley

Tornadoes are certainly the most fearsome aspects of hazards of supercells. Many of the storms that produce tornadoes also produce hail. However, the hail and its damage are commonly overlooked in the reports because the tornado was so much more devastating. Hail, however, can be both deadly and devastating. It is reported that approximately 1000 English soldiers were killed by hail in France on Black Monday in 1360 during the Hundred Years' War and more than 10 000 were killed by hail in China in 1490. These may be exaggerations, but hail can certainly be damaging.

Hail forms by raindrops being carried upward by updrafts in a thunderstorm into extremely cold levels of the atmosphere where they freeze (Figure 12.14). The hail then drops back down into the level above freezing and moisture condenses on the surface of the hail. If the hail is heavy enough or the updraft is weak, the hail continues to drop to the ground. Otherwise, the updraft carries the hail back into the extremely cold level and the condensed water freezes to the hail causing it to grow. If the cycle continues, the hail can keep growing and become quite large. The updraft wind speeds can be up to 110 mph (180 kmph).

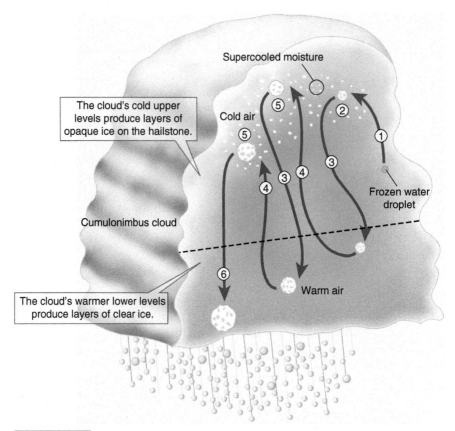

FIGURE 12.14 Diagram showing the formation of hail by circulation between cold and warm sections of the cloud. *Source*: Adapted from EverythingLubbock.com.

The sweeping of hail to the top of its circuit may take as much as 30 minutes based on the force of the updrafts in the thunderstorm. The top of supercell thunderstorms is usually greater than 6 mi (10 km) high. Hail is commonly at elevations above 20 000 ft (6100 m) within the storm. Between 20 000 and 10 000 ft (6100–3000 m), 60% of hail is still within the thunderstorm, but 40% has fallen clear of the convection cell.

The layers of ice on the hail are not simply from up and down cycles inside a thunderstorm. The winds may be horizontal and help to form a rotating updraft, especially in supercell thunderstorms or get caught in downdrafts or wind shear. Strong surface winds can make hail fall at an angle or even essentially sideways. Wind-driven hail is very dangerous and can shear siding from houses, break windows, break side windows on cars, and inflict severe injury or even death to people and animals.

The speed of descent of the hail depends on its size. The fall speed for hail of <1 in. (2.5 cm) is between 9 and 25 mph (14.5–40 kmph). In a typical severe thunderstorm, hail is between 1 and 1.75 in. (2.5–4 cm) and fall speed ranges from 25 to 40 mph (40–64 kmph). In the strongest supercells that produce some of the largest hail, it is commonly 2–4 in. (5.1–10.2 cm) and the fall speed is between 44 and 72 mph (71–116 kmph). These sizes and speeds can vary considerably due to variability in the hail shape, degree of melting, fall orientation, and local conditions. Hailstones of 4 in. (10.2 cm) or more have been known to fall at speeds over 100 mph (161 kmph). The largest hailstone on record in the United States was on 23 June 2010 in Vivian, South Dakota. It had a diameter of 8 in. (20.4 cm), a circumference of 18.62 in. (47.3 cm), and weighed 1 lb 15 oz (0.88 kg).

The various specific sizes of hail have associated names as follows:

Pea = 0.25 in. (0.64 cm).
Mothball = 0.5 in. (1.3 cm).
Penny = 0.75 in. (1.9 cm).
Nickel = 0.88 in. (2.3 cm).
Quarter = 1 in. - hail quarter size or larger is considered severe (2.5 cm).
Ping–pong ball = 1.5 in. (3.8 cm).
Golf ball = 1.75 in. (4.4 cm).
Tennis ball = 2.5 in. (6.4 cm).
Baseball = 2.75 in. (7 cm).
Tea cup = 3 in. (7.6 cm).
Softball = 4 in. (10.2 cm).
Grapefruit = 4.5 in. (11.5 cm).

The High Plains of Northeast Colorado to the east of the Rocky Mountains are known for destructive hail. In terms of states, Nebraska, Colorado, and Wyoming have the most hailstorms. The area around where these three states meet has long been nicknamed "Hail Alley" as a result (Figure 12.14A). However, Texas usually has the most damage from hailstorms.

Warm, moist air from the Gulf of Mexico moved into Colorado by moving through the Rio Grande Basin on 8 May 2017. A cold front exhibiting upslope flow along the Colorado Front Range of the Rocky Mountains near Denver and Fort Collins moved into the unstable air (Figure 12.14B). The forecast warned of damaging winds but included the potential for hail as well. This weak frontal boundary was located

Generation of Severe Weather 239

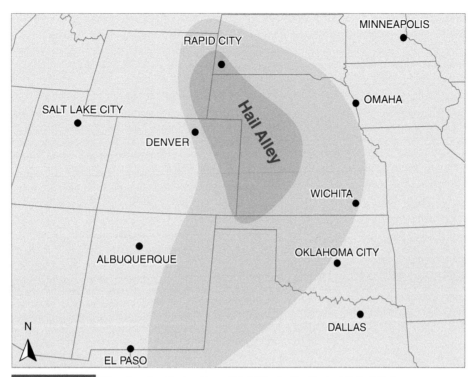

FIGURE 12.14A Map showing the location of Hail Alley in the Central United States.

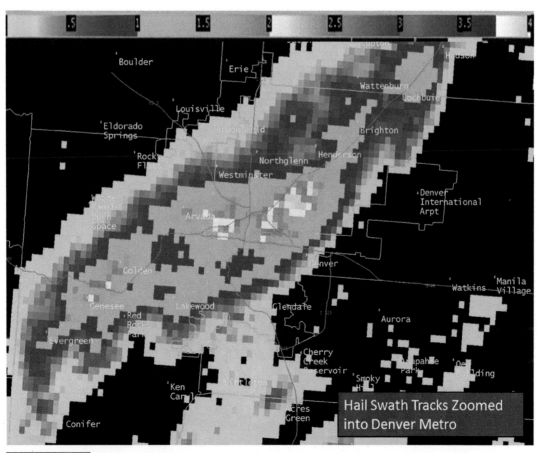

FIGURE 12.14B Radar image of the Denver hailstorm. *Source*: Image courtesy of the US National Weather Service.

over the Palmer Divide to the south of Denver. It initiated discrete highly convective supercell thunderstorms exhibiting rotation south and east of Denver around 1:30 p.m.

The slow-moving thunderstorms produced large hail between 2:00 and 3:00 p.m. The storms first headed north-northeast into the West Denver area at 22 mph (kmph), but abruptly shifted to an east-northeast direction. It was at this point, at 2:40 p.m., that Doppler radar detected large hail and the first severe hail was reported from the west of Denver. It was followed by numerous reports of hail above 2 in. (5.1 cm) and up to 2.75 in. (7 cm) (Figure 12.14C). In all, there were 60 reports of damaging hail. This area is densely populated and includes numerous automobile dealerships setting up the potential for extensive damage.

The storm was moving so slowly that hail piled up several inches deep, causing travel problems throughout the metro area (Figure 12.14D). After passing through Denver over an approximate three-hour period, the storm swung back to traveling in a north-northeast direction. It later merged with other storms in the area into a large mesoscale convective system that produced more than 2 in. (5.1 cm) of rain in the Greeley, Colorado area.

The hailstorm caused extensive damage. In most areas, the hail was golf-ball-sized which was large enough to break windshields and the windows of some businesses (Figure 12.14E). However, in Wheat Ridge to the northwest of downtown Denver, the hail was baseball-sized (2.75 in. [7 cm]). The hospital building and medical offices of the Lutheran Medical Center sustained damage to windows. In this area, power lines were downed and there were multiple collisions on the slick and hail-covered roads. Because the storm struck during the evening rush hour in the Denver metropolitan area, automobile damage was extensive. As the storms struck the Denver International Airport, many flight delays occurred.

FIGURE 12.14C Photo of golf-ball-sized hail from the 2017 Denver storm. *Source*: spxChrome/Getty Images.

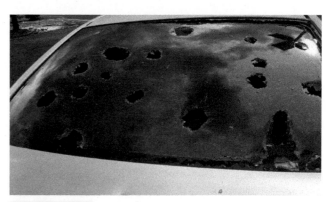

FIGURE 12.14E Photo of hail damage to a car window. *Source*: National Weather Service, NWS.

FIGURE 12.14D Photo of hail covering Denver city streets. *Source*: Len Edgerly, Flickr, Inc., CC BY-SA 2.0.

FIGURE 12.14F Photo of hail damage to a house. *Source*: Claudia Weinmann/Alamy Stock Photo.

The 8 May 2017 hailstorm was the costliest natural disaster in Colorado history. It resulted in 167 000 auto hail damage claims filed. There were 56 409 buildings damaged (Figure 12.14F). The Colorado Mills Mall in Lakewood experienced such extensive hail damage that it was forced to close for eight months. The storm hammered the roof with large hail for 15 minutes, causing parts to break allowing rain to flood the mall. As much as 3 in. (7.6 cm) of water filled the mall. The total cost of the damage was $2.3 billion dollars which is far greater than the vast majority of tornadoes and even some hurricanes.

CASE STUDY 12.4 2020 Midwest Derecho

Great Inland Hurricane

The 2020 Midwest derecho was a severe weather occurrence on 10 and 11 August 2020. It struck Iowa, Illinois, Nebraska, Wisconsin, and Indiana in the Midwestern United States where it caused exceptional damage (Figure 12.15). This extensive storm system has been described as an "inland hurricane." Unlike most other weather disasters in this book which are the result of single storms, a derecho is a weather occurrence of multiple connected storms. Derechos are rare and almost never as strong as the 2020 disaster.

A derecho is a large-scale and long-lived windstorm that is connected to a rapidly moving band or squall line of showers and thunderstorms that forms a bow echo on weather radar (Figure 12.15A). A derecho can be as destructive as a weak to moderate tornado, but the damage is mainly the result of straight-line rather than rotating winds. However, tornadoes can also form along the bow of a derecho. Derechos can sometimes be recognized as a bank of ominous-looking "shelf" clouds that appear darker than normal clouds.

Derechos include wind gusts of a minimum of 58 mph (93 kmph) along most of their length. The wind gusts are typically less than 100 mph (161 kmph), but, in rare cases, they can be as much as 130 mph (209 kmph) or more. Derechos are large features that are at least 250 mi (400 km) long and 50 mi (80 km) wide. However, the largest derechos can be more than 1000 mi (1609 km) long and more than 200 mi (320 km) wide. They are also long lasting with durations of more than 12 hours and high-speed forward movement.

Derechos are zones of organized storms that form along cold fronts, but they commonly outrun them. They develop a self-perpetuating mesoscale convective system at the point where the upper-level divergence is strongest and low-level inflow and convergence are greatest. This system is driven by downbursts that occur where wet air in a thunderstorm evaporates when it meets the drier air surrounding it. This evaporation cools the air around it by latent heat transfer and the cool dense air sinks rapidly to the ground creating the downburst. The downburst can draw more dry air into the storm which forms even stronger downbursts or groups of downbursts. This positive feedback system drives the instability that intensifies the wind and storms.

Derechos commonly appear as squall lines marking the cold front that form bow- or spearhead-shaped features on weather radar

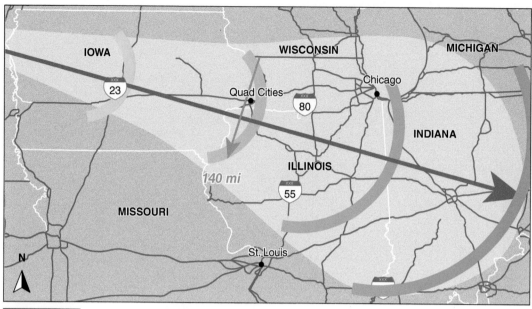

FIGURE 12.15 Illustration of the movement of the 2020 Derecho across the Northern United States.

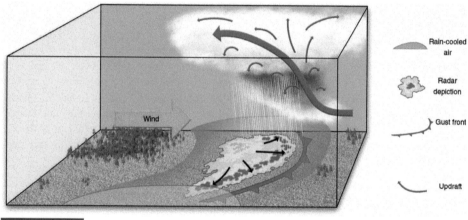

FIGURE 12.15A Illustration of a derecho including radar image and winds.

(Figure 12.15a). These are called bow echoes or spearhead echoes and they occur because a mesoscale high-pressure system forms behind the initial convective line driving it forward at high speed. This high-pressure area is the result of the downbursts behind the squall line.

There are several types of derechos:

Serial derecho: A derecho that is associated with a deep low-pressure system.

Single-bow derecho: This is a single large bow echo that is at least 250 mi (400 km) long.

Multibow derecho: This is where multiple bow derechos form along a squall line that is at least 250 mi (400 km) long.

Progressive derecho: A line of thunderstorms that forms a bow shape and travels several hundred miles along a stationary front.

Hybrid derecho: A mixed serial and progressive derecho.

Low dew-point derecho: A derecho in an area with limited low-level/-altitude moisture, but appreciable mid-level moisture.

The 2020 Midwest derecho was among the most powerful ever witnessed and established new concerns for these large weather events. The storm system that spawned the derecho developed on the Southeast South Dakota–Northeast Nebraska border in the morning of 10 August 2020 (Figure 12.15B). It rapidly intensified to a severe storm that produced damaging winds as it moved. In Western Iowa, the storm became extreme, producing widespread severe wind damage. High velocity wind gusts lasted for as long as 30–45 minutes each.

The storm moved at great speed, entering Illinois and Southern Wisconsin by 2 p.m. on 10 August and Northern Indiana by 5 p.m. In this region, the winds decreased slightly, but the intensifying thunderstorms produced 17 confirmed tornadoes including 12 in Illinois, three in Indiana, and two in Wisconsin. Seven of these tornadoes were rated EF0 and 10 were rated EF1. The derecho struck Southeastern Michigan by 6 p.m. and Western Ohio by 7:30 p.m. It progressively weakened during this time and ultimately dissipated overnight.

The 2020 Midwest derecho traveled a distance of 770 mi (1240 km) from Southeastern South Dakota to Ohio in about 14 hours. It moved west to east at an average speed of a very quick pace of 55 mph (88.5 kmph) evolving primarily from a single-bow to a multibow derecho as it traveled. The length increased from about 150 mi to more than 250 mi along the route.

The maximum estimated winds for the derecho were an amazing 140 mph (225 kmph) in Southwest Cedar Rapids, Iowa. This wind speed is equivalent to an EF3 tornado or a category 4 hurricane. The maximum measured wind speed was 126 mph (203 kmph) at Atkins, Iowa, which is equivalent wind speed to an EF 2 tornado or a category 3 hurricane. In all, there were 16 reports of winds of 90 mph (140 kmph) or greater in Iowa and Illinois.

The 2020 Midwest derecho produced severe damage especially in Eastern Iowa and Northwestern Illinois, where most of the tornadoes occurred. Between 10 and 13 August, there were more than 1.9 million power outages in the region with a maximum of 1.4 million outages at one time. In Linn County, Iowa, 95% of residents experienced power outages, with Cedar Rapids experiencing a 98% power loss at one point. In total, there were 759 000 power outages in Illinois, 585 000 in Iowa, 283 000 in Indiana, and 345 000 in the other states.

Damage to buildings and other structures was also severe (Figure 12.15C). More than 8000 homes were severely damaged or destroyed in Iowa and $23.6 million worth damage was inflicted on public infrastructure. Essentially, every structure within 75 mi^2 (190 km^2) of Cedar Rapids was damaged. The cost of cleaning up the debris left by the storm alone exceeds $21.6 million.

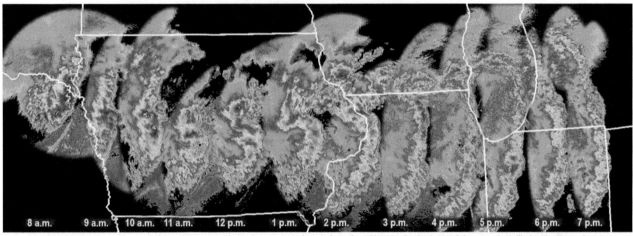

FIGURE 12.15B Sequential radar images of the migration of the 2020 Derecho across the Northern United States. *Source*: National Weather Service -NOAA.

FIGURE 12.15C Photograph of damage caused by the 2020 Derecho. *Source*: National Weather Service-NOAA.

The 2020 Midwest derecho caused tremendous damage to agriculture. Nearly 66% of the 21.3 million acres (86 200 km^2) that were planted with corn and soybeans in Iowa in 2020 were damaged or destroyed (Figure 12.15D). This accounts for 45% of the state's total 30.6 million acres (124 000 km^2) of agricultural land.

The total cost of damage to homes, farms, businesses, livestock, and crops is still being calculated (Figure 12.15E). The current estimate is $6.5 billion with nearly $4 billion in Iowa alone. This exceeds most tornadoes and many hurricanes. It was also responsible for four deaths and several hundred injuries.

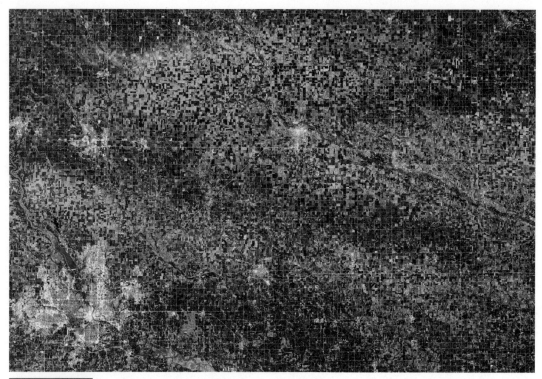

FIGURE 12.15D Satellite image of Iowa after passage of the 2020 Derecho. The brighter green areas are produced by flattened corn plants. *Source*: Image courtesy of NASA.

FIGURE 12.15E Photograph of damage caused by the 2020 Derecho. *Source*: Courtesy of NWS.

CASE STUDY 12.5 1993 Storm of the Century

When Disasters Coincide

On very rare occasions, all of the components come together to produce a weather disaster that spans the full range of severe conditions. This was the case with the 1993 Storm of the Century (Figure 12.16). It was perhaps the most intense mid-latitude cyclone ever to have formed over the eastern part of the United States. This extratropical cyclone included a nor'easter, a blizzard, a tornado outbreak, a derecho, and a major ice storm among other dangers. This extensive storm struck 26 states and Southern Canada and impacted 40% of the United States population with devastating effects despite being the best forecasted event to that point in United States history.

The Storm of the Century began as a low-pressure system that developed in the Gulf of Mexico during the day of 12 March 1993. Farther north, a strong arctic high-pressure system was building over the middle of the country, providing a strong contrast in air masses. The low-pressure system strengthened rapidly as it crossed the Gulf of Mexico during the afternoon of 12 March. The deteriorating conditions there sank several ships and caused danger for all other ship traffic in the area.

The storm crossed the coast along the Florida Panhandle about midnight on 13 March. It generated a storm surge up to 12 ft (3.7 m) in Apalachee Bay on the Florida coast. A line of severe thunderstorms developed along a front to the south of the storm center and extended across Florida during subsequent hours of 13 March. This squall line included embedded bow echoes that swept ahead of the system's cold front in a serial derecho across the Florida peninsula. The squall line raced off the coast of Florida at 5:00 a.m. on 13 March and crossed Central Cuba after sunrise.

The squall line produced damaging straight-line winds of the derecho and 11 confirmed tornadoes and possibly 15 tornadoes across Florida. Straight-line wind gusts reached 96 mph (154 kmph) in the Tampa Bay area and reported wind speeds of 120 mph (193 kmph) in Cuba. The first recorded tornado touched down in Chiefland at 12:38 a.m. at F2 strength. At 1:30 a.m., a tornadic waterspout at F0 strength struck Tampa. The tornado outbreak in Florida lasted 2 hours and 32 minutes. Of the 11 confirmed tornadoes, four were F0, four were F1, and three were F2, but only two had track lengths longer than 1 mi (1.6 km). These and the other thunderstorms of the system produced at least 60 000 lightning strikes over a 72-hour period.

In the meantime, the low-pressure system moved quickly toward the northeast where it encountered the very cold air from the Arctic blast. The system moved inland through Southern Georgia and continued into the Appalachians. By early afternoon, the center of the low-pressure system had lower barometric pressure than had ever been observed for any winter storm or even hurricane in the Southeastern United States. By the mid to late afternoon on 13 March, cold air wrapped in and the storm center moved north into Northeastern North Carolina, but it continued to spread north. By the time it reached its full extent on 14 March, the storm stretched from Canada to Honduras. The storm finally dissipated 15 March 1993.

The low pressure and cold combined to produce widespread heavy snow and blizzard conditions from Alabama to the Western Carolinas and Virginia (Figure 12.16A). All-time record snowfall blanketed the entire southeast and later spread north through the Central and Northern Appalachians. The Appalachian Mountains experienced

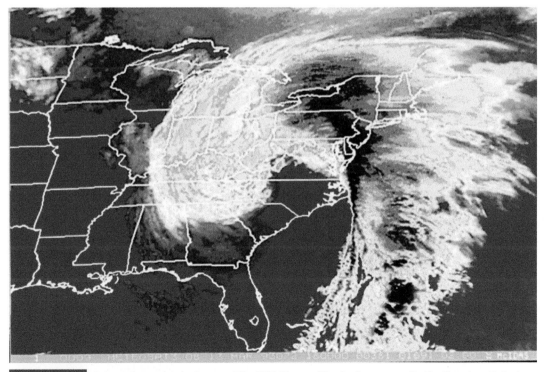

FIGURE 12.16 Color-enhanced radar image of the 1993 Storm of the Century across the Northeastern United States. *Source*: Image courtesy of NOAA.

FIGURE 12.16A Photograph of the snowstorm of the 1993 Storm of the Century. *Source*: Courtesy of the US National Weather Service.

FIGURE 12.16B Photo of snow on cars in New York. *Source*: New York Daily News Archive/New York Daily News/Getty Images.

up to 5 ft (1.5 m) of snow with drifts up to 35 ft (11 m) in Tennessee. There was even 6 in. (15 cm) of snow in the Florida Panhandle. The snow spread northward with totals of 44 in. (112 cm) in West Virginia, 43 in. (110 cm) in Syracuse, New York, 36-in. (91 cm) (Figure 12.16B) and 10-ft (3 m) drifts in Western Pennsylvania, 35 in. (89 cm) in New Hampshire, and 19 in. (48 cm) in Portland, Maine as examples (Figure 12.16C).

The drifting of the snow was caused by powerful winds driven by the strong pressure gradient all along the storm. Highest wind gusts during this storm included 144 mph (232 kmph) on Mount Washington, New Hampshire, 131 mph (232 kmph) in Nova Scotia, Canada, 101 mph (163 kmph) on Flattop Mountain, North Carolina, and 98 mph (157 kmph) in South Timbalier, Louisiana among others. These were accompanied by frigid temperatures such as −12 °F (−24 °C)

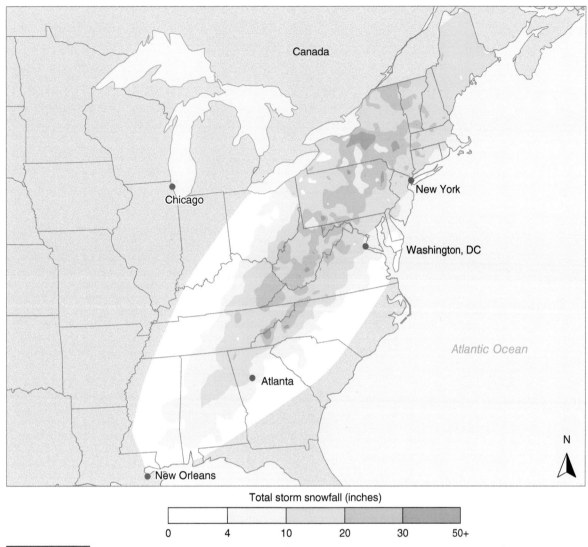

FIGURE 12.16C Map showing the distribution of snow across the Eastern United States caused by the 1993 Storm of the Century.

in Burlington, Vermont, −11 °F (−24 °C) in Syracuse, New York, −10 °F (−23 °C) on Mount Le Conte, Tennessee, and −4 °F (−20 °C) in Waynesville, North Carolina among others which were record-breaking for such a late-season storm.

The 1993 Storm of the Century affected 26 states and much of Eastern Canada impacting 40% of the population of the United States. It shut down every major airport on the East Coast of the United States, at one point. More than 10 million customers lost electric power as a result of the storm. The damage from the entire storm was in excess of $2 billion in 1993. A total of 318 people lost their lives in the storm including 49 from Pennsylvania, 44 from Florida, 23 from New York, and 19 from North Carolina.

Thanks to significant advances in technology at the time, the National Weather Service forecasters predicted this huge storm five days ahead of time. This was the first time in history that this was possible. They provided a five-day lead time to 100 000 000 people in the eastern third of the United States of the impending storm. Further, they were able to provide relatively accurate storm and blizzard warnings some two days in advance. This was unprecedented at the time.

Reference

Kasparian, Jerome & Ackermann, Roland & Andre, Yves-Bernard & Mechain, Gregoire & Méjean, Guillaume & Prade, Bernard & Rohwetter, Philipp & Salmon, Estelle & Stelmaszczyk, Kamil & Yu, Jin & Mysyrowicz, André & Sauerbrey, Roland. (2008). Progress towards lightning control using lasers. Journal of The European Optical Society-rapid Publications - J EUR OPT SOC-RAPID PUBL. 3. 10.2971/jeos.2008.08035.

CHAPTER 13

Ocean Circulation and Coastal Systems

CHAPTER OUTLINE

13.1 Shape, Size, and Location of Oceans 251

13.2 Ocean Processes 255

13.3 Ocean Circulation 260

Earth's Fury: The Science of Natural Disasters, First Edition. Alexander Gates.
© 2022 John Wiley & Sons Ltd. Published 2022 by John Wiley & Sons Ltd.
Companion website: www.wiley.com/go/gates/earthsfury

Words You Should Know:

Barrier islands – Long, thin islands of sand that lie very close to coastlines.
Bathymetry – The topography of the seafloor beneath the oceans.
Beaches – Strips of land consisting primarily of sand and lying along coastlines.
Delta – Land built at the mouth of a river entering the ocean.
Estuary – A river flooded by the ocean.
Fetch – Size of the ocean surface across which the wind can act to produce waves.
Harbors – Protected areas along coastlines, typically up rivers, where coastal communities are commonly located.
Longshore currents – Ocean currents that move parallel to the shoreline.
Tides – Periodic flow of water in and away from the shore driven by the gravity of the moon and the sun.
Waves – Regular wind-driven perturbations on the surface of the ocean whose size is a reflection of the size of the basin and strength of the wind.

13.1 Shape, Size, and Location of Oceans

Oceans cover 71% of the earth's surface (Figure 13.1). They impose strong control on the weather systems including facilitation of most of the coastal and near-inland weather-generated natural disasters including hurricanes, typhoons, and cyclones. The shape of their seafloors and coastlines also greatly controls the impact of these storms as well as tsunamis. Many of the most destructive natural disasters in history strike from the oceans. It is important to understand the basic ocean processes and the architecture of the basins to understand and appreciate those natural disasters.

Ocean basins are floored by ocean crust everywhere except along some coastlines. The reason is that ocean crust is denser and thinner than continental crust. It is drawn deeper into the planet by gravity than continental crust, so it sits at a lower elevation. Because water also goes to the lowest elevation as a result of gravitational pull, ocean crust is flooded with water. The bathymetry of ocean floors has a distinct architecture depending on if the margin is tectonically active or not. If the continental coast is not a plate margin, it is a passive margin. A good example of a passive margin is the East Coast of North America. The submarine area closest to

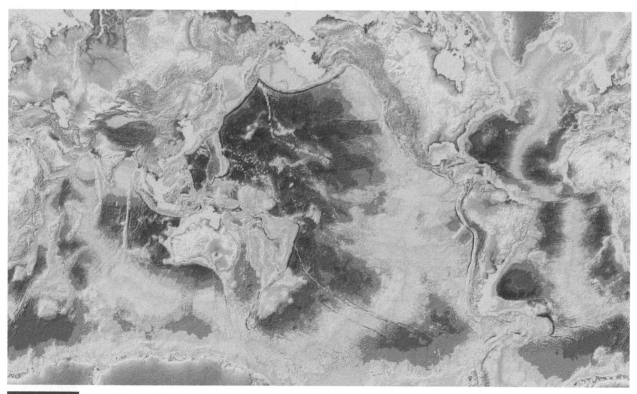

FIGURE 13.1 Map of the world showing the bathymetry of the ocean basins. *Source*: Image courtesy of NOAA.

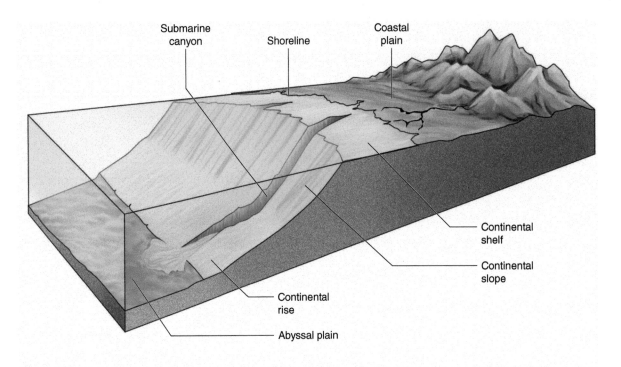

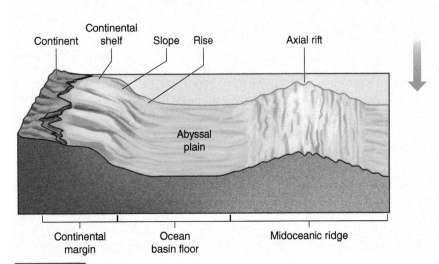

FIGURE 13.2 Block diagram of a typical ocean basin with a passive margin showing the different parts of the continent to deep ocean transition. *Source*: Adapted from Deep-sea trench. Encyclopædia Britannica, Inc.

the shore is the continental shelf which is flooded thin continental crust (Figure 13.2). Depending on the continent, the shelf can be up to several hundred miles wide and the slope of the shelf is a very shallow 0.5°. Storms and tsunamis impact this seafloor.

The continental slope is on the oceanward side of the shelf and it is relatively narrow. It is floored by transitional continent–ocean crust and is much steeper with slopes up to 2°. The rise sits at the base of the slope and on ocean crust. It is also narrow. Oceanward of the rise is the abyssal plain which is very wide. It is floored by ocean crust and it is flat. The center of the ocean basin is marked by an undersea mountain range called the mid-ocean ridge, the most rugged part of the ocean floor.

In active margins where there is subduction of ocean crust, the slope goes directly from the shore into the slope and finally into the trench. The trenches are the deepest points on the ocean floor and occur all around the Pacific Ocean as well as locally in the Atlantic Ocean and Indian Ocean. Oceanward of the trench, the seafloor is abyssal plain and mid-ocean ridge at various distances.

The coastline, where the ocean meets land, can have various geometries that amplify or subdue the energy of natural disaster from the ocean. In general, concave shorelines tend to funnel the wave energy and increase their size and therefore the impact of the disaster (Figure 13.3). Convex shorelines tend to dissipate the wave energy, reducing the impact of the disaster. Shorelines can also be smooth or jagged. In smooth shorelines, the waves are generally the same height, so impact of the disaster tends to be about the same regardless of location (Figure 13.4). Jagged shorelines can funnel or dissipate the wave energy from place to place making the impact of the disaster quite variable. Tsunamis, for example, can be 75 ft (23 m) high in one place and 20 ft (6.1 m) high within 0.5 mi (0.8 km) of each other.

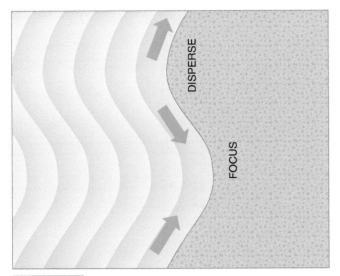

FIGURE 13.3 Map view of an uneven coastline showing where wave energy would be dispersed (diverging arrows) or focused or amplified (converging arrows).

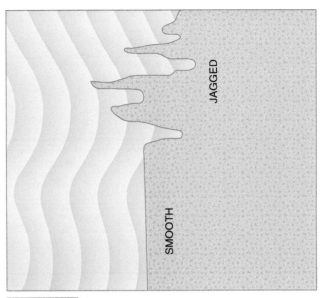

FIGURE 13.4 Map view of a coastline showing a smooth segment versus a jagged segment.

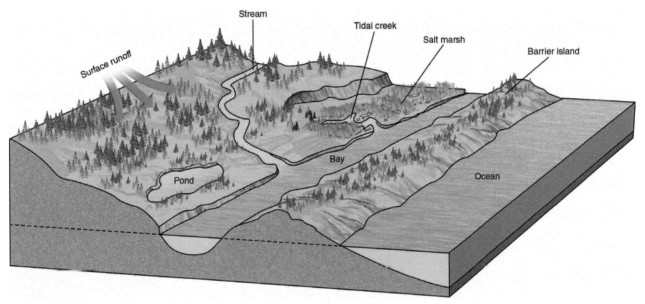

FIGURE 13.5 Block diagram of a coast showing the placement of a barrier island separated from the mainland by a bay.

There are also specific features of the shoreline that affect the impact of the disaster. Most shorelines are lined with beaches which are bands of sand and possibly gravel. Beaches typically include dunes behind them which can protect the shoreline from strong waves. However, the sand is easily removed by strong waves, so they can be modified in a disaster. During periods of rapidly rising sea level, ocean water can fill behind the beach and form a lagoon or bay on the landward side of the beach (Figure 13.5). This leaves the elevated strip of beach as an island along the coast. These are called barrier islands and they form a semi-continuous strip from New York to Florida along the Atlantic Coast of the United States. Because these barrier islands are separated from the land, residents can be stranded during disasters. This can be especially dangerous because the islands can be breached by powerful waves.

One of the main shoreline areas of danger during a disaster is where rivers enter the ocean. In areas where the ocean has low wave and tide energy, land can be built into the ocean by successive river deposits. These are called deltas and the best example is at the end of the Mississippi River in the Gulf of Mexico (Figure 13.6). There are many types of deltas depending upon the wave energy, tide energy, and ruggedness of the shore area. If the sea level is rising, the sea can flood back up the river removing the land along the banks. This is called an estuary and most rivers along the mid-Atlantic and southeast coast of the United States fall into this classification (Figure 13.7). Low-lying areas around rivers and anywhere else along the coast that flood during high tide are called *tidal flats*. All low-lying areas in any of these categories are subject to severe flooding during natural disasters and must be carefully monitored.

FIGURE 13.6 Satellite image of the end of the Mississippi River Delta in the Gulf of Mexico. The green area is land that was built into the gulf by the river. *Source*: Image courtesy of NASA.

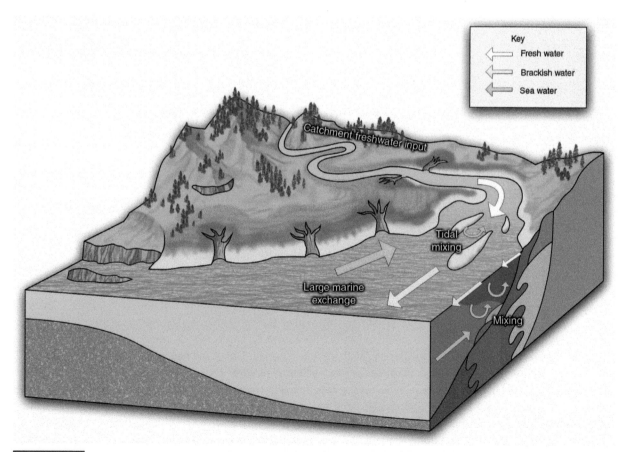

FIGURE 13.7 Block diagram of the mouth of an estuary with blue arrow showing saltwater incursion during high tide and the white arrow showing freshwater input from the river. *Source*: Adapted from OzCoasts: Australian Online Coastal Information.

13.2 Ocean Processes

13.2.1 Tides

Tides are the periodic rise and fall of the surface of any water body on the surface of the earth. Tides are caused by response of oceans or lakes to the gravity of the moon with input from the sun. Ocean tides are the most important and typically have two high and two low tide cycles per day separated by 12 hours (Figure 13.8). High tides occur when the moon is directly overhead or over the direct opposite side of the earth. Low tides occur when the moon is directly on the horizon. If there is a high tide at midnight, there will be a low tide at 6 a.m., a high tide at noon, and a low tide at 6 p.m.

The elevation between the high and low tides is called the tidal range (Figure 13.9). The distance the tides penetrate upstream in a river or inland during high tide is called the tidal reach. The tidal height of high tides and low tides and consequently the tidal reach vary throughout the month and year. This small variability is primarily caused by gravity of the sun. If the sun, the moon, and the earth align (with the sun on either side of the earth), high tides are at their highest and low tides are at their lowest (Figure 13.10). The reason for the extremes is because the gravities of the sun and the moon are additive in this arrangement. This is referred to as spring tide and it generally corresponds to a full moon or a new moon. If the alignment of the sun and the earth is at a right angle to the alignment of the moon and the earth, it corresponds to neap tide. Neap tides result in lower high tides and higher low tides than normal. Neap tides typically occur at half moon. The distance between the earth and the moon and the earth and the sun also varies. This has a lower impact than the alignment, but it can have a small effect.

The main controlling factor in the tidal height, range, or reach at a particular location is the shape of the ocean basin that contains it. Small, shallow ocean basins or lakes produce a small tidal range and reach and may even have only one tidal cycle per day. Larger deeper basins generally produce larger tides. The size and bathymetry of the basin and especially the shape of the margins also play a big role in tidal range and reach. In New Jersey, the typical tidal range averages about 6 ft (1.85 m), whereas in the Bay of Fundy, Nova Scotia, Canada, the tides are an amazing 32 ft (9.85 m), yet both face the Atlantic Ocean. The difference is the shape of

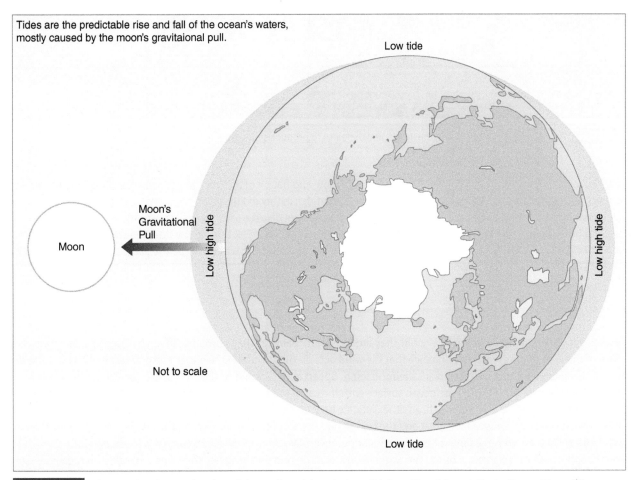

FIGURE 13.8 Diagram showing a polar view of the earth and the relation of high and low tides relative to the position of the moon. The position of the tides relative to the moon remains constant, but the earth rotates, which shifts the position of the tides on the earth.

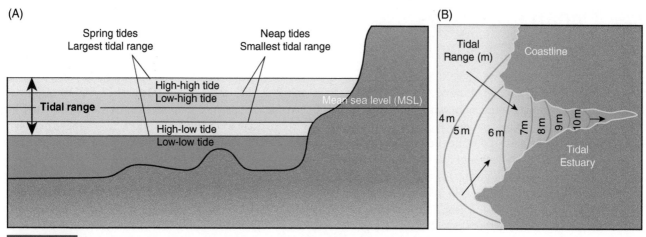

FIGURE 13.9 Diagrams showing tidal ranges. (A) Cross-section view of tidal range by height relative to mean sea level. (B) Map showing positions of sea-level height contours in an estuary which move upstream as height increases. This is tidal reach.

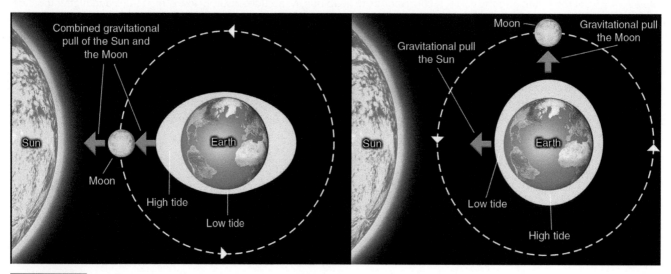

FIGURE 13.10 Relative positions of the sun, the moon, and the earth to produce spring tide (left) and neap tide (right).

the coastline, concave or convex, and that the ocean floor is shallow and flat off of the New Jersey shore but deep off of Nova Scotia. There is more water available and better funneling to make larger tides in Bay of Fundy.

The reason that an understanding of the tides is important in natural disasters is that they contribute to the size and impact of the event. If a hurricane strikes the New Jersey shore with a storm surge of 20 ft (6.1 m) at low tide, the water will inundate everything up to 17 ft (5.2 m) above sea level. If it strikes at high tide, the height of the water will be the surge height plus the high tide height or 23 ft (7 m). At spring tide, it would be even higher.

13.2.2 Waves

Ocean waves occur all around an ocean basin, but they are most noticeable at the shoreline. Driven by winds, waves travel across the surface of the water. The greater the distance that the wind acts on the surface of the water, the greater the height of the resulting waves. This distance of this interaction is called the fetch. Speed and duration of the winds also determine the wave heights. Ocean waves form sinusoidal waves on the ocean surface with the same parameters of waves in physics (Figure 13.11). The peak of the wave is the crest, the bottom is the trough, and the distance between peaks or troughs is the wavelength. The height of the wave is the difference in elevation between the crest and the trough and the amplitude is one-half of the height. The speed of the wave is measured in its period, which is the time it takes for similar points on two adjacent waves to pass a reference point.

Even though waves move across a basin, the water in any location remains basically the same regardless of the wave size and speed. As a wave moves through open water, the water particles move in a circular motion (Figure 13.12). As the wave approaches an area, the water is first pulled out to sea before the lifting, falling, and returning to its original position as the wave passes by. This circular motion of the water particles is called an orbital and they are larger nearer to the surface. The depth to which the wave

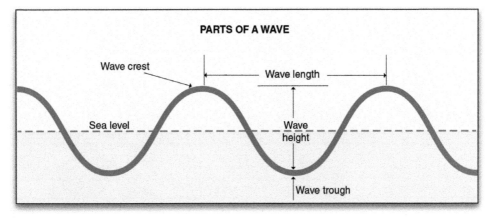

FIGURE 13.11 Cross-section diagram showing the parts and parameters of a wave relative to sea level.

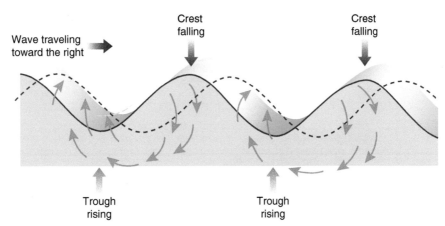

FIGURE 13.12 Cross-section diagram showing the movement of water particles as a wave passes through. The diagram shows the current state of the wave and the dashed line shows the future position of the wave. The red arrows show relative movement of water particles in each area as the wave moves to its future position.

penetrates is equal to one-half of the wavelength and is called the wave base. Below this, the water is still. Wave base constantly shifts up and down with the tides, being deeper in the basin at low tide and shallower at high tide. It has a bigger range during spring tide and the least range during neap tide. Wave base also varies seasonally with larger winter waves penetrating much deeper than gentler summer waves. Storm waves also produce a much deeper wave base.

As the wave approaches the shore, the wave base starts to make contact with the seafloor. The dragging of the wave on the seafloor slows the forward motion, especially at the base. The wave starts to tilt forward and it rises up with respect to sea level (Figure 13.13). Successive waves progressively crowd together as the result of this friction and slowing. The orbitals of water movement progressively flatten and the wave becomes more asymmetric. It falls over in the breaker zone and the momentum of the wave pushes the water up the shoreface into the swash zone. The interaction of the waves with the seafloor moves sediment. Gentler waves have shallower wave bases and do not have enough energy to hold sediment in suspension. Sand can then deposit closer to the shore and beaches grow wider in the summer. In winter, the waves are more energetic and have deeper wave bases. They scour out beach and near-shore sand and keep it in suspension. Energy is not low enough for deposition for up to several hundred yards offshore. In winter, sand is consequently removed from the beach and stored in offshore sand bars. There is a constant cycle of sand being slowly moved to the beach in summer and removed to the offshore during the winter.

Waves often do not approach the shore straight ahead, but instead approach it at an angle. As a result of the obliquity, there is movement of water parallel to the shore in the direction of the acute angle of the wave crests with the shore (Figure 13.14). This lateral movement of water is called the longshore current, and it varies with the angle of approach. The more oblique the wave, the stronger the longshore current. This angle of approach typically only changes during storms, but in local areas it may vary with tides as well. As the wave approaches the shore, it drags on the seafloor and slows down. The leading edge of oblique waves slows more than the wave that angles into deeper water. This difference in speed causes the wave to bend as it approaches the shore, a phenomenon called wave refraction. For shorelines that are not straight, the bending of the wave as it approaches the shore can

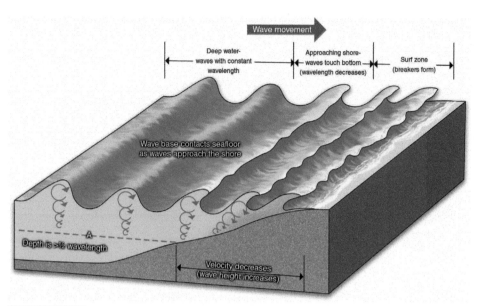

FIGURE 13.13 Block diagram showing the changes that a wave undergoes both in terms of surface shape and the shape of the water particle orbitals at depth as the wave approaches a shoreline.

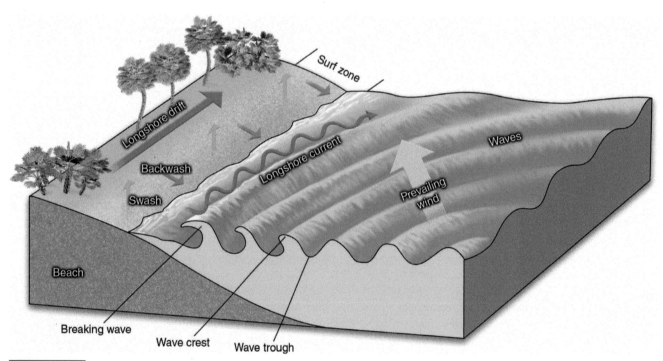

FIGURE 13.14 Block diagram showing waves approaching a shoreline at an angle and the subsequent generation and path of a longshore current and the resulting path of sand particles in littoral drift.

be quite complex. In some cases, longshore currents can be driven toward each other forming a buildup of water. This buildup is gravitationally unstable and escapes seaward in a rip current which is dangerous to swimmers.

The longshore current transports sediment along the coastline as littoral drift. The sediment particles move toward and away from the shore with each successive wave in addition to moving laterally along the beach. In this way, net transport of sand is along the beach, which has been referred to as *rivers of sand* as a result. Littoral drift extends the length of barrier islands if the angle of the waves to the shore remains relatively constant. This growth can create problems if it crosses a channel or shipping lane. If the angle of the shoreline varies or there is an obstruction, the sand will pile up in one area and be starved from another (Figure 13.15). Uneven coastlines have interspersed areas of sand buildup and starvation. This problem also happens if beach residents build jetties, piers, and breakwaters in front of their property. The flowing sand is blocked by jetties and builds up.

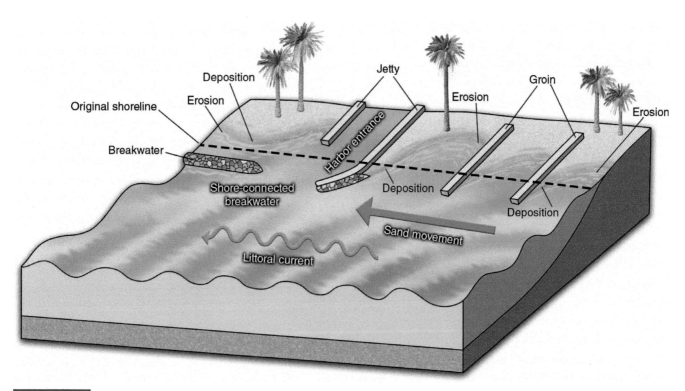

FIGURE 13.15 Block diagram showing the scalloped shape of a shoreline that undergoes littoral drift and has groins, jetties, and breakwaters constructed along it.

FIGURE 13.15A Photograph of a shoreline with groins. *Source*: Scenics & Science/Alamy Stock.

FIGURE 13.15B Photograph of a shoreline with breakwaters. *Source*: Hardaway & Gunn 2015/with permission of Spinger Nature.

The longshore current removes sand from the downflow side of the jetty, producing a sand deficit and giving shorelines a scalloped appearance (Figure 13.15A). Breakwaters reduce the energy of the waves reaching the shore and allow sand to deposit behind them. The beach slowly advances out to the breakwater creating a spit of land (Figure 13.15B). In simple cases, the littoral drift continues around the spit, but, in complicated cases, the longshore currents change, which alters the whole sediment budget.

13.3 Ocean Circulation

The water in ocean basins circulates similar to the atmosphere. The main driver of ocean circulation is the temperature difference between the equator and the poles. Heated ocean water is less dense and rises to the surface where it spreads out. Cold water at the poles is denser and sinks. This temperature variation drives a convection cell that circulates the ocean water with warm surface waters moving toward the poles and deep cold waters moving toward the equator. The circulation of these currents is also generally controlled by the Coriolis Effect, but the bathymetry of the seafloor has much stronger control on the circulation patterns. The currents are controlled by the shapes of the ocean basins and continents. They also must go around islands and are funneled by the mid-ocean ridges (Figure 13.16).

Surface currents, such as the Gulf Stream, can have great impacts on weather. Because hurricanes, typhoons, and cyclones require warm surface waters to proliferate, the warm surface currents can impact the areas of generation and paths of these storms. In addition to laterally moving currents, there are also areas of upwelling of cold deep ocean waters. These tend to be areas of high productivity. There are also areas of downwelling. These currents can strengthen, weaken, and even shift locations over time. The currents exert significant control on weather patterns. El Niño and La Niña weather phenomena cause significant changes in temperature, precipitation, and severe weather patterns including natural disasters. The cause of these phenomena is shifts in Pacific Ocean circulation and they can affect areas worldwide as can other shifts in ocean currents. Climate change is causing especially rapid and unpredictable changes in atmospheric and ocean circulation.

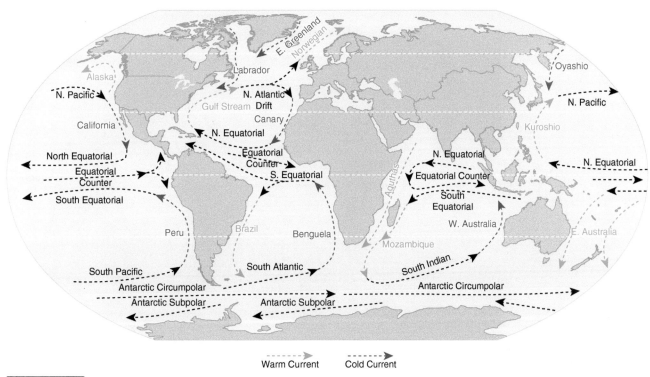

FIGURE 13.16 Map of the earth showing the major warm and cold ocean currents.

CASE STUDY 13.1 1982–1983 El Niño

Ocean Current Disaster

An El Niño event is defined by unusually warm surface ocean temperatures in the equatorial band of the Pacific Ocean. This contrasts with a La Niña event which is marked by unusually cold surface ocean temperatures in the same band. The El Niño is an ocean-atmosphere system oscillation in the Pacific Ocean that impacts weather worldwide. Probably the greatest El Niño in history was from 1876 to 1878 and caused record-breaking droughts that helped trigger disastrous famines in Asia, Africa, and South America. It appears to have killed more than 50 million people globally or up to 3% of the world population. This was an extreme rarity. Recently, El Niños are getting stronger and more frequent including major events 1972–1973, 1982–1983, 1997–1998, and 2015–2016 possibly in response to global warming.

In normal times, trade winds blow to the west along the equator in the Pacific Ocean from South America toward Asia (Figure 13.17). The winds push the warm surface waters to an area off the Asian coast. As a result, the sea level near Indonesia is about 1.5 ft (0.5 m) higher in elevation than near Ecuador. In addition, the sea surface temperature near Indonesia is about 14 °F (8 °C) warmer than it is near Ecuador, partly because currents drive cold water from deeper levels toward the surface along South America. This upwelling of cold water is nutrient-rich, and supports diverse marine ecosystems, and the major fisheries. Unlike the weather around the cold South American waters, clouds and rainfall are common in the rising air over the warm water near Indonesia.

During an El Niño, the trade winds decrease in the Pacific Ocean stopping the pile up of warm water near Asia and reducing the upwelling near South America (Figure 13.17). This causes a rise in sea surface temperature near South America and cuts off the supply of nutrient-rich water to the euphotic or main production zone/water level. The result is a sharp decline in ocean productivity. The weakening of the easterly trade winds shifts weather patterns so that rain moves eastward with the warm water and it results in flooding in Peru and droughts in Indonesia and Australia. It also results in major changes in global atmospheric circulation causing changes in weather far from the Pacific Ocean.

When the 1982–1983 El Niño struck, it was considered the strongest and most devastating of the century. It was so strong that the trade winds actually reversed direction. The thermocline off South America dropped to about 500 ft (124 m) depth. On 24 September 1982, the sea surface temperature at Paita, Peru increased 7.2 °F (4 °C) in just 24 hours. The event was so impactful that waves of the warm water it produced were still detectable in the Pacific Ocean 12 years later. It produced weather-related natural disasters all around the world. In Australia, Africa, and Indonesia, there were record droughts, dust storms, and widespread wildfires. In Peru, the heaviest rainfall in recorded history fell. Some areas that had average rainfalls of 6 in. (15 cm) received 11 ft (3.4 m) of rain. As a result, local rivers were measured at 1000 times their normal flow. The 1982–1983 El Niño caused about 1300–2000 deaths and more than $13 billion (US) in damage.

The 1982–1983 El Niño also induced other problems. For example, the warm, wet spring on the East Coast of North America increased mosquito populations, and, as a result, there were encephalitis outbreaks. In New Mexico, the cool, wet spring increased flea-ridden rodent populations and there was a rise in bubonic plague. Above-normal temperatures in Alaska and Northwestern Canada reduced

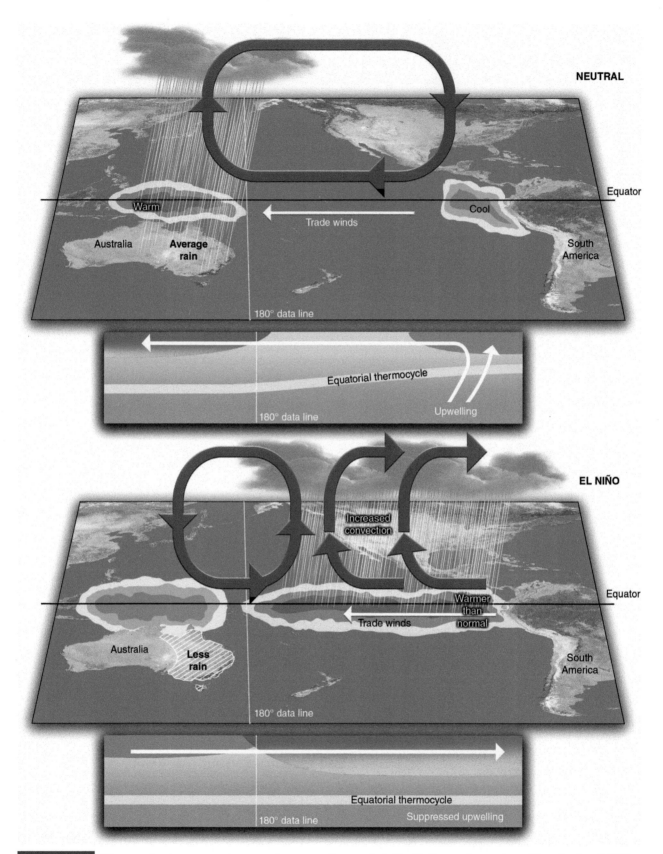

FIGURE 13.17 Diagrams showing maps and cross sections across the equatorial Pacific Ocean of normal conditions (top) and El Niño conditions (bottom).

salmon populations. One of the strongest typhoons on record struck Hawaii, but the Atlantic hurricane season was relatively quiet.

Even though the 1982–1983 El Niño was considered an anomaly at the time, just 14 years later, the 1997–1998 El Niño was even more powerful. It temporarily increased the global air temperature by 2.5 °F (1.5 °C) and killed off approximately 16% of the world's reef systems. At the time, it was the warmest year in recorded history. It too caused widespread droughts, flooding, and other related natural disasters around the world. Extreme rainfall in Kenya and Somalia led to a severe outbreak of Rift Valley fever and Indonesia suffered one of its most devastating droughts on record. Surface sea water temperatures off the coast of Peru reached 19.8 °F (11 °C) above average. Perhaps the biggest impact was that in 1997–1998, the western Pacific experienced a record of 11 super typhoons, with 10 of them reaching Category 5 intensity. The global estimates of 1997–1998 El Niño damage range from $32 to $96 billion (US).

Reference

Hardaway, C. S., & Gunn, J. R. (2015). Headland Breakwaters. Encyclopedia of Estuaries, 350–355. https://doi.org/10.1007/978-94-017-8801-4_313.

CHAPTER 14

Hurricanes, Cyclones, and Typhoons

CHAPTER OUTLINE

14.1 The Most Dangerous Storms 268

14.2 Surprise! They Have Nothing to Do with Fronts 268

Earth's Fury: The Science of Natural Disasters, First Edition. Alexander Gates.
© 2022 John Wiley & Sons Ltd. Published 2022 by John Wiley & Sons Ltd.
Companion website: www.wiley.com/go/gates/earthsfury

Eye. At the center of well-developed storms, a large area of falling air creates calm and clear conditions that can trick residents into thinking the storm is past even though significant danger remains.

Storm Surge. The strong winds around the eyewall create a bulge on the ocean surface that comes ashore as a large and powerful wave that batters and floods shore communities.

HURRICANE, TYPHOON AND CYCLONE
These fragile, enigmatic storms only develop within a stable tropical air mass rather than along a front and only intensify if no other wind or weather affects them. Yet, they are the most powerful and devastating storms on the planet.

Words You Should Know:

Extratropical cyclone – A cyclone formed through the temperature contrast between warm and cold air masses.

Eye – The center of rotation of a hurricane. They are typically clear.

Eyewall – The wall of storm surrounding the hurricane eye and the most powerful part of the storm.

Hurricane – A tropical cyclone with maximum sustained surface winds of 74 mph (118 kmph) or greater, also called a typhoon or a cyclone.

Post-tropical cyclone – A cyclone that no longer exhibits tropical cyclone characteristics.

Rainbands – Bands of rain forming a whorl on the outer part of a hurricane and separated by clear areas.

Saffir–Simpson Scale – The accepted scale for classifying hurricane strength.

Storm surge – The bulge of seawater plowed ashore by a hurricane.

Tropical cyclone – A tropical and sometimes subtropical low-pressure system not associated with a front with organized convection and closed wind circulation around a well-defined core including depressions, storms and hurricanes, and typhoons.

Tropical depression – A tropical cyclone with maximum sustained winds of 38 mph (61 kmph) or less.

Tropical disturbance – A tropical weather system with organized convection and not associated with a front persisting for 24 hours or longer.

Tropical storm – A tropical cyclone with maximum sustained winds ranging from 39 to 73 mph (62–117 kmph).

Tropical wave – A diffuse area of low pressure moving east to west across the tropics that can form a tropical cyclone.

14.1 The Most Dangerous Storms

Hurricanes and their equivalent cyclones and typhoons are among the deadliest natural disasters. The greatest natural disaster in the United States was the Galveston hurricane of 1900, as result of which more than 8000 people lost their lives. Officially, the greatest natural disaster of the twentieth century was the Bhola cyclone of 1970 in which more than 300 000 and up to 500 000 people perished. Even with the modern, sophisticated weather prediction technology, hurricanes such as Katrina in 2005 in New Orleans and Sandy in 2012 in New York and New Jersey can still devastate even the best-prepared areas.

Hurricanes are storms that originate in the tropics and are the most powerful meteorological events on Earth. A fully developed hurricane produces enough energy per day to power the needs of the United States for an entire year. These huge storms are called hurricanes only in the Atlantic Ocean and adjacent related waters. They are cyclones in the Indian Ocean and typhoons in the Pacific Ocean. All are produced through the same processes and are equally powerful.

14.2 Surprise! They Have Nothing to Do with Fronts

Unlike most severe weather such as tornadoes and blizzards, hurricanes do not form at boundaries between air masses (fronts). They begin as innocuous weakly developed clusters of thunderstorms off the coast of West Africa in the Atlantic Ocean. Hurricanes only develop within large stable air masses that are more typically associated with pleasant weather. They also require a surface ocean water temperature of 79 °F (26 °C) or greater. At that temperature, the humidity is elevated in the air above the ocean surface. The warm, moist air is swept into the updrafts of the thunderstorms. As the air rises, the moisture in it condenses to form rain, and the condensation from vapor to liquid releases latent heat energy into the storm. This added energy causes more vigorous circulation and greater updraft. The updraft lowers the pressure in the storm. The decreasing pressure draws ever stronger winds into the storm. The winds whip up the ocean surface into waves that in turn expose greater surface area on the water from which more humidity can be emitted. More humidity results in more condensation and more latent heat transfer which quickly powers up the storm.

The storm rotates counterclockwise in the Northern Hemisphere as the result of the Coriolis Effect, driven by the earth's rotation. Storms rotate clockwise in the Southern Hemisphere. These massive spinning storms cannot cross the equator due to their momentum. As the storm grows into a hurricane, it forms an "eye" at its center which is an area that is calm and clear. The eye typically ranges from 20 to 40 mi (32–64 km) in diameter. The most dangerous part of the storm is the eyewall which surrounds the eye. It contains the strongest winds and heaviest rains of the hurricane. Spiral rainbands of 5–10 mi (8–16 km) wide and several hundred miles long encircle the storm outward from the eyewall. Hurricanes, cyclones, and typhoons are commonly about 300 mi (480 km) wide but range from 200 to over 500 mi (320–800 km) in diameter. There is strong wind between the rainbands but little to no rain. The rain and heavy clouds in a hurricane occur where air is rising and therefore lower pressure. Where there is no rain, air is falling, and pressure is higher.

Hurricanes are largely unguided because they are not connected to other weather systems. This makes the speeds and directions that a hurricane can travel very difficult to predict and why they can follow such unusual paths. They commonly travel at about 15–20 mph (24–32 kmph), but they can sit still or achieve speeds of 60 mph (96 kmph) in extreme situations. Hurricanes are actually fragile storms that can be easily degraded. If they make landfall, encounter ocean with surface temperatures below 79 °F (85 °C), encounter a front, or high-level winds, they become weaker.

14.2.1 Classification of Hurricanes

There are two classification systems for tropical disturbances. The main system classifies them as tropical depressions, tropical storms, or hurricanes. Tropical depressions are organized systems of thunderstorms that evolve from a tropical wave with surface circulations and maximum sustained winds of 38 mph (61 kmph). Tropical storms are organized systems of strong thunderstorms with surface circulations and maximum sustained winds of 39–73 mph (62–117 kmph). Hurricanes are intense tropical weather systems with strong thunderstorms and well-defined circulations and with maximum sustained winds of 74 mph (118 kmph) or more.

The Saffir–Simpson Scale is the accepted system for classifying hurricanes. It has five categories of hurricanes and they are based solely on wind speed:

- Category 1 hurricanes have sustained winds of 74–95 mph (119–153 kmph). They produce damage to unanchored mobile homes, shrubbery, and trees, but little damage to building structures.
- Category 2 hurricanes have sustained winds of 96–110 mph (154–177 kmph). They produce damage to mobile homes, shrubbery, and trees with some trees uprooted. There is also coastal flooding and damage to roofing, doors, and windows of buildings.
- Category 3 hurricanes have sustained winds of 111–130 mph (178–209 kmph). They produce structural damage to small residences and utility buildings and damage to shrubbery and trees which are typically defoliated and can be uprooted. Mobile homes are destroyed. Coastal flooding can extend inland 8 mi (13 km) or more.
- Category 4 hurricanes have sustained winds of 131–155 mph (210–249 kmph). They produce destruction of mobile homes, roof removal on small residences, and extensive damage to doors and windows. Areas with elevations below 10 ft (3 m) above sea level may flood requiring evacuation of inland residential areas up to 6 mi (10 km) from the coast.
- Category 5 hurricanes have sustained winds greater than 155 mph (249 kmph). They are massive storms that remove roofs from many residences and industrial buildings and can produce complete building failures. Small utility buildings are typically blown away. Flooding causes major damage to lower floors of structures less than 15 ft (4.6 m) above sea level and within 1500 ft (462 m) of the shoreline. Residential areas on low ground within 5–10 mi (8–16 km) of the coast are evacuated, in most cases.

CASE STUDY 14.1 1900 Galveston Hurricane

Storm Surge and the Worst US Disaster

The great Galveston hurricane of 1900 was the worst natural disaster in United States history. At that time, there was no centralized system of hurricane tracking, so the early history of the storm can only be reconstructed where it was spotted. The first indication of the storm was on 27 August 1900, when a passing ship detected a tropical storm east of the Windward Islands. It appears that the storm tracked west-northwestward and entered the Caribbean Sea on 30 August (Figure 14.1). It struck the Dominican Republic as a weak tropical storm on 2 September and by 3 September, it was off the southern coast of Cuba. It was on 4 September that the Galveston weather station first received notice that a storm was moving northward from Cuba. It entered the Gulf of Mexico on 6 September and as a result of movement over the calm, warm waters, it quickly strengthened into a hurricane. The track is largely a matter of speculation, but rough surf and high tides in Mississippi and Louisiana indicate that it moved westward across the Gulf of Mexico. In addition, it appears to have strengthened to a Category 4 hurricane by 7 September.

On the morning of 8 September, the sky was blue, but the tide flooded several blocks of downtown Galveston, Texas nearest to the gulf. Residents went about their regular business without paying much attention to it. Galveston is located on an island of sand and at that time had an elevation of less than 9 ft (2.7 m), so it was common for some flooding. There were no warnings issued from the weather station, so there was no alarm. By 8 a.m., the sky had become overcast with patchy drizzle and the wind had become brisk. By 10 a.m., the waves had grown in size and power and the head of the weather station issued verbal warnings to the crowds that had gathered to watch the waves. He contacted the United States Weather Bureau (predecessor of the National Weather Service) in Washington, D.C. but at 11:15 a.m. was told that the wind shifted and the storm he was witnessing would stay out to sea. However, the storm would continue to intensify.

By early afternoon, the conditions had deteriorated and all residents were seeking shelter. At about 2:30 p.m., the rain gauge was blown off of its perch. At about 5:15 p.m., the anemometer was blown away as well but not before recording average wind speeds of 84 mph (135 kmph) with gusts exceeding 100 mph (161 kmph). The wind caused a great amount of damage to buildings and other structures, but the real danger in this hurricane was the storm surge. The storm surge is seawater that is plowed by the wind and forward motion of the storm forming a bulge of water on the ocean surface. As the storm

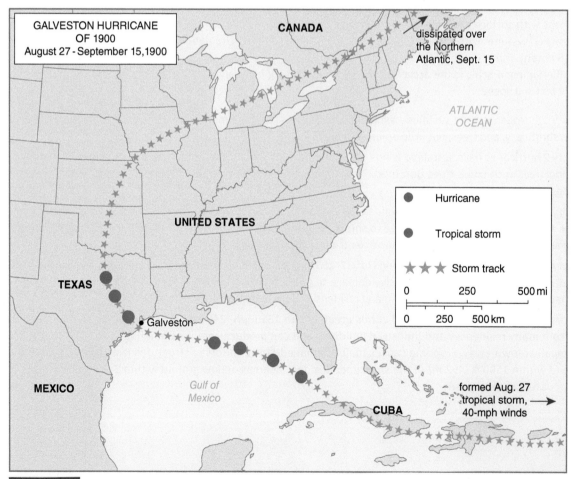

FIGURE 14.1 Map of the track of the 1900 Galveston hurricane.

makes landfall, sea level is raised by this surge until it passes. In this case, the storm surge was 8 to locally 15 ft (2.4–4.6 m) high which was higher than Galveston Island and the entire Texas coastline was inundated. The intensity of the hurricane continued to increase as night descended and the residents had to face these storm perils in the pitch darkness. The storm finally abated during the early morning hours of the next day.

Many stories of survival and death emerged from the deadly evening. More than 700 people sought shelter in the town hall, but the roof collapsed, and many were killed. Other people sheltered in a major downtown hotel and were forced to climb the stairs en masse just ahead of the rising flood waters. The same story came from a tower where survivors were crammed into a winding staircase (Figure 14.1A).

In the morning when it became light, survivors witnessed the horrific devastation and carnage (Figure 14.1B). Debris and corpses were everywhere. Nearly 7000 buildings were badly damaged or destroyed, including 3636 homes (Figure 14.1C). All bridges to the mainland were washed out, so the survivors were stranded. The strong waves had washed all of the sand out of the town cemetery and the coffins had been swept into the city and out to sea. Local authorities declared martial law and forced men at bayonet point to gather the bodies and pile them on barges. They were then weighted with rocks and dumped into the Gulf of Mexico. Unfortunately, many of them washed back onshore and they had to be burned. They were decaying and liable to spread disease.

In all, there were at least 6000 fatalities and as many as 12 000 people killed. The most cited number in official reports is 8000 fatalities. In addition, 10 000 people were left homeless. Considering that the population of Galveston at the time was 38 000, this was an enormous impact on the residents (Figure 14.1D). The total cost of this disaster in Galveston, the Caribbean, and across the United States and Eastern Canada where it restrengthened was approximately $35.4 million.

There were several major outcomes from this disaster. First, the city of Galveston decided that they would raise the ground level in the city and erect a seawall against large storm waves. The US Army Corps of Engineers constructed a 17-ft (5.2 m) high and 10-mi (16 km) long seawall along the Gulf of Mexico shoreline of Galveston Island. The residents then raised the buildings in the city of Galveston by as much as 17 ft (5.2 m) and pumped sand underneath their foundations. This also raised the height of the surface in the city. Unfortunately, in 1915, another major hurricane impacted Galveston and the surge breached the seawall. As a result, the height of the seawall was raised to 19.2 ft (5.85 m). To date, no storm surge has breached the new wall.

On a larger scale, after Galveston, the US Weather Bureau began to track all Atlantic tropical storms and established a department to study hurricanes. They also began to open up communication channels both internationally and within the country. Information about severe weather conditions were now disseminated broadly and warnings were issued when human populations were threatened.

Finally, the disaster shifted the political influence of the western Gulf of Mexico. Prior to this, Galveston had established itself as the major metropolis and trade center of the region. The city was dubbed the "New York of the South" by residents and businessmen alike. The devastation caused by the storm convinced the shippers in the area to move their operations north to Houston's safer harbor. As a result, Houston has been the major port in the region ever since.

FIGURE 14.1A Photograph of the destroyed center of the city as a result of the Galveston hurricane. *Source*: Courtesy of NOAA/US Department of Commerce.

272 CHAPTER 14 Hurricanes, Cyclones, and Typhoons

FIGURE 14.1B Photograph of the debris field of destroyed houses as a result of the Galveston hurricane. Inset shows the area prior to the hurricane. *Source*: Courtesy of NOAA/US Department of Commerce.

FIGURE 14.1C Photograph of the destroyed houses as a result of the Galveston hurricane. *Source*: Courtesy of US Library of Congress.

FIGURE 14.1D Photograph of a destroyed house as a result of the Galveston hurricane. *Source*: Courtesy of US Library of Congress.

CASE STUDY 14.2 1938 Long Island Express Hurricane

Headwinds and the Dangerous Side of a Hurricane

The most devastating hurricane to ever strike the Northeastern United States occurred in 1938. It has been called the Long Island Express, the Yankee Clipper, and the Great New England hurricane because hurricanes were not named in 1938. Because there was no radar or other reliable way to track hurricanes in 1938, the United States Weather Bureau was unaware of the storm until 17 September when a Brazilian ship reported hurricane-force winds and extremely low barometric pressures near the Bahamas. It took a reanalysis project in 2014 to interpret the full track of the 1938 hurricane.

The tropical depression that would become the deadly hurricane formed off the coast of West Africa on 9 September 1938 (Figure 14.2). It moved westward into the Sargasso Sea and achieved hurricane strength on 15 September. It continued to track westward at about 20 mph (32 kmph) and continued to strengthen to peak wind speed of 160 mph (260 kmph) near the Bahamas four days later, making it a Category 5 hurricane.

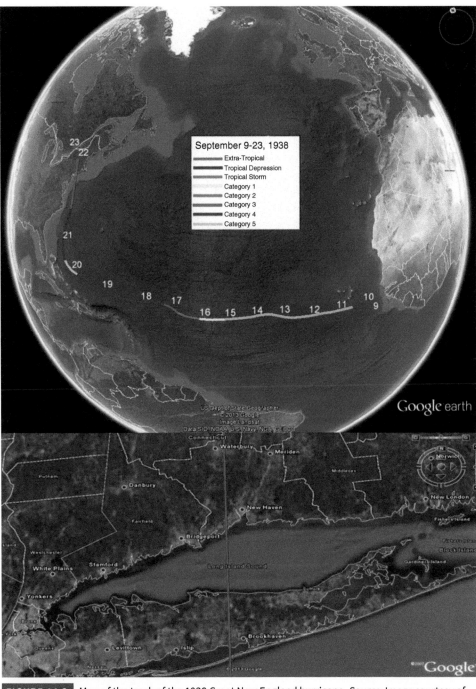

FIGURE 14.2 Map of the track of the 1938 Great New England hurricane. *Source*: Image courtesy of the US National Weather Service.

At this same time, a sharp cold front moved eastward and resulted in a channel or tongue of moist, tropical air along the East Coast and into New England. This steered and propelled the storm at a very high speed northward and parallel to the coastline. On 21 September, it was detected 75 mi (120 km) east of Cape Hatteras, North Carolina at a strong Category 3. At 9:00 a.m. on 21 September, the US Weather Bureau issued storm warnings north of Atlantic City, New Jersey and south of Block Island, Rhode Island. However, they anticipated that the storm would go out to sea and spare the Northeast. As a result, at 10:00 a.m., they downgraded it to tropical storm warnings. The 11:30 a.m. advisory predicted gale-force winds for the area, but no more mention of a tropical storm or hurricane.

It was a deadly mistake to underestimate the intensity and speed of the storm. At 2:00 p.m., while still under the 11:30 a.m. advisory, hurricane-force gusts were occurring on the south shore of Long Island, New York and near hurricane-force gusts were impacting the coast of Connecticut. At 3:45 p.m., the hurricane made landfall at Bellport, Long Island with maximum sustained winds of 120 mph (195 kmph).

At the point that the hurricane impacted Long Island, it was moving rapidly northward at a minimum speed of 47 mph (76 kmph), but some estimates place it as high as 60 mph (97 kmph) or even 70 mph (113 kmph). This is the fastest forward speed of a hurricane ever recorded. Because hurricane winds rotate counterclockwise, the winds on the east of the eye, or headwinds, moved from south to north and parallel to the forward motion. In this case, the forward speed of the storm is added to its circulating wind speed on the east side and the measured wind speed is increased accordingly (Figure 14.2A). On the other hand, on the west side of the storm, the forward speed is subtracted from the circulating wind speed and the total wind speed is reduced. These are the tail winds.

At the time of landfall, the eye of the hurricane was about 50 mi (80 km) across and the whole storm was 500 mi (800 km) wide. Toward the west, the wind speed was 60 mph (97 kmph) at Central Park, New York City, and Battery Park recorded winds of 70 mph (110 kmph) with gusts to 80 mph (130 kmph). In this area, storm surge was around 14 ft (4.3 m). On the eastern end of Long Island, there were no weather stations, but the damage was extreme, and there were many more deaths than to the west despite the paucity of population compared to that in the west. It is estimated that winds reached up to 150 mph (240 kmph), with waves surging to around 25–35 ft (7.6–10.7 m) high based upon damage.

A similar relationship occurred as the storm made its second landfall in New England. The Blue Hill Observatory in Eastern Massachusetts recorded sustained winds of 121 mph (195 kmph) and a peak gust of 186 mph (300 kmph) at 6:11 and 6:16 p.m. Sustained winds of 91 mph (146 kmph) and gusts up to 121 mph (195 kmph) were recorded on Block Island, Rhode Island. Similarly, a station in downtown Providence measured sustained winds of 100 mph (161 kmph) and gusts up to 125 mph (201 kmph). The storm surge was 14 ft (4.3 m) across most of the western Connecticut coast, but increased 18–25 ft (5.5–7.6 m) from New London, Connecticut east to Cape Cod, Massachusetts. With this distribution, Rhode Island was ground zero for maximum impact. Water in the streets of downtown Providence, Rhode Island was just a few inches deep as the hurricane struck and increased to waist deep within minutes (Figure 14.2B). It continued to rise to nearly 20 ft (6.1 m). Sections of Falmouth and New Bedford, Massachusetts were submerged under as much as 8 ft (2.4 m) of water. All three locations had a very rapid rise in flood waters, achieving the highest water within 1.5 hours of landfall.

The storm continued across New England. It produced heavy rainfall of 10–17 in. (25.4–43.2 cm) across most of the Connecticut River Valley. The heavy rain flooded small streams which fed into the Connecticut River. It reached a height of 35.4 ft (10.8 m) in Hartford which was 19.4 ft (5.9 m) above flood stage (Figure 14.2C). Winds continued as well. On Mount Washington, New Hampshire, winds gusted to 163 mph (262 kmph).

Death and destruction from this storm were extreme (Figure 14.2D). There were more than 700 deaths, with 390 in Rhode Island alone, and a total of 708 injured. Some 4500 homes, cottages, and farms were destroyed and more than 15 000 were damaged. There were 26 000 automobiles destroyed, and $2 610 000 worth of fishing boats, equipment, docks, and shore plants damaged or destroyed. More than 20 000 mi (32 187 km) of electric and telephone lines were downed. There were 1700 livestock and up to 750 000 chickens killed and half the entire apple crop was destroyed at a cost of $2 million. It is estimated that 2 billion trees were lost. The cost of this tremendous damage was $308 million which is the equivalent of $5.1 billion adjusted for inflation in 2016 dollars. However, it is estimated that an identical hurricane striking in 2005 would have caused $39.2 billion in damage.

The only benefit of the 1938 Hurricane was that it effectively ended the unemployment experienced as a result of The Great Depression. Clean up operations employed all persons interested.

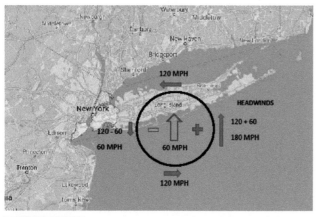

FIGURE 14.2A Diagram showing the dangerous versus less-dangerous side of the 1938 Great New England hurricane as it crosses Long Island, NY.

FIGURE 14.2B Photograph of street flooding as a result of the 1938 Great New England hurricane.
Source: Courtesy of the US National Archives and Records Administration.

FIGURE 14.2C Photograph of river flooding as a result of the 1938 Great New England hurricane.
Source: Courtesy of the US National Weather Service.

FIGURE 14.2D Photograph of the debris field of destroyed houses as a result of the 1938 Great New England hurricane on Long Island, NY. *Source*: Frank Markus/National Weather Service.

CASE STUDY 14.3 1998 Hurricane Mitch

Orographic Precipitation

The origin of Hurricane Mitch was a tropical wave that moved from Africa to the Atlantic Ocean on 10 October 1998 (Figure 14.3). Once it reached the Caribbean Sea on 19 October, it began to organize, and by 22 October, it was a tropical depression located south of Jamaica. It quickly intensified to a hurricane two days later and was a Category 5 hurricane two days after that with sustained winds of 180 mph (290 kmph) and a record low pressure of 905 mb. It quickly crossed the Gulf of Mexico and battered the Swan Islands off the coast of Honduras with waves 22 ft (6.7 m) high (Figure 14.3A). It made landfall on 29 October in Honduras in a much weakened state at a Category 2 hurricane with winds of 98 mph (158 kmph) and a storm surge of 12 ft (3.6 m). It would seem that Central America had avoided a catastrophe,

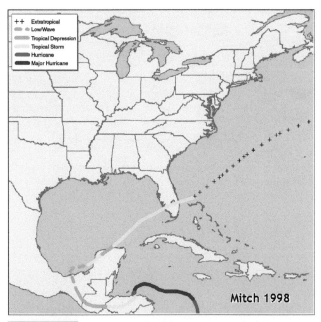

FIGURE 14.3 Map of the track of the Hurricane Mitch in 1998.
Source: Image courtesy of NOAA.

FIGURE 14.3A Satellite image of Hurricane Mitch over Central America. *Source*: Courtesy of NOAA.

but instead the storm became the second most devastating Atlantic hurricane ever.

There were two main reasons that it was so destructive despite being a weaker storm. The storm moved very slowly at an average of 4.6 mph (7.4 kmph), so it took a week for it to move across Honduras, Nicaragua, Guatemala, Belize, and El Salvador allowing it to produce enormous amounts of precipitation. It moved up the Isabella Mountains (Central American Arc or Cordillera) which also greatly increased the precipitation. When weather moves up a mountain slope, it has the same effect as a low-pressure system where the air is rising. The rising air or storm, in this case, carries warmer, moister air to higher altitude and necessarily colder levels of the atmosphere. Because colder air cannot carry as much moisture, it condenses and precipitates. This raining out of storms moving up slopes is called orographic precipitation and it can make even weakened tropical storms devastating.

Hurricane Mitch lumbered across the impacted Central American countries dumping record amounts of rain in a very short time. The highest official rainfall total was 36.5 in. (92.8 cm) in Honduras, 18.4 in. (46.7 cm) of which fell in a single 24-hour period. However, there were unofficial rainfall totals as high as 75 in. (190 cm). The volcanic mountain slopes were saturated and gave way producing enormous mudflows and debris flows that buried entire villages (Figure 14.3B).

FIGURE 14.3B Photograph of mudflows down mountain slopes in Central America. *Source*: Courtesy of the US Geological Survey.

The active Casita Volcano in Nicaragua produced a lahar from excessive rainfall. The runoff flooded all rivers to record levels. One river in Honduras officially reached 33 ft (10 m) above flood stage but unofficially they were even higher. The swollen rivers caused huge inland floods that swept away structures and even towns (Figure 14.3C).

Honduras suffered the greatest damage from Hurricane Mitch with Nicaragua almost as bad. Across the two, 70–80% of the transportation network was wiped out, 70% of the crops were destroyed, and more than 100 000 cattle perished as did 60% of fowl among other losses. Total damage was $6 billion in 1998 dollars which was extreme for the economically challenged countries. Loss of human life was also extreme. There were more than 9000 confirmed deaths and 11 000 people missing and presumed dead. So many people were buried in mud and swept away that they will never be found (Figure 14.3D). The death toll can therefore reliably be estimated at 20 000. There were also 2.3 million people left homeless by the event.

FIGURE 14.3C Photograph of broadly flooded areas in Central America as a result of Hurricane Mitch. *Source*: Canada.ca.

FIGURE 14.3D Photograph of debris from river flooding in Central America as a result of Hurricane Mitch. *Source*: Courtesy of NOAA.

CASE STUDY 14.4 2005 Hurricane Katrina

Tail Winds Can Still Be Dangerous

The costliest natural disaster in the history of the United States was Hurricane Katrina in August 2005. It originated from the merger of a tropical wave and the remnants of tropical depression in the Atlantic Ocean. The storm strengthened from a tropical depression over the Southeastern Bahamas on 23 August to a tropical storm on 24 August as it headed toward Florida (Figure 14.4). It became a Category 1 hurricane with 80 mph (129 kmph) winds just before it made landfall in South Florida. It crossed Florida and entered the warm waters of the Gulf of Mexico on 26 August where it quickly strengthened. It grew from Category 3–5 strength in just nine hours on the morning of 28 August. It reached its peak strength that day, with maximum sustained winds of 175 mph (280 kmph). Katrina defied predictions of making landfall in Florida or Alabama and swung from its westward track to a northward track in Louisiana, heading just east of New Orleans (Figure 14.4A). This was the killer track that residents of New Orleans had been warned about for many decades. Katrina weakened to a Category 3 before making landfall first in Southeast Louisiana with sustained winds of 125 mph (201 kmph) on 29 August. It made landfall once more to the north along the Mississippi Gulf Coast with sustained winds of 120 mph (193 kmph).

New Orleans, however, was not unprepared for Katrina. At 10 a.m. on 28 August, when Katrina was upgraded to a Category 5 storm, the mayor of New Orleans ordered the first ever mandatory evacuation of the city. A strongly worded warning was released that day predicting that the area would experience devastating damage and be uninhabitable for weeks. The federal government issued hurricane evacuation plans requiring areas along and near the Gulf Coast to evacuate in three phases, beginning with the immediate coast 50 hours before the onset of tropical storm winds. Inland phase II areas were to begin evacuation 40 hours before tropical storm winds, and phase III areas, which included New Orleans, were to evacuate 30 hours before. On 29 August, models suggested that Katrina's storm surge might breach the city's levees and flood walls. These warnings were heeded as 80% of the 1.3 million residents of Greater New Orleans evacuated ahead of the storm.

The perfect killer track moved the storm northward and just east of New Orleans. The powerful right-front quadrant head winds pushed a huge storm surge up to 27 ft (8.2 m) northward onto the Gulf Coast of Mississippi (Figure 14.4B). It penetrated 6 mi (10 km) inland in many areas and up to 12 mi (19 km) locally. The circulating winds on the leading edge of the storm pushed the swelled Gulf of Mexico water westward through the Rigolets, a narrow opening in a thin strip of land, and into Lake Pontchartrain, Louisiana. This huge lake is 35 mi (56 km) long and 25 mi (40 km) wide immediately north of New Orleans and it quickly overfilled with this water. There are a series of earthen levees along the south shore of the lake to protect the city from storms on the lake. Hurricane Katrina passed to the east of the lake at Category 3 strength. The weaker left-side tail winds blew due south creating a

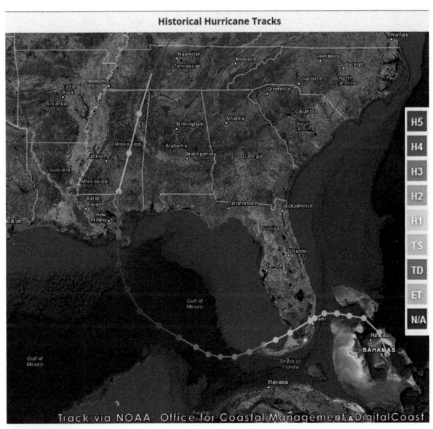

FIGURE 14.4 Map of the track of the Hurricane Katrina in 2005. *Source*: Courtesy of the US National Weather Service.

FIGURE 14.4A Satellite image of Hurricane Katrina over Louisiana. *Source*: Courtesy of NOAA.

FIGURE 14.4B Photograph of houses that were damaged and destroyed by the storm surge of Hurricane Katrina on the Mississippi Gulf coast. *Source*: Courtesy of NOAA.

second storm surge on the now overfilled lake estimated at 13–16 ft (4.0–4.9 m) high that swamped the levees and flowed southward flooding 80% of the city (Figure 14.4C).

This tail wind surge occurred while New Orleans was still experiencing the brunt of the storm. Much of the city was experiencing sustained winds of Category 1 or 2 hurricane strength. There was accompanying heavy rain of 8–10 in. (20–25 cm) and locally as much as 15 in. (38 cm). But it was the surge that caused the most damage producing 53 levee breaches in the federally built levee system that protected New Orleans and the failure of the 40 Arpent Canal levee which was breached in approximately 20 places (Figure 14.4D).

The damage totals from Katrina were staggering. At least 1836 people lost their lives either directly or in the subsequent flooding, making it among the deadliest hurricanes in US history. Federal disaster declarations were issued for 90 000 mi^2 (230 000 km^2) of land and up to three million people were without electricity. The storm is estimated to have caused $125 billion (2005 US dollars) in damage, making it the costliest natural disaster in US history. The federal government was

FIGURE 14.4C Photograph of broadly flooded areas in New Orleans, Louisiana as a result of Hurricane Katrina. *Source*: Courtesy of NOAA.

FIGURE 14.4D Photograph of flooded houses in New Orleans, Louisiana with only the roofs showing. *Source*: Courtesy of NOAA.

slow in responding to the disaster and was roundly condemned. Many people who had not evacuated, sheltered in the Superdome, which sustained significant damage during the storm. There were inadequate facilities for these people with no food or electricity and significant civil unrest ensued. New Orleans would still not have recovered from the destruction many years later despite extensive rebuilding efforts. Primarily as the result of Katrina, the state of Louisiana experienced a population decline of 219 563 or 4.87%.

CASE STUDY 14.5 2013 Super Typhoon Haiyan

Meteotsunami and Coastal Shape

On 2 November 2013, an area of low pressure developed several hundred miles east-southeast of Pohnpei, Micronesia (Figure 14.5). The system tracked westward and strengthened. It achieved tropical storm status on 4 November and was named Haiyan. It rapidly intensified to typhoon intensity on 5 November and the next day it became a super typhoon with the equivalent of a Category 5 on the Saffir–Simpson Hurricane Scale. On 7 November, Haiyan's maximum 10-minute sustained winds were measured at 145 mph (230 km), but another source estimated them at 180 mph (285 km). The one-minute maximum sustained winds were measured at 195 mph (315 km), making Haiyan the strongest typhoon and indeed tropical cyclone on record at the time (Figure 14.5A). Early the next morning at 5:40 a.m. local time, Haiyan made its first landfall in Guiuan, Eastern Samar, Philippines at peak intensity. It would make five other landfalls in the Philippines before emerging in the South China Sea on 9 November, local time, where it turned northwestward and weakened.

Typhoon Haiyan, also known as Super Typhoon Yolanda in the Philippines, resulted in the death of 6300 people with 1061 missing and 28 689 injured making it the second most devastating typhoon to strike the area. It had ferocious winds that caused great damage, but the storm surge at first landfall made it unique.

As Haiyan made landfall, it crossed small islands off southeast Samar Island, and then continued westward across the Leyte Gulf to Tolosa on Leyte Island. It moved north up the funnel-shaped Bay of San Pedro off Leyte Gulf. The shape of coastlines can greatly affect how waves come ashore. Where the coastline is convex, wave energy is dispersed, but funnel-shaped coastlines focus the energy similar to

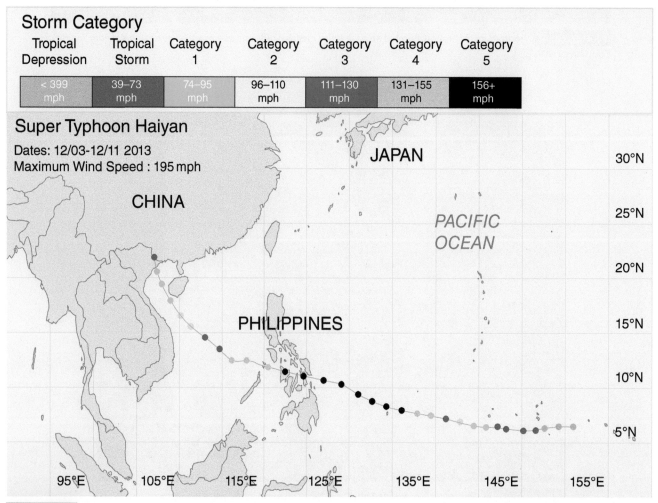

FIGURE 14.5 Map of the track of the 2013 Super Typhoon Haiyan.

FIGURE 14.5A Satellite image of Super Typhoon Haiyan in the Philippines. *Source*: Courtesy of NOAA.

the Venturi Effect in engineering applications (Figure 14.5B). Along the shores of Leyte Gulf, the storm surge was 10–14 ft (3–4 m). However, at the north end of Bay of San Pedro, at 5:30 a.m., tide gauges recorded a drop in sea level of about 3.3 ft (1 m). Within seconds, the storm surge flowed up the bay and into the cities at its head at tsunami speeds (Figure 14.5C). The storm surge heights were 16 ft (5 m) along the shores of the bay, 23–26 ft (7–8 m) in the city of Tacloban, and a maximum of 46 ft (14 m) run-up height at the city of Guiuan. Videos of the event show the tsunami striking the cities and causing comparable damage.

Because the Haiyan wave was not caused by an earthquake or submarine landslide, it is not a true tsunami. Tsunami-like sea waves produced by meteorological phenomena are termed meteotsunamis or meteorological tsunamis. It was these storm surge waves that caused most of the deaths and damage (Figure 14.5D). The damage was estimated at $2.2 billion making it the costliest typhoon disaster in Philippine history (Figure 14.5E). The storm was further made infamous by the poor response of the government to the disaster. A documentary on Haiyan was made by Discovery Channel entitled "*Megastorm: World's Biggest Typhoon.*"

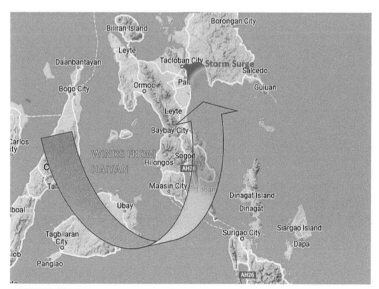

FIGURE 14.5B Diagram showing the Super Typhoon Haiyan over the Philippines causing the meteotsunami in Tacloban. *Source*: Base map from Google Maps.

FIGURE 14.5C Photograph of the Philippines coastline showing the storm surge. *Source*: SunStar Cebu.

FIGURE 14.5D Photograph of broad damage in the Philippines from Super Typhoon Haiyan. *Source*: Erik De Castro/Reuters.

FIGURE 14.5E Photograph of storm surge damage in the Philippines from Super Typhoon Haiyan. *Source*: Associated Press.

CASE STUDY 14.6 2017 Hurricane Maria

Early and Rapid Intensification

A tropical wave formed on 12 September 2017 off the western coast of Africa. It slowly organized as it moved westward across the Atlantic Ocean. By 16 September and about 665 mi (1070 km) east of Barbados, it achieved tropical depression status. The warm ocean surface temperatures of 84 °F (29 °C) encouraged its rapid intensification to Tropical Storm Maria just six hours later. By late in the day on 17 September, Maria achieved hurricane status. Maria then underwent among the most rapid intensifications ever recorded, doubling its wind speed from 85 mph (140 kmph) to 165 mph (270 kmph) in just 24 hours. This is an increase from Category 1–5 in one day, at which point, it was located 15 mi (25 km) east-southeast of Dominica on 18 September. Maria made landfall on Dominica on 19 September at Category 5; the first time the island experienced such a strong storm in its history. Maria weakened to Category 4 as the result of interaction with land, but restrengthened to Category 5 as it entered the Caribbean Sea. There, the sustained wind speed increased to 175 mph (280 kmph), the peak speed for Maria.

Fortunately, Maria underwent an eyewall replacement cycle that weakened it to Category 4 strength just before it made landfall near Yabucoa, Puerto Rico (Figure 14.6). Landfall was at 6:15 a.m. on 20 September and the sustained wind speed was 155 mph (250 kmph) making it the most intense storm to strike Puerto Rico since 1928 (Figure 14.6A). The island was devastated by Maria, which weakened significantly as it crossed. It then began to strengthen to a major hurricane once again when it was over the warm Atlantic Ocean during the afternoon. It reached a second peak with wind speed of 125 mph (205 kmph) on 22 September, just north of Hispaniola. There, it threatened several island nations before weakening.

When Maria underwent eyewall replacement, it caused the eye to triple in size from 10 mi (16 km) to 32 mi (51.5 km) across, and, as a result, the powerful eyewall covered the entire island. Heavy rainfall soaked Puerto Rico, with one location reporting nearly 38 in. (96.5 cm) of rain in one day (Figure 14.6B). These rains caused serious flooding. In some areas, flood waters were reported to have risen more than 6 ft (1.8 m) in 30 minutes and reached a depth of 15 ft (4.6 m). The rains caused landslides on hillslopes across the island at densities as high as 25 landslides per square mile (10 landslides/km^2). Maria also produced a storm surge along the coast with maximum inundation of 6–9 ft (1.8–2.7 m) (Figure 14.6C).

The damage that Hurricane Maria imposed on Puerto Rico was catastrophic. This island was already struggling economically. The Puerto Rico Electric Power Authority suffered budget cuts as well as 30% loss of its workforce since 2012. The age of the power plants averaged 44 years, the electric grid was poor, and the infrastructure was aging across island. Most of the water supply also did not meet environmental standards. Further, Puerto Rico had yet to recover from the damage that Hurricane Irma had imposed on it just two weeks before. It used up all available rescue and relief resources in the area in that natural disaster. This set the stage for a catastrophe.

Forecasts warned of Maria and evacuations were ordered in advance of the storm resulting in a 20% increase in travel from Puerto Rico the day before it struck. Officials opened 450 shelters in the afternoon of 18 September, and by 19 September, at least 2000 people had sought shelter.

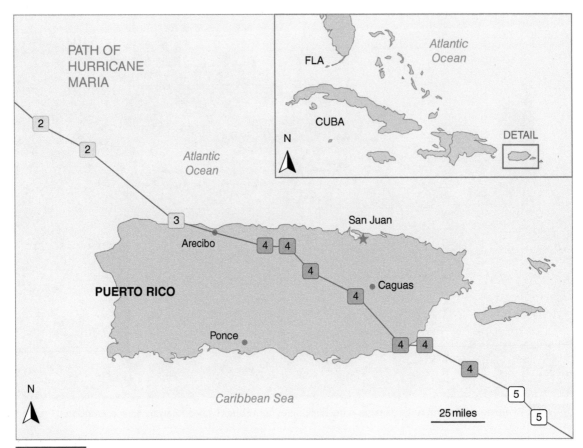

FIGURE 14.6 Map of the track of the Hurricane Maria in 2017.

FIGURE 14.6A Satellite image of Hurricane Maria over Puerto Rico. *Source*: Courtesy of NOAA.

When the storm struck, it destroyed Puerto Rico's entire power grid, leaving 100% of the 3.4 million residents without electricity. The cell networks were 95% destroyed and the above-ground phone and Internet cables were 85% destroyed. The weather radar antenna of Puerto Rico that was designed to withstand winds of more than 130 mph (209 kmph) was destroyed. It took nine months to replace it. More than 80% of the territory's agriculture was destroyed including the entire coffee industry.

The recovery from Maria was very slow. By 26 September, 95% of Puerto Rico was still without power, less than half the population had tap water and 95% still had no cell phone service. By 6 October, 89% still had no power, 44% had no tap water service, and 58% had no cell phone service. By 20 October, one month after the storm, 88% were still without power (about 3 million people), 29% had no tap water (about 1 million people), and 40% had no cell service. By 17 November, nearly two months after, only 392 mi (631 km) of Puerto Rico's 5073 mi (8164 km) of road were open, about 50% of the sewage plants were operational, and 60 000 homes still lacked roofs.

Three months after Maria, 45% of Puerto Rico (1.5 million people) was still without power, 14% had no tap water, but over 90% of cell service was restored and 86% of cell towers were functioning. By the end of January 2018, four month after the storm, approximately 450 000 customers were still without power and regular blackouts increased that number. For example, on 1 March 2018, two power stations shut

FIGURE 14.6B Photograph of wind and rain of Hurricane Maria. *Source*: AFP Contributor/Getty Images.

FIGURE 14.6C Photograph of flooded area in Puerto Rico during Hurricane Maria. *Source*: The Washington Post/Getty Images.

down and a major blackout occurred in which more than 970 000 people lost electricity.

Many residents of Puerto Rico abandoned the island. By 5 November, more than 100 000 people had left Puerto Rico for mainland United States. By 17 December, 600 people still remained in shelters and 130 000 had left the island for the mainland United States.

Maria was the deadliest hurricane in Dominica since 1834 and the deadliest in Puerto Rico since 1899. The total death toll is 3059 with an estimated 2975 killed in Puerto Rico making it the second worst natural disaster in the United States. The total damage was $90 billion in Puerto Rico alone which makes Maria the third costliest hurricane in US history. The total estimated cost of the storm is $92 billion.

CASE STUDY 14.7 2008 Cyclone Nargis

Impact of Cyclones on Low-Lying Areas

Among the deadliest tropical cyclones of all time, the 2008 Cyclone Nargis struck Myanmar, the former country of Burma. Myanmar is the largest country in Southeast Asia containing an area of 261 228 mi² (676 578 km²) and a population of 51.5 million people. It has 1243 mi (2000 km) of coastline on the east coast of the Bay of Bengal. This area is largely the Irrawaddy Delta which forms a large tide-dominated alluvial floodplain. The delta extends 180 mi (290 km) south of the river mouth, covers an area of more than 13 500 mi² (235 000 km²), and was once home to an extensive tract of mangrove forests. The delta region is densely populated with some 7.7 million people, and very low-lying with most of the area between 9.8 and 16 ft (3–5 m) above sea level. These factors leave this area of Myanmar at grave danger for disasters from the sea.

The Irrawaddy Delta area of Myanmar is often flooded from June to August during the mid-monsoon season. The removal of the mangrove forests made the flooding worse. As a result, there are numerous structures to address and alleviate the regular flooding. However, there are large areas of rice paddies that despite being so low-lying are fed by freshwater from rain.

The system that would become Cyclone Nargis developed as a low-pressure wave in the Bay of Bengal in the last week of April 2008. On 27 April, the system strengthened into a tropical depression located 465 mi (748 km) east-southeast of Chennai, India. The depression moved north-northwest at about 7 mph (11 kmph) and intensified. On 28 April, the system was 340 mi (547 km) east of Chennai and a cyclone equivalent to a Category 1 hurricane on the Saffir–Simpson Scale. High pressure ridges to both the northwest and southeast held the cyclone stationary over the warm Bay of Bengal. By the morning of 29 April, Nargis had winds of 100 mph (161 kmph), making it equivalent to a Category 2 hurricane. Later in the day, however, drier air was infused decreasing the convection of the cyclone and it weakened. By the time it began to move northeastward, it was a tropical storm once again.

On 1 May, the now eastward-tracking Nargis underwent explosive intensification over a 24-hour period from a weak Category 1 storm to a Category 4 storm. It is thought that the intensification was due to warm, upper ocean waters in the Bay of Bengal extending deeper than normal. This abnormally deep layer of warm water increased the available energy to the cyclone, thereby fueling its growth by 300%. By 2 May, Cyclone Nargis attained its peak strength with one-minute sustained winds of 130 mph (209 kmph). It was at its peak intensity that Cyclone Nargis made landfall on 2 May, in the Irrawaddy Division of Southwest Myanmar (Figure 14.7). Further enhancing the impact, this

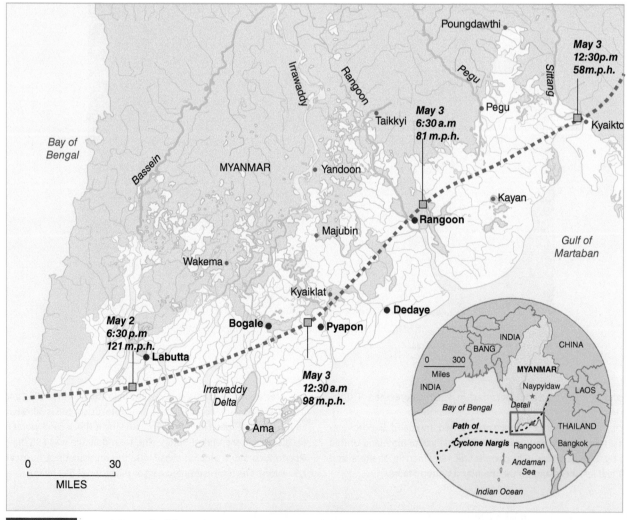

FIGURE 14.7 Map of the track of the Cyclone Nargis in 2015. *Source*: Adapted from Dartmouth Flood Observatory.

huge storm only moved slightly inland, remaining primarily along the coast. This prevented the usual weakening that cyclones undergo as they move over land.

The storm skirted the coast with winds of up to 124 mph (200 kmph) and was accompanied by heavy rain and a deadly storm surge (Figure 14.7A). Although the storm passed 155 mi (250 km) southwest of the largest city and former capital, Yangon (Rangoon), winds there were 80 mph (129 kmph) causing appreciable structural damage. The extreme winds were exacerbated by a 12-ft (3.6 m) storm surge which penetrated 25–30 mi (40–50 km) up the low-lying and densely populated Irrawaddy Delta, reaching a peak about midnight (Figure 14.7B). Unfortunately, the massive deforestation (75% of the mangrove forests), low elevation, dense population, and poorly constructed housing maximized the damage of the surge. The storm surge impacted 90 mi (150 km) of coastline and peaked at more than 16 ft (5 m) height at Pyinsalu. The weakening storm finally turned northeast and encountered a hilly terrain along Myanmar's border with Thailand. At that point, the system dissipated on 3 May.

Cyclone Nargis impacted more than 50 townships, primarily in the Yangon and Irrawaddy Divisions, severely affecting at least 2.4 million people. The heavy rains and storm surge caused flooding that destroyed 90–95% of the buildings in the low-lying delta, leaving more than one million people homeless (Figure 14.7C). Immediately following the cyclone, some 260 000 people were living in camps. The low-lying areas experienced catastrophic fatality rates exceeding 80%, in some cases. The death toll may never be known, but official death toll estimates exceed 138 000–146 000, the seventh deadliest cyclone ever recorded worldwide and the worst natural disaster in the history of Myanmar. However, it is feared that the true death toll is between hundreds of thousands and a million people.

The storm surge surprised residents capsizing small boats and canoes. Some people tied themselves to trees and weathered the storm waves all night. These accounted for the majority of the survivors. Residents ignored warnings because they lacked cyclone awareness and evacuation plans. There was also an absence of high ground or shelters in the area, and there was no indigenous knowledge of comparable prior storm surge flooding because the mangrove forests usually protected them.

Farmland, livestock, and fisheries were all destroyed. It is estimated that 65% of the country's rice paddies were flooded with salt water of the surge (Figure 14.7D). Subsequent rain and flooding flushed some of the salt, but 50 000 acres were unfit and abandoned. More than 93 900 acres (38 000 ha) of natural and replanted mangroves were destroyed. The saltwater also damaged 43% of freshwater ponds and flooded most drinking water wells. Erosion caused the loss of more than 3.25 ft (1 m) of soil in some areas and at various locations 330 ft (100 m) of coast was lost to the ocean. A golden Buddhist stupa that had been on dry land was seen sticking out of the water 500 ft (150 m) offshore. Damage estimates totaled as much as $13 billion (US 2008), which made Nargis the most destructive cyclone ever recorded in the Indian Ocean to that point.

This cyclone became infamous for the government's resistance to international aid. Military leaders banned foreign personnel, including relief workers, from the disaster area. On 8 May, the Myanmar representative to the United States finally asked the United Nations for aid. However, the Myanmar government continued to

FIGURE 14.7A Satellite image of Cyclone Nargis in the Bay of Bengal. *Source:* Courtesy of NOAA.

FIGURE 14.7B Photograph of a flooded area in Myanmar with only the house roofs showing. *Source*: Associated Press/Shutterstock.com.

FIGURE 14.7C Photograph of broadly flooded areas in Myanmar as a result of Cyclone Nargis.
Source: STR/AFP via Getty Images.

FIGURE 14.7D Photograph of a flooded area in Myanmar as a result of Cyclone Nargis. *Source*: AP Photo.

refuse international aid and placed harsh restrictions on even the most basic forms of assistance. Finally, on 9 May, a week after the disaster, Myanmar finally bowed to international pressure and accepted outside aid, but limited it to food, medicine, and basic supplies. With 1.5 million people facing death from starvation, exposure, and disease, world leaders pleaded with Myanmar to allow in additional assistance. It was not until 19 May that members of the Association of Southeast Asian Nations were admitted into the country to deliver aid, but it was not until 23 May that other international aid workers were let into the country.

CASE STUDY 14.8 1970 Cyclone Bhola

Storm that Changed History

Some natural disasters are so impactful that they change history in several aspects. This was the case with the 1970 Bhola cyclone. Some of the changes were precipitated by developing problems in the country. The area that was impacted by the cyclone was East Pakistan at the time. The relationship between East and West Pakistan was uneasy. The seat of the government was in West Pakistan and the people of East Pakistan regarded themselves as poorly served in the distribution of resources and services. Pakistan had been under military rule for the previous decade. Just prior to the Bhola cyclone on 23 October 1970, another cyclone threatened the area, and the Pakistani government evacuated a large area of East Pakistan at great expense. The storm never materialized. It was for this reason that the government was reluctant to make the same mistake twice.

When the Bhola storm formed as a tropical depression over the central Bay of Bengal on 8 November 1970, it was monitored by Pakistan as a potential threat as it traveled northward (Figure 14.8). However, it quickly intensified to a cyclone with winds reaching 115 mph (185 kmph) by 11 November and it abruptly swung northeastward directly toward East Pakistan. By the time it impacted the coast on 12 November, the average wind speed was 130 mph (205 kmph) with one-minute sustained winds of 150 mph (240 kmph). Even more devastating was the storm surge which averaged 20 ft (6 m) but locally reached heights of 35 ft (10.5 m). These numbers are not extraordinary for such storms and make it equivalent to a Category 3 hurricane on the Saffir–Simpson Scale. Once it made landfall, it quickly degraded to a tropical storm in less than one day. It was the area that it struck and the lack of warning and preparation that made it such a disaster.

The coastal area of East Pakistan sits in the Ganges River Delta which is such a low energy system that it has a very low average elevation. Thirty-five percent of this area is less than 20 ft (6 m) above mean sea level. More than 20% of the country, on average, is flooded during annual river floods. When the storm surge struck, it came without warning because none had been issued, let alone the evacuation orders, and it was much higher than the land surface. Whole islands and coastal areas were completely submerged by the storm surge with only the tops of trees sticking out of the ocean waters (Figure 14.8A). Residents climbed the trees and hung on through the battering winds. After the storm, government officials would conclude that the majority of the dead were women and children because they were not strong enough to hold onto the trees through the storm.

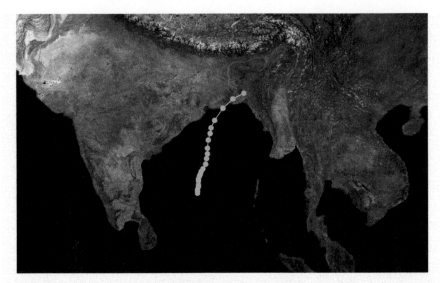

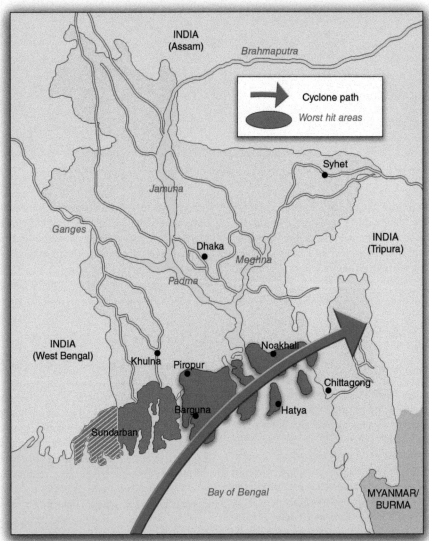

FIGURE 14.8 Map of the track of the Cyclone Bhola in Bangladesh in 1970. Inset shows track of Cyclone Bhola in the Bay of Bengal. *Source*: Nilfanion/Wikimedia Commons/Public Domain.

FIGURE 14.8A Photograph of broadly flooded areas in Bangladesh as a result of Cyclone Bhola. *Source*: AP Images/Harry Koundakjian.

Because so many victims were washed out to sea and census was so poor in the area, the actual death toll will never be known, but it was at least 300 000 making it among the worst natural disasters in history. Some estimates place it closer to 500 000, but these appear to be exaggerations. These deaths were largely from the two million residents who lived in the most damaged area. In total, more than 4.8 million people were directly affected by the cyclone. Approximately 85% of homes in this area were destroyed or severely damaged including 400 000 houses and 3500 schools. More than 280 000 cattle and half a million poultry were killed, and one million acres of crop were destroyed or badly damaged. The worst damage was to the fishing industry where 9000 sea boats and 90 000 river boats were destroyed accounting for 65% of Pakistan's total fishing capacity.

Results of the disaster appeared as flyovers in the evening news around the world. They showed bloated human and cattle carcasses floating in flooded areas with the occasional survivor marooned on a small island madly waving at the helicopter (Figure 14.8B). Aid poured into Pakistan from around the world but, the central Pakistani government was slow to respond and waterborne disease swept through the area. Offers of aid from India were even originally declined. Four months later, by March 1971, 600 000 people still lacked adequate shelter. The central government was globally criticized for this disregard.

The lack of warning, lack of evacuation, and slow response to the disaster so enraged the people of East Pakistan that they overwhelmed the elections just a few weeks later, helping the opposition party to victory. However, this was still not enough and armed conflicts broke out across East Pakistan. This situation escalated and in March 1971, East Pakistan declared independence as the new nation of Bangladesh and war began with West Pakistan. This conflict broadened into the Indo-Pakistani War by December. The war ended on 16 December 1971, with Bangladesh being recognized as a sovereign nation by the Pakistani government.

The second major historical change brought about by the Bhola cyclone was more cultural. George Harrison, from the recently disbanded Beatles rock band, had long been interested in the culture of the India–Pakistan region. He and Bengali musician Ravi Shankar organized the first major benefit concert for a social cause. The Concert for Bangladesh was held between 2:30 and 8:00 p.m. on Sunday, 1 August 1971 in Madison Square Garden, New York City (Figure 14.8C). The performers featured such superstars as Ringo Starr, Bob Dylan, Eric Clapton, Billy Preston, Leon Russell, and the band Badfinger. The effort was so successful that a motion picture and album of it were released the following year. It established a new practice for concerts and for raising money for social causes. There have been numerous benefit concerts since this event. It also raised more than $12 million for Bangladesh relief.

FIGURE 14.8B Photograph of a relief worker dragging a body of a child out of the mud. *Source*: Larry Burrows/The LIFE Images Collection/Getty Images.

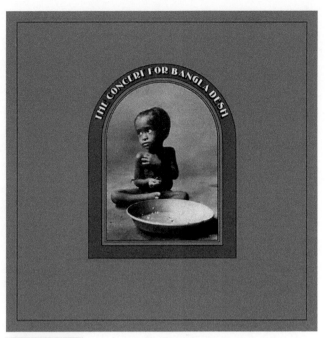

FIGURE 14.8C Album cover for the Concert for Bangladesh. *Source*: The Concert for Bangladesh (album) / Wikipedia.

CHAPTER 15

Tornadoes and Supercells: Terrors of the Plains

CHAPTER OUTLINE

15.1 It Starts with a Cold Front 297

15.2 How Powerful, How Far? 298

Earth's Fury: The Science of Natural Disasters, First Edition. Alexander Gates.
© 2022 John Wiley & Sons Ltd. Published 2022 by John Wiley & Sons Ltd.
Companion website: www.wiley.com/go/gates/earthsfury

Words You Should Know:

Enhanced Fujita Scale – The new scale to measure the power of a tornado ranging from EF0 as the weakest to EF5 as the strongest.
Fujita Scale – The old scale to measure the power of a tornado ranging from F0 as the weakest to F5 as the strongest.
Funnel cloud – A tornado that has not touched down on the ground.
Radar hook – A hook visible on radar images that indicates tornadic circulation.
Supercell – An exceptionally vigorously convective and large thunderstorm cell.
Wall cloud – Disk-shaped rotating cloud that hangs below the cloud deck. It is a precursor to a funnel cloud.

15.1 It Starts with a Cold Front

Tornadoes are the most ferocious type of storm on Earth by virtue of their high speed and fast winds. They can appear suddenly and demolish an entire town within minutes. Only a few human structures can withstand tornadoes. Fortunately, they are relatively small, short-lived, and the more destructive tornadoes are restricted to specific geographic areas. These areas, however, can experience repeated tornado activity.

Tornadoes form from active weather system under very specific conditions that contain thunderstorms. The thunderstorms form in unstable conditions of hot, moist air near the ground surface interacting with cooler air aloft. These conditions produce vigorous vertical circulation of air to high levels. The circulation occurs in structures called convection cells that may be individual in single-cell storms, multicell with multiple single cells, or supercell if the circulation is especially large and intense. Tornadoes can form in multicell storms, but supercell storms are, by far, the primary tornado producers.

The dominant direction of circulation in the storm determines its nature. Storms that are dominated by upward movement bring warm moist air to elevated and cooler levels causing high amounts of condensation. These storms are characterized by high amounts of precipitation but minor amounts of wind. Storms that are dominated by downward movement of air produce strong straight-line winds and microbursts but minor precipitation. Supercell storms produce intense upward and downward movement of air. The upward movement causes the storms to grow to great height. They form a characteristic anvil shape with thick, cumulus clouds and an abrupt flat top that extends behind the storm (Figure 15.1). The intense circulation both upward and downward can sweep freshly condensed rain to high altitudes where it freezes into hail, falls, gathers more moisture, and is swept back upward to have a new coat of ice added. A repeated number of these cycles produce large hail, whose size depends on the number and intensity of the cycles traveled. This same strong upward and downward motion can also produce a tornado. This is the reason that hail is commonly associated with tornadoes.

Surprisingly, storms that produce tornadoes primarily begin with horizontal rather than vertical motion (Figure 15.2). Cool dry air at high altitude that moves in one direction rides over warm moist air at ground level that moves in the opposite direction or at some angle. The cool air is denser and sinks as the warm, less dense air rises. This creates a horizontal rotation or rolling effect in the unstable atmosphere. As the result of instability in the vertical flow, there are variations along the length of the horizontal rotation and drag on the surface, and the developing storm flips up on end to produce a vertically rotating or spinning cloud. The cloud

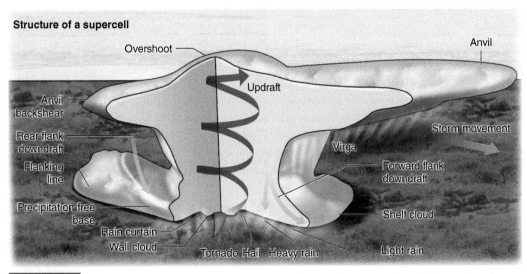

FIGURE 15.1 Diagram of a supercell showing the location of a tornado. *Source*: Adapted from Supercell.svg, Wikimedia Commons.

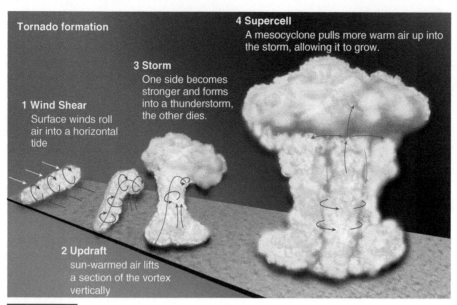

FIGURE 15.2 Diagrams showing the formation of a tornado in four steps starting with horizontal rotation to updraft and uplift of the rotation to a vertical mesocyclone. *Source*: Science History Images/Alamy Stock Photo.

primarily spins counterclockwise, the common direction for all major storms in the Northern Hemisphere, or it can spin clockwise. The vigorously spinning cloud is a mesocyclone, and it develops into a supercell. The upward motion makes the storm grow tall and stronger and the downward motion forms the tornado. The first sign of downward motion is the development of bulbous, mammatus clouds. The next step is development of a wall cloud extending below the spinning storm cloud. The wall cloud comprises a circular band of clouds that extend below the main cloud deck. The funnel cloud drops downward from the center of the wall cloud. This funnel cloud only becomes a tornado if it reaches the ground surface. The tornado spins very fast producing ferocious winds because it is so thin. Just like a figure skater who spins slowly with their arms outstretched but very quickly if they pull their arms in, the fast-spinning tornado is a thin extension of the slowly spinning, large mesocyclone or supercell above.

Supercell storms with tornadoes are not symmetric. One part of the storm is dominated by updraft and another by downdraft. The area of updraft is characterized by hail and heavy rain, whereas the area of downdraft is relatively dry but can contain a tornado. Tornadoes are more commonly on the dry side of a storm.

15.2 How Powerful, How Far?

The classification of tornado strength is based purely on wind speed. The standard classification scheme until 2007 was the Fujita Scale. It normally ranged from F0 to F5, but F6 tornadoes were possible in very rare cases. The characterization of the levels is as follows:

- *F0 tornadoes*: Also known as *gale tornadoes*, F0 is characterized by wind speeds of 40–72 mph (64–116 kmph). F0 produces damage to chimneys, breaks off branches of trees, pushes over shallow-rooted trees, and damages sign boards. The frequency of F0 storms is about 29%.
- *F1 tornadoes*: F1 is classified as a moderate tornado with wind speeds of 73–112 mph (117–180 kmph). The damage from F1 includes peeling off the surface layer of roofs, the pushing off the mobile homes from their foundations or overturning them, the pushing off moving cars from roads, and possible destruction of attached garages. The frequency of F1 tornadoes is about 40%.
- *F2 tornadoes*: F2 is classified as a significant tornado with wind speeds of 113–157 mph (181–252 kmph). F2 inflicts considerable damage including tearing off roofs of frame houses, demolishing mobile homes, pushing boxcars over, uprooting or breaking large trees, and turning light objects into missiles. The frequency of F2 tornadoes is 24%.
- *F3 tornadoes*: F3 are severe tornadoes with wind speeds of 158–206 mph (253–330 kmph). F3 damage includes tearing off the roof and some walls of well-constructed houses, overturning trains including the engine, and uprooting most trees. The frequency of F3 tornadoes is about 6%.
- *F4 tornadoes*: F4 are devastating tornadoes that have wind speeds of 207–260 mph (331–417 kmph). F4 can level well-constructed houses and propel structures with weak foundations for a considerable distance. Cars and other large objects can become missiles. About 1% or 2% of storms can reach this strength.

- *F5 tornadoes*: F5 are incredible tornadoes with wind speeds of 261–318 mph (418–509 kmph). F5 can lift strong frame houses off foundations and carry them considerable distances before they disintegrate. Massive missiles are thrown distances in excess of 325 ft (100 m). They remove bark from trees and severely damage steel-reinforced concrete structures. Far less than 1% of tornadoes attain F5 intensity.
- *F6 tornadoes*: Only one or two F6 inconceivable tornadoes with wind speeds of 319–379 mph (510–606 kmph) have been proposed under this category, but much less documented. If F6 is ever achieved, evidence for it might only be found in ground swirl patterns. The damage will be so extreme that it might not be identifiable through engineering analysis.

It was found that the Fujita Scale has several weaknesses. It is subjective being based solely on the damage caused by the tornado. The damage among different types of construction is not distinguished. If specific damage indicators are absent, it is difficult to apply. If a 3/4-mi (1.2-km)-wide tornado does not strike any buildings, designating a Fujita Scale rank cannot be done. Fujita Scale designation is based on the opinion of observers and on the worst damage observed even if it is only one building or house that shows damage. It overestimates wind speeds in tornadoes that are greater than F3. For these and several other reasons, the meteorological profession improved the Fujita Scale to create the Enhanced Fujita (EF) Scale.

On 1 February 2007, the EF Scale was implemented in the United States. The EF Scale considers 28 damage indicators to determine the rating of the tornado. Each one of the indicators includes a description of the construction for that category. Degree of damage is determined by observation of each indicator that is present. An estimate of wind speed is determined by comparing the observed damage with a calibrated guide. The observed damage is constrained to a lower and upper wind speed range. There are 28 features including many specific types of buildings, towers, and trees from which wind speed is estimated. Ultimately, wind speed is estimated using as many features as can be identified. The advantage of the EF system is the multiple sources of estimation including vegetation. This is a much more inclusive and accurate system than the Fujita Scale. The wind speed is interpreted for winds lasting three seconds or more rather than just peak speed. The scale is:

EF0: 65–85 mph (104–137 kmph).
EF1: 86–110 mph (138–177 kmph).
EF2: 111–135 mph (178–217 kmph).
EF3: 136–165 mph (218–264 kmph).
EF4: 166–200 mph (265–320 kmph).
EF5: >200 mph (320 kmph).

15.2.1 Path of a Tornado

A vast majority of funnel clouds that descend from a storm do not reach the ground and become tornadoes. Tornadoes typically begin as low-rank storms (EF0) and lift and touch down a few times before they become established. Once established on the ground, they may intensify. Typically, they begin as short, thin funnels. With intensification, the funnel widens and lengthens. The tornado path direction and length are variable. Some travel in straight paths, but others sweep side to side as they move. The longer funnels tend to sweep more. The forward speed of a tornado typically ranges from 20 mph (32 kmph) to upwards of 60 mph (96 kmph) depending on the storm. The forward speed affects the circulating wind speed. On the side of the tornado where the wind direction is the same as the movement direction, the wind speed and forward speed are added to produce a higher-velocity side. In most cases, this is the right side of the tornado. On the left side, the forward motion is subtracted from the circulating wind speed which produces lower total wind speeds.

Toward the end of the life of a well-established tornado, it can become ropey and form unusual patterns before it ends. Tornadoes primarily stay on the ground for a few seconds to a few minutes especially for storms that are EF0 or EF1. Larger storms with high EF ratings tend to remain on the ground for longer periods of time. They can be as short as 30 minutes or less, or as much as three hours in the worst cases.

15.2.2 Debris

The first indication that a tornado is on the ground is commonly the debris cloud emanating from the base. Tornadoes can generate several tons of debris and spread it in all directions. The debris cloud can include fragments from all kinds of construction materials, vegetation, and soil. The winds pulverize much of the material and produce a huge amount of dust or particulate that leaves a trail marking the path of the tornado. It may also circulate around the affected area for days depending upon the amount of accompanying rain in the storm.

CASE STUDY 15.1 2011 Joplin Tornado

EF5 Through a City

In the most active year for tornadoes on record, the Joplin tornado was probably the most devastating of all. Tornado activity in 2011 was the most intense on record both by number and strength. The 2011 Super Outbreak alone produced more and stronger tornadoes than most entire seasons. Less than one month after this devastating Super Outbreak, the EF5 Joplin tornado struck Missouri as the strongest tornado in the state history and the seventh deadliest in United States history. Missouri is typically not known for strong tornadoes.

On 22 May 2011, a powerful supercell moved from Southeast Kansas to Southwest Missouri as part of a late-May tornado outbreak (Figure 15.3). Tornado sirens were sounded all along the path. The National Weather Service issued a tornado warning at 5:17 p.m. for Joplin, Missouri and the sirens sounded 20 minutes before the tornado struck (Figure 15.3A). At 5:34 p.m., the EF0 tornado touched down just outside of the city. It was observed to have multiple vortices rotating around the main center of circulation.

As the tornado moved east-northeast toward Joplin, the quarter-mile (0.4 km) wide funnel advanced from EF1 to EF2 strength. The forward speed of the tornado was about 20–25 mph (32–40 kmph) as it entered Joplin and it remained at that speed through the rest of its path. As it began to travel through Joplin, it developed into a three-quarters of a mile (1.2 km) wide massive wedge tornado at full strength (Figure 15.3B).

Based upon the phenomenal damage evaluated after the event, the tornado is estimated to have varied from EF4 to EF5 across the entire city of Joplin. In a parking lot west of St. John's Hospital, 200–300 lb (91–136 kg) concrete parking stops that were fixed to the asphalt using rebar were lifted and thrown 90–180 ft (27–54 m). Adjacent buildings had their steel trusses rolled up like paper and the concrete walls were destroyed.

Perhaps the most phenomenal damage was to St. John's Hospital itself. This huge structure was lifted from its foundation and deformed (Figure 15.3C). It had to be completely demolished and rebuilt. It had every window blown out on three sides of the building. Inside the building, the ferocious wind caused destruction of interior walls and ceilings on every floor. Most of the roof was removed or heavily damaged. It was calculated that the wind speeds necessary to cause such damage must have exceeded 200 mph (322 kmph).

The tornado continued to move across the city where it destroyed the newer section of the high school and severely damaged the outer walls (Figure 15.3D). The Greenbriar Nursing Home, the Franklin Technology Center, and St. Mary's Catholic Church and School were completely demolished and swept off their foundations. There were 21 deaths at the nursing home.

After this, the width of the wedge increased to nearly 1 m (1.6 km) and had winds estimated at more than 200 mph (320 kmph) with gusts at approximately 225–250 mph (360–400 kmph) based on observed damage. At this point, the tornado became rain-wrapped, so it could not even be seen as it approached. It completely demolished a Walmart Supercenter, a Home Depot, and many businesses and restaurants in this area. In the Home Depot parking lot, the asphalt was peeled from its base and thrown hundreds of yards. It also lifted 100 lb (45.4 kg) manhole covers from their fixtures in the roads and threw them long distances. This requires a tornado of enormous power and was part of the reason for the calculation of such extreme wind speeds.

The tornado then traveled on an east to east-southeast trajectory where it began to weaken. In this last part of the track toward the

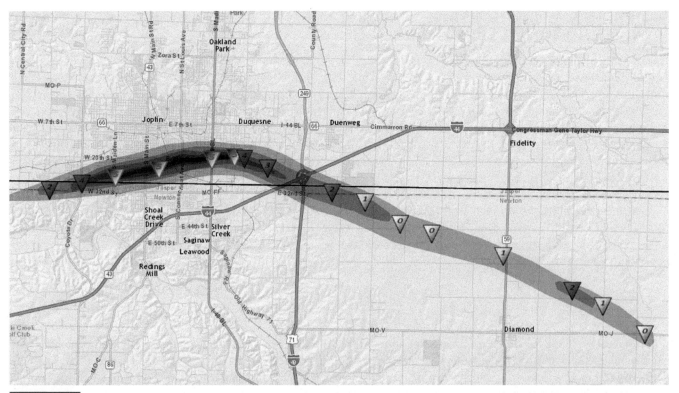

FIGURE 15.3 Map showing the path of the 2011 Joplin tornado across Missouri with Enhanced Fujita Scale levels shown. *Source*: Image courtesy of the US National Weather Service.

FIGURE 15.3A Satellite image of the storm that produced the Joplin tornado. *Source*: Courtesy of NOAA.

FIGURE 15.3B Photo of the Joplin tornado. *Source*: Courtesy of the US National Weather Service.

outskirts of Joplin, the tornado was about 1500 ft (457 m) wide and produced EF1 to EF0 damage. The tornado finally terminated at 6:12 p.m. just outside of the city limits.

The path length of the Joplin tornado was about 22.1 mi (35.6 km) long (Figure 15.3E). The segment within the city of Joplin was about 6 mi (9.7 km) and was by far the most devastating. It caused a total of 158 deaths and 1150 injuries. This makes it the deadliest tornado in the United States since 1947. It also destroyed 25% of Joplin and damaged about 75% of the city. The ferocious winds destroyed 6954 homes and another 359 homes had major damage. In addition, 516 homes had minor damage. More than 15 000 vehicles of various sizes and weights were lifted, thrown, and destroyed. In total, the Joplin tornado caused $2.8 billion in damage making it the costliest single tornado in United States history.

302 CHAPTER 1 Tornadoes and Supercells: Terrors of the Plains

FIGURE 15.3C Aerial before and after photos of the St. John's Hospital in Joplin destroyed by the tornado. *Source*: Courtesy of the US National Weather Service.

FIGURE 15.3D Photo of damage done by the Joplin tornado. *Source*: Courtesy of the US National Weather Service.

FIGURE 15.3E Satellite image of the path of stripped vegetation produced by the Joplin tornado. *Source*: Courtesy of NOAA.

CASE STUDY 15.2 2011 Super Outbreak

The Largest Outbreak

The 2011 Super Outbreak was, by far, the largest outbreak of tornadoes in United States history producing 360 confirmed tornadoes from 25 to 28 April. On 27 April alone, there were 216 tornadoes, the largest single-day total ever. These tornadoes struck the Southern, Midwestern, and Northeastern United States through 21 states from Texas to New York and into Southern Canada (Figure 15.4). The great advances made in forecasting, warning systems, and preparedness as a result of previous outbreaks were no match for this massive event. As a result, it was not only deadly but also the costliest tornado outbreak in history.

The event began with a deep upper-level low-pressure trough that moved into the Southern Plains on 25 April. The imbedded storms moved northeastward along the trough (Figure 15.4A). The atmospheric conditions were similar on 26 April. The forecast was for severe thunderstorms and a strong threat of intense tornadoes throughout the afternoon and evening. The prediction was that discrete tornado-bearing supercells in the afternoon and early evening would give way to a mesoscale convective complex with damaging winds and hail overnight. On 27 April, the storm systems intensified and shifted eastward into the Ohio, Mississippi, and Tennessee Valleys. It moved offshore into the Atlantic Ocean on 28 April. The Storm Prediction Center issued 56 major severe weather watches over the four days including 41 tornado watches. Of these, 10 watches were designated "particularly dangerous situation" watches and there were 15 severe thunderstorm watches in them.

At 3:25 p.m. on 25 April, a "particularly dangerous situation" tornado watch was issued for most of Arkansas and parts of Missouri, Oklahoma, Texas, and Louisiana. Several EF2 and one EF3 tornadoes struck Arkansas causing destruction, but most tornadoes were minor. In all, 42 tornadoes struck that day causing 5 deaths and 59 injuries.

26 April was a similar situation but with two areas of primary activity. Tornado watches were issued for the Great Lakes in the afternoon. Supercells across Central Michigan produced thunderstorms and two tornadoes. As it moved east, severe thunderstorms across Pennsylvania and New York produced 2-in. hail and an EF1 tornado. The southern center included stronger supercells that produced numerous tornadoes across Texas, Louisiana, and Arkansas. Several of the tornadoes were strong at EF2 and even EF3 in Kentucky, but the vast majority was weak. There were 55 tornadoes on 26 April.

27 April proved to be, by far, the most catastrophic day of the outbreak. There was actually no break in activity between 26 and 27 April as continuous severe thunderstorms produced tornadoes from the evening of 26 April into the late morning of 27 April. A line of morning storms moved across Louisiana, Mississippi, Alabama, and Middle and East Tennessee in the early part of the day, strengthening eastward in Alabama. Most of the tornadoes in this line were weak, especially earlier in the day. However, several deadly EF2, EF3, and even one EF4 tornado struck Eastern Mississippi and Alabama.

A second and more powerful line of supercells developed in the late morning and early afternoon. The line moved across Mississippi producing a deadly EF5 tornado northeast of Philadelphia, Mississippi. It was so powerful that it tore pavement off roads and scoured off the top 2 ft (0.61 m) of soil, locally. This was one of four EF5 tornadoes produced that day by this line. An EF4 tornado developed in this early group in Eastern Mississippi crossing into Alabama. By 1:45 p.m., a "particularly dangerous situation" tornado watch was issued for a large part of Alabama and parts of Mississippi, Tennessee, and Georgia, and a large number of powerful tornadoes were produced, particularly in Alabama. Many of them were deadly including the multivortex tornado in Cullman, Alabama which began at 3 p.m. and was documented by several television news cameras.

Perhaps the most famous tornado of the 2011 Super Outbreak was the Tuscaloosa to Birmingham EF4 tornado (Figure 15.4B). This huge tornado began at 5:10 p.m. and stayed on the ground for 90 minutes while it carved an 80-mi (129 km) long path of extreme destruction (Figure 15.4C). It was shown on several television networks in the area and rebroadcasted throughout the United States where horrified viewers watched the huge storm destroy parts of large cities. Radar showed an impressive ball of debris around the base of the tornado. This tornado claimed 65 lives, the most of any single storm.

Powerful tornadoes continued to be generated eastward into Tennessee through the evening where they caused great destruction. A separate area of severe weather developed from Central and Northern

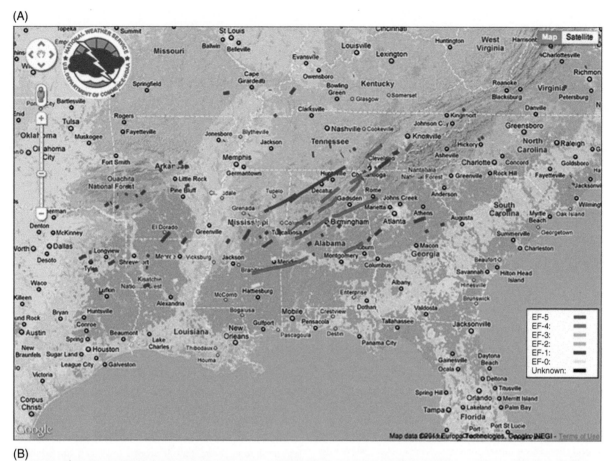
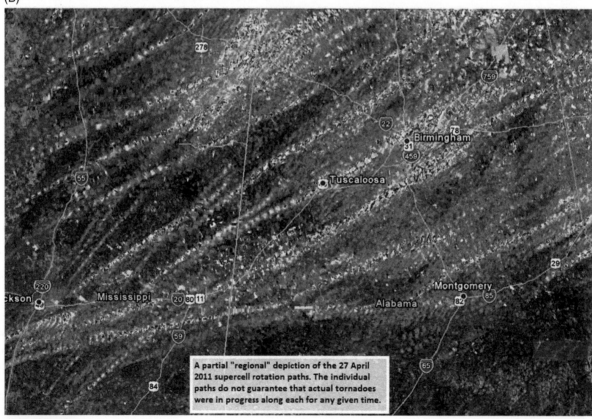

FIGURE 15.4 Maps showing (A) the paths of the 2011 Super Outbreak across Southeastern United States and (B) the supercell paths shown by radar returns of the 2011 Super Outbreak across Mississippi and Alabama. *Source*: US National Weather Service.

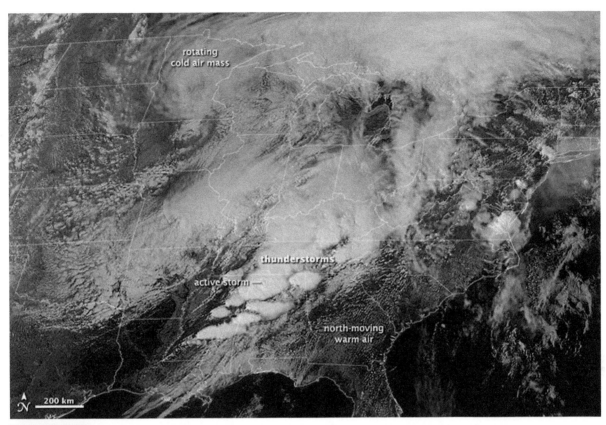

FIGURE 15.4A Satellite image of the storm that produced the 2011 Super Outbreak. *Source*: Courtesy of NOAA.

FIGURE 15.4B Photo of an EF5 tornado devastating Tuscaloosa, Alabama. *Source*: David Mabe/Alamy Stock Photo.

Virginia northward through Maryland, Pennsylvania, and New York that afternoon and evening. There were even tornado watches into Southern Canada. Tornadoes developed all along the East Coast from North Carolina to New York, continuing into early 28 April. Additional tornadoes broke out all along the East Coast from Florida through Pennsylvania and Southern New York during the day of 28 April, some of which were strong and deadly. The entire system then moved into the Atlantic Ocean.

By the time the 2011 Super Outbreak was over, it set numerous records as the greatest outbreak of tornadoes ever. A total of 360

FIGURE 15.4C Aerial photo of a damage path done by a tornado in the 2011 Super Outbreak. *Source*: AP Images/Dave Martin.

tornadoes were produced over a four-day period with 216 on 27 April alone, the highest single-day total. Of these, 238 tornadoes were in Alabama alone. The six most impacted National Weather Service offices issued 303 tornado warnings on 27 April with the Alabama office issuing 90 of the warnings. Tornadoes were on the ground for a total of 3200 path miles (5150 km) for the outbreak. A single EF5 tornado was on the ground for 132 continuous miles (212 km) (Figure 15.4D). A total of 348 people died across six states in the outbreak including

FIGURE 15.4D Infrared satellite image of a path of stripped vegetation produced by a tornado from the 2011 Super Outbreak. *Source*: Courtesy of NOAA.

FIGURE 15.4E Aerial photo of damage done by a tornado in the 2011 Super Outbreak. *Source*: Courtesy of NOAA.

324 deaths directly from tornadoes making it the deadliest outbreak of the decade and longer. It also caused more than 3100 injuries. The outbreak caused a total of $12 billion in damage, which makes it the costliest in US history (Figure 15.4E).

A study of the debris that was lofted and dispersed by tornadoes on 27 April 2011 was conducted in 2012. It utilized social media to track the debris by user observation and found that it had been dispersed by as far as 220 mi (354 km). This was the first such study to utilize social media.

CASE STUDY 15.3 1974 Super Outbreak

Impetus for Better Tornado Monitoring

The 1974 Super Outbreak of tornadoes was the second largest outbreak for a 24-hour period in history. It was the first outbreak to produce more than 100 tornadoes in less than 24 hours. Between 3 and 4 April 1974, there were 148 confirmed tornadoes in 13 states and Ontario, Canada. The storms stretched from the Gulf Coast to Canada including Alabama, Georgia, Illinois, Indiana, Kentucky, Michigan, Mississippi, New York, North Carolina, Ohio, Tennessee, Virginia, and West Virginia (Figure 15.5). As a result of this catastrophic event, numerous changes were made to the system of how tornadoes were forecast and studied.

The stage was set for the outbreak when on 1 April, a strong low-pressure system developed across the Central Plains of the United States. It moved into the Mississippi and Ohio River Valley areas and collided with a surge of very warm, moist air which increased the instability. As a result, on 1 and 2 April, a few F2- and F3-rated tornadoes struck parts of the Ohio River Valley and Southern United States. Early on 3 April, the National Weather Service had already forecast severe weather and an unrelated tornado struck prior to the outbreak. A large low-pressure trough had developed over much of Central United States, with rotation around its base. At 11:30 a.m., an F3 tornado struck Tennessee. Baseball-sized hail was reported in Illinois at 12:20 p.m. By 1 p.m., three more tornado-producing supercells had developed in the Tennessee and Ohio River Valleys and were moving northeastward at 69 mph (111 kmph).

Severe activity moved into Illinois, Indiana, and Northern Kentucky where between 2:20 and 3:20 p.m. three major supercells produced F5 tornadoes that struck Depauw, Indiana, Xenia, Ohio, and Brandenburg, Kentucky among numerous other smaller tornadoes. By 3:50 p.m., supercells in Illinois and Indiana produced other tornadoes up to F3, one of which traveled a distance of 109 mi (180 km). By 6:00 p.m., the southern extension of the system struck Tennessee, Mississippi, and Alabama where some of the strongest tornadoes developed. In Alabama, this included two F5 tornadoes near Tanner and an extremely powerful F5 tornado that destroyed the town of Guin. Michigan was struck by weaker tornadoes but also by snow. During the morning of 4 April, the southern activity which had moved toward the Appalachians overnight produced the final tornadoes of the outbreak across Southeastern United States. The final tornado of the outbreak ended in Caldwell County, North Carolina at 7:00 a.m.

CHAPTER 1 Tornadoes and Supercells: Terrors of the Plains

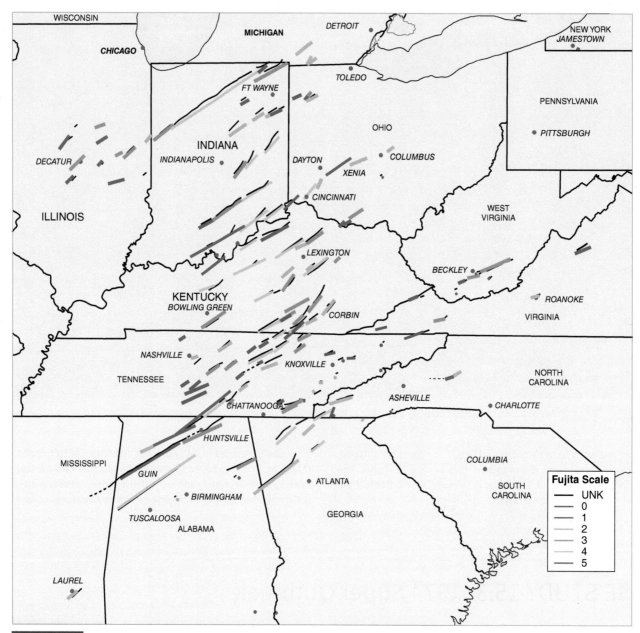

FIGURE 15.5 Map showing the paths of the 1974 Super Outbreak across Southeastern and Central United States.

The 1974 Super Outbreak was characterized by 30 tornado-producing supercells that were responsible for 148 tornadoes. There were seven F5 tornadoes in Indiana, Ohio, and Kentucky, and three in Alabama (Figure 15.5A). There were 23 F4 tornadoes including one in Indiana that had a path length of 121 mi (195 km) which was the longest in the outbreak. At one point in the event, there were 16 tornadoes on the ground simultaneously. There were so many tornadoes on the ground and favorable conditions for tornado development at one time in Indiana that the National Weather Service issued a blanket tornado warning for the entire state. This was the only time in history that an entire state was under a tornado warning.

The 1974 Super Outbreak produced 18 hours of continuous tornadic activity in the most vigorous convective system in history. A total of 319 people were killed directly by tornadoes and 5484 people were injured. The 148 tornadoes had a combined path length of about 2600 mi (4184 km) and damaged an area of more than 900 mi^2 (2331 km^2). The damage caused the declaration of 10 states as federal disaster areas (Figure 15.5B). The cost of the damage was approximately $843 million at the time which is equivalent to $4.58 billion in 2019 dollars.

One of the most devastated cities was Xenia, Ohio (Figure 15.5C). The F5 tornado struck the city traveling at 50 mph (80 kmph) with multiple vortices and a 0.5-mi (0.8 km) width. It is considered to be the worst of the entire outbreak. The death toll for Xenia was 32 people, and about 1150 were injured. About 1400 buildings were heavily damaged or destroyed, which is approximately half of the city (Figure 15.5D). Damage was estimated at $100 million which translates to $471.7 million in 2013 dollars. President Richard Nixon visited only Xenia of the many cities impacted by the outbreak and declared it the worst disaster he had ever seen. He then provided federal aid.

Dr. Ted Fujita of Fujita Scale and Enhanced Fujita Scale fame and his team of researchers performed a 10-month research study on the

FIGURE 15.5A Photo of a tornado near Cincinnati, Ohio. *Source*: Smithsonian Magazine.

FIGURE 15.5B Aerial photo of damage done by a tornado in the 1974 Super Outbreak. *Source*: Courtesy of the US National Weather Service.

1974 Super Outbreak. His research funding to study tornadoes had essentially run out, but the impact of the outbreak was so powerful that it helped propel the rest of his career.

The 1974 Super Outbreak revealed the terrible inadequacies of the US system of preparation and warning systems for tornadoes. As a result, focused efforts were made to improve communications about storms, early warning systems, emergency preparedness, forecast techniques, and the radar network. The result of these efforts is increased lead times for storm warnings, more accurate forecasts of severe weather, greater public awareness

FIGURE 15.5C Photo of an F5 tornado devastating Xenia, Ohio. *Source*: National Weather Service/ Public Domain.

FIGURE 15.5D Photograph of damage done by a tornado in 1974. *Source*: Courtesy of the US National Weather Service.

FIGURE 15.5E Aerial photo of damage done by a tornado in the 1974 Super Outbreak. *Source*: Courtesy of the US National Weather Service.

of preparedness and sheltering, and a more reliable communications network.

The detailed studies also debunked a number of tornado myths that persisted to that time.

Studies of the damage proved that buildings do not explode from pressure differences as the tornado passes over them (Figure 15.5E). To that point, people had wasted precious time opening windows to equalize pressure as tornadoes approached. There was a myth that tornadoes never climbed hills or mountains making those areas safer. However, the F5 Guin, Alabama tornado climbed and descended the 1640-ft (500 m) high Monte Sano Mountain. There was a myth that rivers somehow mitigated tornadoes. However, the powerful F5 Cincinnati/Sayler Park, Ohio tornado crossed the Ohio River without weakening.

CASE STUDY 15.4 1999 Moore Tornado

Strongest Single Tornado

The 1999 Moore tornado was important both for its power and the revised approach to tornado protection it encouraged. The Moore, Oklahoma tornado was just one in a two-day outbreak in the Central Plains of the United States. The outbreak produced 71 tornadoes on 3 May 1999 across five states and another 25 tornadoes on 4 May stretching eastward across the Mississippi River. This outbreak produced the most powerful tornado on record, extensive damage, and loss of life.

Only a slight risk of severe thunderstorms across the Central Plains was forecast on the morning of 3 May 1999. A minor cold front from Western Kansas to Western Texas was approaching a warm, humid air mass that had settled over the Central Plains. However, a previously undetected shifting of the jet stream to the west greatly enhanced the instability of the atmosphere across the area. Even though computer models still only predicted a minor chance of severe weather, meteorologists became concerned about the developing conditions as early as 7:00 a.m. By 11:15 a.m., the forecast of potential severe weather was issued and reemphasized in the 12:30 p.m. update. By 1:00 p.m., wind shear had increased to the point that the development of tornadoes became a probability. At 3:49 p.m., the Storm Prediction Center concluded that severe storms were imminent over the next few hours and tornado watches were issued for most of Oklahoma and parts of Texas and Kansas.

Even before the latest warning was issued, a strong thunderstorm began at 3:20 p.m. in a massive supercell in Tillman County, Oklahoma. Just after 4:00 p.m., the storm crossed into Comanche County where it produced hail up to 1.75 in. (4.4 cm) in diameter. The National Weather Service issued a severe thunderstorm warning for the county at 4:15 p.m. This was upgraded to a tornado warning at 4:50 p.m. when the first of 14 tornadoes formed from this supercell (labeled "A" by the National Weather Service).

At 6:23 p.m., the ninth tornado formed from supercell A and quickly strengthened to F4 strength (Figure 15.6). The wedge tornado

312 CHAPTER 1 Tornadoes and Supercells: Terrors of the Plains

FIGURE 15.6 Map showing the path of the 1999 Moore tornado across Oklahoma with Enhanced Fujita Scale levels shown. *Source*: Base map courtesy of Google Maps.

had a width between one-quarter and one mile (0.40 and 1.61 km) for 6.5 mi (10.5 km) before striking the town of Bridge Creek. There, it further strengthened to F5, the top rating on the Fujita Scale (Figure 15.6A). It was after this that a mobile Doppler weather radar station recorded winds of 301 mph (484 kmph) from this tornado, making it the highest wind speed ever recorded (Figure 15.6B). It scoured about 1 in. (2.5 cm) of asphalt off roads, destroyed 200 houses and mobile homes, and killed 12 people in Bridge Creek. The tornado attained a width of 1 mi (1.6 km) as it crossed the southern Oklahoma City limits where it reduced a well-built home with anchor bolts to bare slab (Figure 15.6C).

At 6:57 p.m., the National Weather Service issued its first ever tornado emergency for the Oklahoma City metropolitan area. As it did, several people attempted to shelter under a highway overpass as had been seen in a recent tornado video, but they were killed or badly injured. The tornado entered the city of Moore, just outside Oklahoma City where it reached F5 strength for the third time. Among the most severe damage of the entire path took place in Moore, where 11 people were killed and 293 were injured. It then reentered Oklahoma City at an F4 rating where it threw a train freight car, weighing 18 tons (16 000 kg) about 0.75 mi (1.21 km). The tornado dissipated in Oklahoma City 1 hour and 25 minutes after it started, leaving a path of destruction up to 1.3 mi (2.1 km) wide and 17 mi (27.4 km) long. The final tornado produced by supercell A dissipated at 8:25 p.m.

Thirty-six people died as a direct result of the tornado and five died from indirect causes for a death toll of 41. In addition, 583 people were directly injured by the tornado of the 645 total injuries in the event. Some 8132 homes, 1041 apartments, 260 businesses, 11 public buildings, and 7 churches were damaged or destroyed by the Moore tornado producing 220 cubic yards (170 m^3) of debris from the buildings (Figure 15.6D). Damage costs were $1.2 billion, making this the first tornado to exceed $1 billion in damage.

The National Weather Service estimated that the number of deaths from the Moore tornado would have exceeded 600 without the advanced warning through the local media, the area warning sirens, and the safety precautions taken by the area residents.

When the National Weather Service meteorologists saw that the Moore tornado would strike a populated area, they issued a "tornado emergency." This was a rarely used forecasting level only for exceptionally dangerous situations prior to the Moore tornado. In these situations, residents are urged to seek shelter in safe rooms. The city of Moore probably contains the highest number of safe rooms per

FIGURE 15.6A Photo of the 1999 Moore tornado. *Source*: Image courtesy of NOAA.

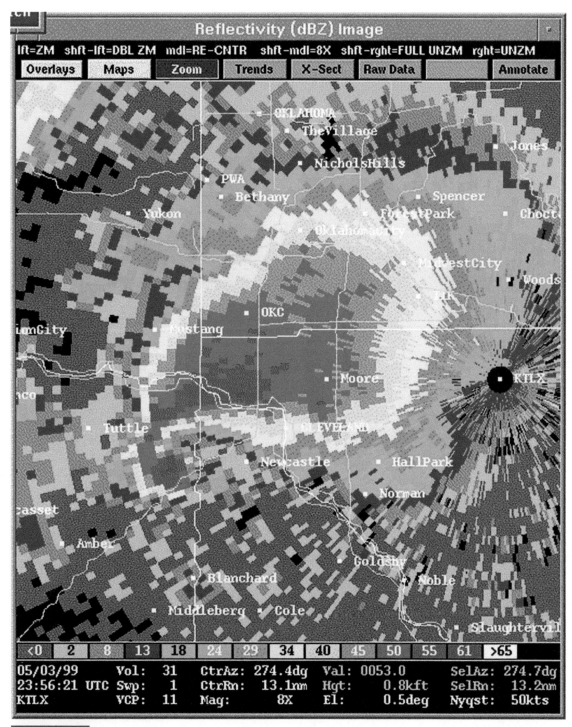

FIGURE 15.6B Radar image of the Moore tornado showing the classic tornado hook. *Source*: Courtesy of the US National Weather Service.

capita in Tornado Alley. In Moore, there were 3170 registered safe room shelters.

Safe rooms are engineered, tested, and certified to withstand the impact of tornadoes. They are armored and anchored to withstand direct winds up to 250 mph (402 kmph) and from debris at speeds up to 100 mph (161 kmph). Safe rooms are inside or outside of a building, above or below ground, and installed in new or existing buildings. They can be installed in a home closet or constructed to protect the students of an entire school.

As a result of the 1999 Moore tornado, Federal Emergency Management Agency (FEMA) and the state of Oklahoma began an incentive program to encourage the installation of more safe rooms area-wide. FEMA provided up to 75% of the costs of these rooms for qualified applicants. Oklahoma and FEMA have provided $57 million through this program and supported construction of 11 386 private safe rooms and 382 safe rooms in public buildings. The protection levels have also been increased in many safe rooms as the result of the Moore tornado.

314 CHAPTER 1 Tornadoes and Supercells: Terrors of the Plains

FIGURE 15.6C Photo of damage of the Moore tornado. *Source*: Image courtesy of the US National Weather Service.

FIGURE 15.6D Aerial photo of damage done by the Moore tornado. *Source*: RICK WILKING/REUTERS/Alamy Stock Photo.

CASE STUDY 15.5 1925 Tri-State Tornado

The Deadliest Ever

The Tri-state tornado was perhaps the deadliest and most destructive tornado in United States history. It struck parts of Missouri, Illinois, and Indiana on the afternoon of Wednesday, 18 March 1925 (Figure 15.7). It was, by far, the most powerful of a major tornado outbreak with at least 12 other strong tornadoes. The path left by the tornado was 151–235 mi (243–378 km) long and extended from Southeastern Missouri to Southwestern Indiana, making it the longest ever recorded. In addition, there were at least 695 deaths, which makes it the deadliest tornado in United States history and the second deadliest in world history.

In 1925, the technology did not exist to do a full meteorological analysis on the outbreak, but using modeling coupled with observational data from the region, an approximate recreation of the situation was developed. A surface low-pressure area in Northeastern Oklahoma faced a warm front moving north and eastward at 7 a.m. on 18 March. The low-pressure area drew in warm, moist air from the Gulf of Mexico. As a result, there were early morning showers and thunderstorms in Kansas, Kentucky, and Indiana.

The low-pressure area moved to Southcentral Missouri by noon and a low-pressure trough developed to its northeast which is where the Tri-state supercell formed. At about 1:00 p.m., the Tri-state tornado was first sighted as a small condensation funnel in the rural area of Moore Township, Shannon County, Missouri. By 2 p.m., the supercell and tornado neared the Mississippi River and other storms developed in the warm air mass by around 3 p.m. These storms also generated destructive tornadoes in Tennessee, Kentucky, and Indiana and powerful tornadoes in Alabama and Kansas.

The Tri-state tornado was attached to a classic supercell for the first couple hours of its life, but it gradually transitioned to a high-precipitation supercell in Southern Illinois. It crossed among rural areas, farmland, cities, and towns as it traveled northeast. Witnesses described the tornado as amorphous rolling fog or boiling clouds on the ground. It first struck a mining town as it steadily grew. Deep scouring of the ground was observed and sheets of iron were lifted and carried as far as 50 mi (80 km). The tornado then developed a double funnel as it strengthened to an interpreted F5 tornado. At least 12 people were killed in Missouri by the Tri-state tornado and 200 were injured.

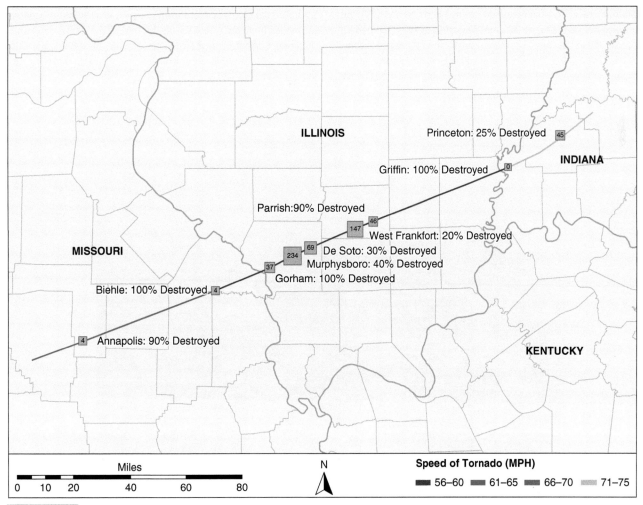

FIGURE 15.7 Map showing the path of the 1925 Tri-state tornado showing the tornado speed and number of deaths in impacted towns. *Source*: Adapted from Wilson and Changnon, 1971; Peter Felknor- Tri state Tornado, 1992; Grazulis, Thoms-P - Significant Tornadoes 1680-1991, 1993.

The tornado crossed the Mississippi River into Southern Illinois where it deeply scoured the ground in rural areas. It was reportedly wrapped in a huge ball of dust and debris at times. At 2:30 p.m., it struck and obliterated the entire town of Gorham, Illinois. Virtually every building was destroyed and railroad tracks were even torn up. At least 36 people were killed and 170 were injured, accounting for more than half the population of Gorham (Figure 15.7A).

The tornado tracked northeast at 62 mph (100 kmph) on average, but locally it achieved forward speeds of 73 mph (117 kmph). By this point, the tornado had a width of about 1 mi (1.6 km). With this awesome power, the huge storm slammed into the city of Murphysboro, Illinois with catastrophic results. While destroying or severely damaging most buildings in the city, it killed 234 people, the greatest loss of life from a tornado in a single city in United States history. In addition, the number of injured was a staggering 623 people.

The village of De Soto was the next community targeted by the Tri-state tornado. The village was essentially obliterated while killing 56 people and injuring 105. The partial collapse of the De Soto School accounted for 33 of the deaths, all of whom were students. It was the worst death toll at a single school as a result of a tornado in United States history. The wind was so strong that 300 ft (91 m) of railroad track was torn up and blown away.

The Tri-state tornado then inflicted the same degree of damage in West Frankfort. At this point, it had transitioned into a high-precipitation, rain, and hail-wrapped storm which means that the tornado was not visible to the residents or victims. It killed 102 people and injured 389 in the town. Then the small town of Parrish was completely destroyed, killing 33 people and injuring 60. The destruction of Parrish was so devastating that many survivors and businesses chose to relocate, and the town was abandoned. The death toll in Franklin County, Illinois was 192.

The tornado then entered Hamilton County where it widened to 1.5 mi (2.4 km) which appears to have been its widest. In rural and sparsely populated Hamilton and White counties, 65 additional people were killed and another 120 were injured.

The powerful tornado next crossed into Indiana where it demolished the town of Griffin. It is reported that every single building in the town was damaged or destroyed. The death toll was 46 in Griffin and surrounding areas and there were 202 injuries. It next destroyed much of the south side of the town of Princeton where 44 people died and 152 were injured. After Princeton, the tornado traveled about 10 mi (16 km) northeast into Pike County before it dissipated at about 4:38 p.m. At least 95 people were killed in Indiana.

This horrific tornado killed at least 695 people including 12 in Missouri, 95 in Indiana, and 588 in Illinois. Some studies estimate that deaths exceeded 747. The number of people injured was at least 2298. It impacted at least 19 communities across 14 counties. Four of these communities were completely abandoned because damage was so extreme and other rural areas never recovered (Figure 15.7B). It changed the history of the region. At least 15 000 homes were destroyed and total damage was $16.5 million in 1925 dollars, but $1.4 billion in 1997 dollars (Figure 15.7C).

This tornado had an extremely long path of 151–235 mi (243–378 km) and it lasted from around 12:40 p.m. to 4:40 p.m., approximately four hours. Although it is reported as a single continuous Tri-state tornado, it is more likely a series of multiple independent tornadoes in a continuous family. Studies of the damage in some of the areas estimate that the wind speeds may have been in excess of 300 mph (480 kmph) in some locations, which would make it perhaps the strongest in history.

FIGURE 15.7A Photo of damage from the Tri-state tornado. *Source*: Topical Press Agency/Hulton Archive/Getty Images.

FIGURE 15.7B Photo of damage from the 1925 Tri-state tornado. *Source*: Image courtesy of NOAA.

FIGURE 15.7C Photo of damage from the Tri-state tornado. *Source*: Courtesy of the US Library of Congress.

CASE STUDY 15.6 1970 Venice Waterspout

Tornadoes on Open Water

A very strong and deadly waterspout struck Venice, Italy on the evening of 11 September 1970 causing significant damage. Although such tornadic waterspouts are rare, in some areas, such as the Adriatic, Aegean, and Ionian Seas, they can account for half of the total number of tornadoes. Tornadic waterspouts are really tornadoes over water. They can be formed by supercells just like land-based tornadoes and produce impressive damage, in some cases. If a waterspout moves onshore, a tornado warning is issued just like a land-based tornado as some of them can cause significant damage and injuries to people. Typically, fair-weather, non-tornadic waterspouts quickly dissipate if they make landfall.

Waterspouts mostly form in the tropics and subtropics, but other areas also report waterspouts, including Europe, Australia, New Zealand, and even the Great Lakes. Some also occur on the East Coast of the United States and the California coast. The deadliest tornado in Greek history occurred from a waterspout on Megdovas Lake on 17 December 1959, when a boat was destroyed killing 21 workers. In Japan, 25% of typhoon-associated tornadoes originated as powerful waterspouts. These tornadoes of waterspout origin are stronger than those originating over land. Italy ranks sixth in Europe in terms of tornadoes with an average of 12–18 per year.

A powerful tornado formed at 8:45 p.m., about 39 mi (63 km) from Venice in the countryside beyond Padua, Italy (Figure 15.8). It damaged or destroyed 300 houses and injured a number of people while onshore. At 9:32 p.m., the tornado reached the Venice Lagoon and became a waterspout. It tore the tiles off the roof of the hospital on La Grazia Island. At 9:35 p.m., waterspout struck the 400-ton Venice Public Transport Company lagoon passenger ferry "Aquileia." The 60 passengers were hurled to the floor as the boat was lifted out of the water and spun. Some of the passengers were thrown from the boat as it dropped to the water. It sank in 30 seconds drowning the remaining 36 people. At 9:36, the waterspout turned toward Sant'Elena island in the city of Venice. It encountered another ship carrying 50 passengers. The ship slowed down to stop and dock at Sant'Elena, but the waves and wind caused it to take on water and it sunk, drowning 21 of the passengers. At 9:37, the waterspout crossed Sant'Elena uprooting trees and tearing roofs from houses. It struck a soccer stadium and it partially collapsed (Figure 15.8A). Power was disrupted in Venice and several shops and houses were damaged. The waterspout moved northeast near the inlet to the lagoon at San Nicolò, damaging farms and beach villages around Punta Sabbioni, Ca' Savio, and Cavallino. The waterspout came ashore at Jésolo and destroyed a camp site, killing 11 additional people and injuring several hundred before dissipating.

In all, the tornado and waterspout lasted about 58 minutes. During that time, it traveled 43 mi (70 km) northeastward at a speed of 44 mph (72 kmph). The wind speed was in excess of 136 mph (220 kmph) making it an F2 tornado. It caused about $3.7 million (€2.5 million) in damage. The death toll is in question, but the estimates range from 36 to 69 victims and at least 300 people were injured (Figure 15.8B).

FIGURE 15.8 Map of the Venice, Italy area. *Source*: Image courtesy of Google Maps.

FIGURE 15.8A Photo of stadium destroyed by the 1970 Venice waterspout. *Source*: Istituto Nazionale Geofisica / CC BY 4.0.

FIGURE 15.8B Photo of a monument to the victims of the 1970 Venice waterspout. *Source*: Istituto Nazionale Geofisica / CC BY 4.0.

CHAPTER 16

Devastating Floods and Their Aftermath

CHAPTER OUTLINE

16.1 Drainage Systems 323

16.2 Discharge, Gauging Stations, and Water Height 325

Earth's Fury: The Science of Natural Disasters, First Edition. Alexander Gates.
© 2022 John Wiley & Sons Ltd. Published 2022 by John Wiley & Sons Ltd.
Companion website: www.wiley.com/go/gates/earthsfury

Words You Should Know:

Upstream flooding – Flooding directly from a precipitation event typically in the upper reaches of a stream and commonly in the mountains and flash flooding.

Downstream flooding – Flooding in the lower reaches of a river caused by events far upstream, so it is rarely accompanied by a precipitation event.

Discharge – The amount of water passing a point in a stream over a period of time.

Gauging station – A device that measures and records the height of the water level in a river.

Flash flooding – Flooding caused by heavy precipitation and rapid runoff that overwhelms the drainage systems.

Flood stage – The height of water in a river when it spills over its banks and into the flood plain.

Drainage basin – A system of tributaries and streams that drain an area bounded by hills and ridges forming drainage divides.

Flooding occurs in a number of natural disasters including tsunamis, hurricanes, the melting of glacial ice over volcanoes, impacts into water, and even avalanches and landslides into water. These types of floods are covered in their respective chapters. This chapter focuses on river flooding from precipitation events through spring thaws and other situations that generate exceptional amounts of water, such as dam bursts, which are also included in the concepts.

16.1 Drainage Systems

Water always seeks the lowest elevation as dictated by gravity. This means that all streams and rivers always sit in the lowest part of a valley. Further, they always flow downhill. These simple concepts are the bases for understanding hydrologic and hydrogeologic systems. The most basic form of this concept is the drainage basin (Figure 16.1). Any valley or even area surrounded by mountains or ridges drains downhill toward a single destination or related destinations. This is a drainage basin. The ridges, hills, or other high

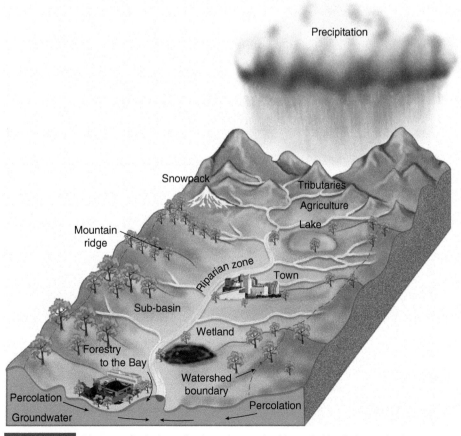

FIGURE 16.1 Diagram of a drainage basin and watershed bounded by drainage divides.

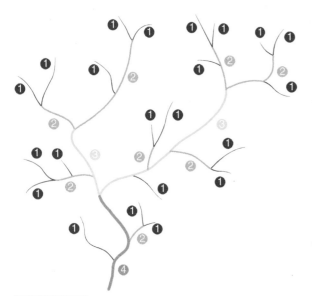

FIGURE 16.2 System for the ordering of tributary streams and rivers in a drainage system.

grounds surrounding this basin are called drainage divides and water may not flow across them. It must flow down one side or the other. They form the boundary of the basin. In some definitions, the drainage basin only includes the bodies of water flowing in this area and the watershed or catchment area also includes all of the surface area, encompassing the surface runoff and the groundwater system as well. However, these terms have been used interchangeably.

The small streams join together to form larger streams and rivers as they collect at the lowest points in the drainage basin. The joining streams are called tributaries and they join at a confluence, eventually emptying into a distributary which is the primary waterbody in the system. Each step from the distributary is assigned an order—first order, second order, etc. (Figure 16.2)—indicating how many other streams join to assemble the distributary. Several tributaries may join together in a single area bounded by hills, so appearing to form a small basin of its own. This is called a sub-basin within the main drainage basin. There can be multiple sub-basins within a single drainage basin or just the main drainage basin. The banks of the streams and impacted areas around them are called riparian zones or areas.

The streams interact with the groundwater system in the watershed (Figure 16.3). The nature of the interaction depends on the depth of the water table. The water table is an underground surface below which all pore spaces in the rock, sediment, and soil are saturated with groundwater. In other words, it is the top surface of the groundwater. If the water table is beneath the level of the stream, stream water infiltrates the streambed and contributes to the

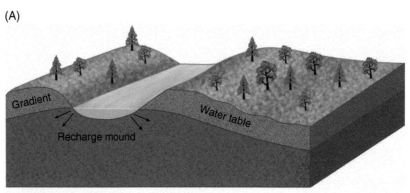

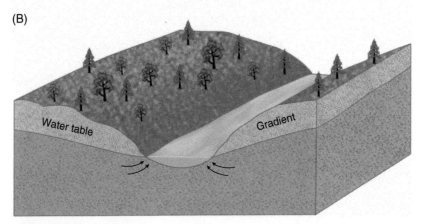

FIGURE 16.3 Block diagrams showing (A) an influent stream and (B) an effluent stream with respect to the water table.

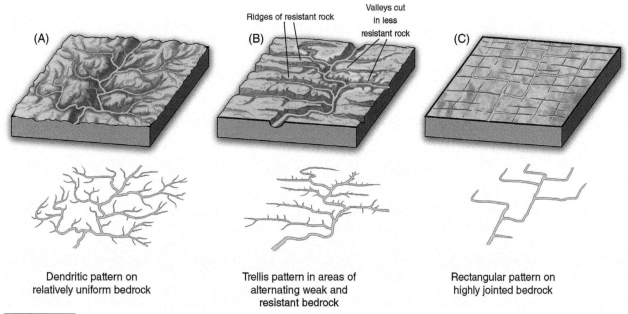

FIGURE 16.4 Drainage patterns for various systems including (A) dendritic pattern, (B) trellis pattern, and (C) rectangular drainage.

groundwater. This is called an influent stream. If the water table intersects the ground surface above the stream, groundwater is contributed to the surface water either through the streambed or as springs along the banks of the stream. This is called an effluent stream. Streams can be effluent or influent at different times depending upon precipitation and time of year among other factors and can vary along the length of the stream. The groundwater, surface water, and surface runoff are all components of the water budget of a watershed.

In map view, the patterns of the streams and tributaries are impacted by the patterns in the bedrock (Figure 16.4). In areas where the layers are flat and even in all directions, the drainage pattern is dendritic, looking like the branches on a tree. If the layers are tilted, streams flow parallel to the layers or cross them at high angles forming a trellis pattern. Orthogonal fracture patterns produce a rectangular drainage pattern. In the streams themselves, the channel holds the water and the flat area around them is the floodplain. This area is dry most of the year, but is underwater during floods.

16.2 Discharge, Gauging Stations, and Water Height

The amount of water that passes through streams is called the discharge. The discharge of a stream is determined by the equation $Q = vA$, where Q = discharge, v = stream velocity, and A = cross-sectional area of the stream.

The cross-sectional area of a channel can be relatively accurately determined, but, in a flood, it tops its banks and spills onto the flood plain where cross-sectional area is nearly impossible to measure. Velocity can also vary considerably across the flow in a flood being much faster in the channel than the floodplain. The discharge varies with time as well as position along the stream. There are a number of components and factors that impact the discharge. The base flow of a stream is contributed by groundwater in effluent parts of the stream. This component is usually seasonal and changes slowly depending on average precipitation and infiltration. The rest of the stream comprises runoff from precipitation, thawing, and sudden events such as dam releases or bursts. This can change catastrophically and is the main cause of flooding.

Most streams and rivers are continuously monitored for their discharge at numerous points along their length in populated areas. However, accurate discharge measurements require expensive equipment that needs constant attention. Instead, much cheaper and self-contained gauging stations are installed (Figure 16.5). These continuously measure the height of the stream. The reason is that the velocity of the stream in the channel is not nearly as important as the height of the stream if it spills over its banks and floods a populated area. Numbers are measured from a baseline, but flood stage is also determined and height above flood stage in feet is commonly reported.

Development and human habitation (urbanization) change the character of discharge events in streams. Vegetation slows runoff, so its removal makes the runoff reach the stream quicker and all at once instead of spreading over a long period.

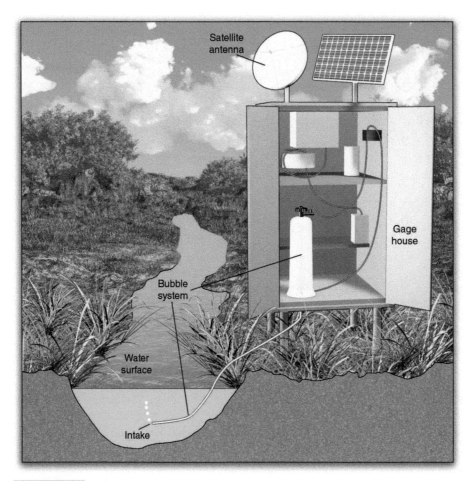

FIGURE 16.5 Diagram of a gauging station measuring the height of water in a river.

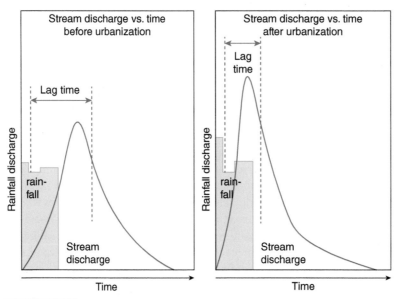

FIGURE 16.6 Time versus discharge curves for precipitation events in (left) a natural drainage area and (right) an urbanized drainage area.

Runoff is even quicker over pavement and buildings which also reduce the amount of infiltration during precipitation events (Figure 16.6). The combined effect of development increases the volume of streams much quicker and to greater levels after a precipitation event than they would naturally. This means that flash flooding is greatly increased in developed areas that are prone to heavy rainstorms.

CASE STUDY 16.1 2013 Uttarakhand Floods, India

Upstream Flooding

Most disastrous floods occur in the lower portions of rivers, far from the source of the flooding. These downstream floods are deceiving because they commonly appear even when the weather is clear (Figure 16.7). The reason is that the precipitation or thaw that generates the flood is in the headwaters of the river which may be quite far away. The flood in this region is termed an upstream flood which is typically in the mountains. Even though these floods can be devastating, the mountainous areas tend to be more sparsely populated. Therefore, there tends to be fewer casualties in these events. This was not the case with the upstream 2013 Uttarakhand floods.

During the week of 7–14 June 2013, the steep mountainsides of the state of Uttarakhand in Northern India were moistened by light rain making them vulnerable to flooding. Uttarakhand lies in the hills at the south side of the Central Himalayan escarpment and includes

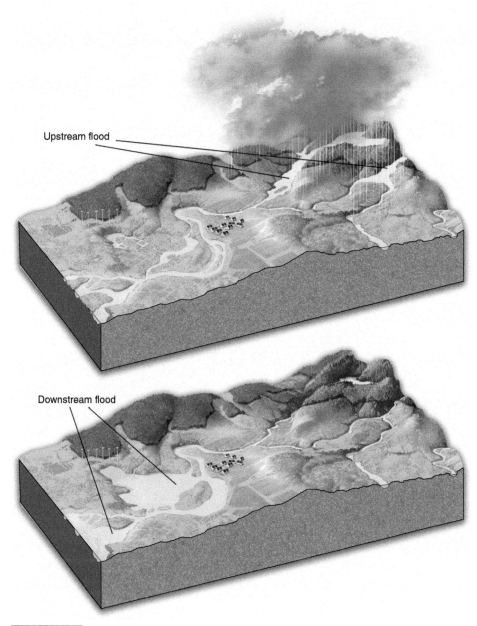

FIGURE 16.7 Diagrams of (top) upstream flooding versus (bottom) downstream flooding.

Himalayan peaks and glaciers (Figure 16.8). The south-facing slopes are naturally prone to flooding and landslides. The following week, an upper-level low-pressure trough extended unusually far southward, and the jet stream reached the Himalayas. There, it merged with a monsoon low-pressure system moving across India producing low-level flow up the Himalayan front that drove large amounts of water vapor from the Bay of Bengal into the Uttarakhand region. Orographic lifting of the moist air produced continuous heavy rain for two to three days as a result.

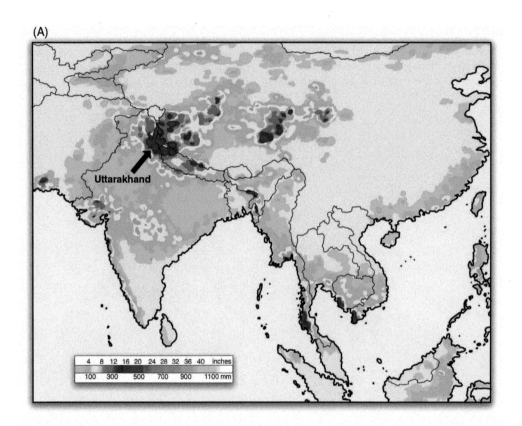

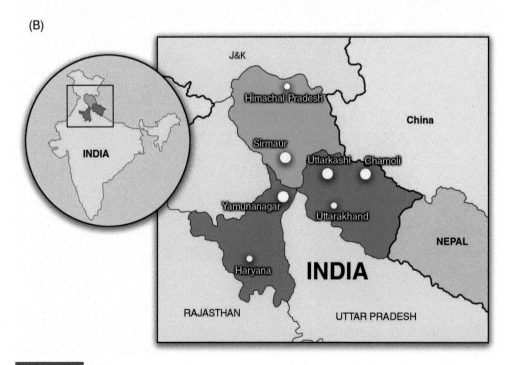

FIGURE 16.8 (A) Maps of areas impacted by the Northern India floods showing the radar map of the storm that produced the flood. *Source*: Image courtesy of NASA. (B) Map of Northern India showing the towns and villages impacted by the floods.

From 15–18 June 2013, some 19.7–39.4 in. (50–100 cm) of precipitation fell over Uttarakhand and neighboring Nepal. This rainfall measured 15.1 in. (38.5 cm) at one station where the normal monsoon rainfall is 2.8 in. (7.1 cm), an increase of 440%. The heavy rain and rising warm air caused melting of the Chorabari Glacier at an altitude of 12 467 ft (3800 m). At higher elevations, there was widespread blinding snow creating hazardous conditions. This extensive precipitation and runoff caused devastating floods and landslides throughout the region and resulted in much destruction and loss of life (Figure 16.8A).

The excessive precipitation over a short period of time resulted in heavy discharge in rivers and streams. The worst damage to human settlements was in the Kedarnath Shrine area, the Mandakini Valley, the Alaknanda Valley, the Pindar Valley, and along the banks of the Kali River in the Dharchula area. The Mandakini River level was 16–23 ft (5–7 m) above flood stage in the wide sections and 32.8–39 ft (10–12 m) high where it is narrow. The force of water gushing from Kedarnath area carried exceptional amounts of sediment including large boulders of 10–32.8 ft (3–10 m) across. This fast-moving debris obliterated everything in its path. The debris caused Kedarnath area to suffer unprecedented devastation with very heavy loss of life and property in two catastrophic discharges at 5:15 p.m. on 16 June and at 6:45 a.m. on 17 June. This second event resulted from the overflow and collapse of Chorabari Lake which released a large volume of water that produced a flash flood in Kedarnath (Figure 16.8B).

More than nine million people were impacted by the flash floods and landslides primarily in five districts (Figure 16.8C). This area is very popular with religious pilgrims who frequently travel to the region. The suddenness of the floods and velocity of sediment-laden flow washed away scores of pilgrims and residents. Landslides also buried many of them under debris flows and rockfalls. The deposits blocked roads stranding survivors in isolated locations that were cut off from rescue and relief operations. Approximately 1 200 000 people were stranded in various locations around the area.

In all, more than 4200 villages were impacted, leaving 2513 houses badly damaged or destroyed. In addition, some 11 091 livestock were destroyed. The estimated death toll, both documented and presumed dead, is 5748, but this is likely not accurate with the number of bodies that were buried or washed away. More than 89% of the casualties apparently occurred in the state of Uttarakhand. Some 69 relief camps were set up to care for 1 510 629 pilgrims and local resident survivors.

This disaster may only be the first in a developing trend. It appears that Northern India has been experiencing ever increasing rainfalls in June since the late 1980s. This increase in rainfall is associated with the same type of regional weather changes that produced the 2013 floods. These were just extreme situations in the trend. Modeling suggests that 60–90% of 2013 rainfall event can be attributed to the trend in weather patterns toward increased rainfall. The trend is modeled to result from increased concentrations of greenhouse gases and aerosols in the atmosphere. In other words, this disaster is yet another effect of climate change.

The other contributing factor in the disaster is the result of local construction projects and practices. The roads in the area are poorly planned and new resorts and hotels are being built in fragile river valleys, in the flood plains (Figure 16.8D). This not only endangers the people inhabiting these areas but the increased buildings and pavement also increase surface runoff causing quick and high discharge during rain events. In addition, there have been in excess of 70 hydroelectric projects including dams in the watersheds of the state of Uttarakhand. Environmental experts report that this activity has resulted in an ecological imbalance in the area, and the restricted river flow and streamside buildings are contributing to an increased number of landslides and more flash flooding.

FIGURE 16.8A Satellite image of a flooded river in the Uttarakhand area. *Source*: Photo courtesy of NASA.

330 CHAPTER 16 Devastating Floods and Their Aftermath

FIGURE 16.8B   Photograph of Northern India flood stranding pilgrims on small islands in rivers. *Source*: MANAN VATSYAYANA/Stringe/Getty Images.

FIGURE 16.8C   Photograph of Northern India floods. *Source*: Xinhua/Alamy Stock Photo.

FIGURE 16.8D Photograph of Northern India flood damage. *Source*: STRDEL/Getty Images.

CASE STUDY 16.2 1931 Yangtze–Huai River Flood

Extended Storm Cycle and Downstream Flooding

There are numerous cases of single storms causing disastrous floods, but in 1931 in Central China, a weather pattern developed that caused repeated flooding for an extended period of time. The resulting inundation severely impacted residents and the entire society for as much as six months and some for much longer. Refugees from flooded out areas packed every even marginally safe town and village throughout the region. Huge numbers of people drowned or died from the epidemics that swept the area as the result of standing water. It is unclear exactly how many people perished during this time, but it this may well have been the worst natural disaster of the twentieth century.

The 1931 Yangtze–Huai river floods were actually a series of interconnected floods primarily from June to September 1931 in China and striking several major cities including Wuhan and Nanjing (Figure 16.9). The 1931 Central China flood had roots long before the flood. The flood area is a major source of agriculture for China. The forests had long been cleared to create gardens and planted fields, which increases erosion rates. Further, water from the rivers was and is regularly used to irrigate the fields and there are a huge number of structures to divert and impound river water. With the addition of expanding urbanized areas, runoff was also increased, and the likelihood of serious flooding had been ever increasing for hundreds of years.

In 1931, the weather conditions would develop that showed just how dangerous the situation had become. The situation began developing during the winter of 1930–1931 which had been particularly harsh. Cold conditions and extensive precipitation produced thick snow and ice deposits in the upper catchments of China's rivers that lasted well into the spring. Melting snow and ice flowed downstream and flooded the middle course of the Yangtze River. These frozen accumulations completely melted in the late spring and added to the unusually heavy rains at the time. This engorged rivers, streams, and lakes and raised the water table.

The already excessive precipitation further increased into the summer. One cause of the precipitation was the extremely powerful East Asian monsoon that China experienced in 1931, possibly the result of an El Niño–Southern Oscillation. This caused excessive rain from June through August. Compounding these rains was an extraordinarily active typhoon season. The Yangtze Basin of China is normally struck by two cyclonic storms per year. In 1931, there were seven to nine typhoon remnants in the month of July alone. These immense storms dumped one and a half times the average annual precipitation in a single month. Four weather stations along the Yangtze River reported more than 24 in. (60 cm) of precipitation for the month of July.

This huge deluge stressed even the best maintained of dams and dikes. However, most of the water control systems that protected human communities alongside the Yangtze and Huai rivers were neglected and stood little chance in the flood. Vast areas along the rivers were inhabited by impoverished communities living in substandard housing. These people were disproportionately impacted by the flooding and were largely displaced. Immense numbers of these displaced people became refugees who sought shelter in the larger cities.

One of the worst impacted cities was Wuhan where residents experienced surreal conditions for the entire summer of 1931. The flooding began early in the spring as river water began pouring into the streets. At first, the flooding was just a nuisance, causing residents

332 CHAPTER 16 Devastating Floods and Their Aftermath

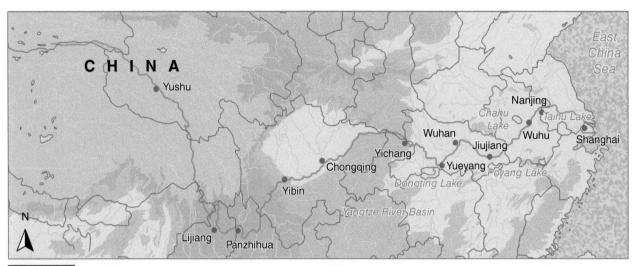

FIGURE 16.9 Map of the Yangtze River in China showing the towns flooded by the 1931 event.

FIGURE 16.9A Photograph of the 1931 Central China river flood inundating normally dry city streets. *Source*: Carter 2020 / with permission of SupChina.com.

to occasionally get their feet wet as they walked on the streets. Then the flood waters grew to ankle deep throughout the city and all of the time. The flood waters mixed with effluent from the overflowing sewers and the whole city endured a horrific stench, which grew worse during the heat of the day. The floodwaters continued to grow even deeper, and the city streets became canals, too deep to wade through (Figure 16.9A). Bottom floors of all buildings were flooded, and residents used all kinds of floatation devices from boats to bathtubs to get around (Figure 16.9B). Compounding the problem were the refugees who had been arriving in the city from the flooded rural areas since the late spring. They numbered as many as 200 000, further straining the resources of the city.

In addition to the overflowing sewers which turned up to 4 ft of water into a microbial soup, it was impossible to bury the dead. They sat in shelters and decayed releasing more dangerous diseases into the water. Wuhan and especially the refugee camps were gripped with rampant smallpox, typhus, cholera, and dysentery, and even malaria. It is estimated that the fatality rate for refugees ranged from 15% to 20%. This epidemic was not unique to Wuhan; all flooded areas experienced it to varying degrees (Figure 16.9C).

FIGURE 16.9B Photograph of a row of houses flooded into the second floor by the river floods. *Source*: Keystone / Stringer/Hulton Archive/Getty images.

FIGURE 16.9C Photograph of the Great China river flood inundating normally dry city streets. *Source*: Image courtesy of Neil Morrison and Historical Photographs of China, University of Bristol.

If Wuhan had not suffered enough, just before 6:00 a.m. on 27 July, there was a catastrophic collapse of the earthen dikes that encircled and protected the city. The water that had been held back poured into the city in a torrent. The coursing flood removed whole neighborhoods. Many thousands drowned or were buried alive. It is estimated that 782 189 urban citizens and rural refugees were left homeless, and the flood covered 32 mi² (83 km²) of the city and approximately 69 000 mi² (180 000 km²) of the surrounding countryside. Wuhan recorded river water levels of 53 ft (16 m) above average, which left the entire city flooded under 4 ft of water for three months.

Unfortunately, there was no relief in August and the flood that had devastated the upstream regions now moved downstream. By early August 1931, perhaps the most populous region of the world was underwater. It is estimated that 150 000 people had already drowned by this time. In the second half of August, the Yangtze River levels were measured at more than 3 ft (1 m) higher than all previous records and it was nearly 10 ft (3 m) above the average for that time of year. Many cities and villages along the Yangtze were completely submerged and barns, cattle, and agricultural machinery had been swept away in the flood.

While the flood was still at peak stage in Wuhan, it was becoming catastrophic downstream. On the evening of 25 August 1931, a torrent of floodwater surged from the Huang He (Yellow) River in the north and down through the ancient imperial Grand Canal near China's capital city of Nanjing. It swept away the dikes that protected Gaoyou Lake north of Suzhou and reportedly drowned nearly 200 000 people in their sleep. The flood spread back upstream on the Yangtze River to Chongqing while isolating Nanjing on an island downstream. The flood filled the agricultural plain of Huguang, ultimately uniting the Yangtze, the Huai, and Yellow rivers across a 41 000 mi² (107 000 km²) developing inland sea. Meanwhile, Nanjing suffered catastrophic flood damage while causing hundreds of thousands of deaths. In addition to drowning, many people died from starvation and waterborne diseases such as cholera and typhus. There are even unconfirmed reports of starving people eating bark off trees and cannibalism.

It is almost impossible to determine the number of people who died in this biblical proportion disaster. The official government report proposed that 140 000 drowned, but claimed that potentially as many as two million people died during the flood event, from drowning, lack of food, or disease. Indeed, probably 70% of the deaths were from disease and famine. Another source calculated the death toll from the event at 422 499. However, this was not the end of the suffering. The following year (1932), a waterborne cholera epidemic ravaged the survivors. The Chinese government officially reported 100 666 cases and 31 974 deaths. However, some other sources from the West propose that death toll for the entire two-year disaster was between 3.7 and 4 million and impacted as many as 53 million people. Others report that 50–80 million people were left homeless. It could be the worst natural disaster in history.

CASE STUDY 16.3 1975 Banqiao Dam Failure, China

Typhoon-Induced Collapse and Flood

One of the worst recent floods resulted from the collapse of Banqiao Dam and 61–63 additional dams in Henan, China in 1975 (Figure 16.10). The Banqiao Dam was supposedly designed to survive a once-in-1000-year flood which would involve 7.6 in. (30 cm) of rainfall per day. However, an unusual meteorological occurrence caused a once-in-2000-year flood in August 1975 causing the dam to collapse. In the Discovery Channel's "The Ultimate 10 Technological Disasters" in the world, the Banqiao Dam collapse was considered the worst, outranking even the Chernobyl nuclear disaster.

Three major reservoir and dam projects began construction in 1951 in the Henan (Zhumadian) area of China including the Banqiao Dam, Shimantan Dam, and the Baisha Dam. The Banqiao Dam was built on the Ru River for water storage, hydroelectric power, and, to a lesser degree, to help manage flooding on the Yellow and Huai river basins. Because the Chinese lacked the expertise, construction was completed under the guidance of engineers from the Soviet Union. By 1953, the three reservoir projects were complete, but a number of settling cracks occurred in the clay core of the earthfill Banqiao Dam. As a result, in 1955, the Banqiao Dam was reinforced with Soviet Union oversight and as a result, it was dubbed "the iron dam." The huge 387-ft (118-m)-high dam held a reservoir with a storage capacity of 398 000 acre-feet (492 million m³) with 304 000 acre-feet (375 million m³) reserved for flood storage.

During Mao Zedong's "Great Leap Forward," from 1957 to 1959, more than 100 dams were built in the Zhumadian region. This race to build resulted in diminished quality of the dams. In addition, during Mao Zedong's "Learn from Dazhai in Agriculture" initiative, a large proportion of the forest cover in the region was stripped away and land degradation was rampant. This severely damaged the ecosystem of the area and increased the impact of flooding. Coupled with the fact that most of the dams, especially Banqiao and Shimantan, were undersized compared to conventional design standards, a disaster was imminent. Also contributing to the danger was a record of poor maintenance and management of the reservoir.

The dam probably would have survived for many decades if not for a very unusual weather event. A disturbance that would become Typhoon Nina was spawned in the Philippine Sea on 29 July 1975. It quickly intensified to a tropical depression and traveled southwestward for the next 36 hours. It then slowed on 31 July and began to intensify to a tropical storm. It attempted to turn to the northwest, but a pressure ridge blocked it and instead it tracked west-northwest. During the late hours of 1 August, Nina underwent explosive development from a marginal typhoon with sustained winds of 75 mph (120 kmph) to a Category 5 monster with sustained winds of 150 mph (240 kmph) the next day. Nina quickly reached peak intensity with wind speeds of 155 mph (250 kmph) making it one of the most powerful ever in the region.

The storm weakened before it made landfall near the coast of Taiwan as a Category 3 typhoon with sustained winds of 115 mph (185 kmph). It made a second landfall near Jinjiang, mainland China. It then moved northwest across Jiangxi and turned north late on 5 August. On 6 August, the remnants of Nina moved to Zhumadian, Henan, where it stalled for three days because it was blocked by a cold front. The stalled system produced phenomenal torrential rainfall, with

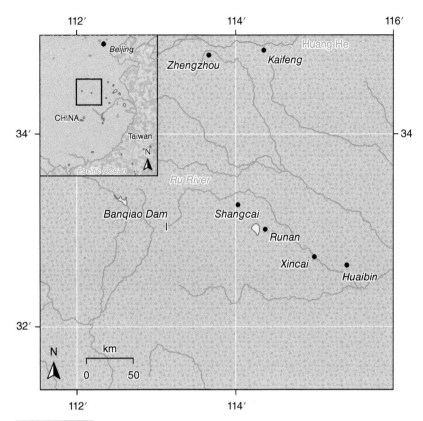

FIGURE 16.10 Map of the Banqiao Dam and downstream towns and villages flooded in Eastern China by the dam burst.

more than 16 in. (40 cm) over a huge area of 7500 mi² (19 410 km²). The heaviest rainfall was near the Banqiao Dam where 64.2 in. (163.1 cm) of rain fell, including 33 in. (83 cm) in just a six-hour period. This precipitation rate is just shy of the world record.

The massive rainfall quickly flooded the Henan area. On 6 August, amid the torrential rain, Banqiao Dam administrators submitted a request to release water from the dam, but it was rejected because of flooding in downstream areas. On 7 August, the request was finally approved, but telegrams did not reach the dam administrators. On 7 August at 9:30 p.m., the People's Liberation Army unit dispatched to the dam issued a dam failure warning by telegraph. They attempted to release water, but several sluice gates for controlling water flow were clogged with silt and just before midnight water began to overtop the dam. Two hours later, on 8 August, 1975 at just past 1:00 a.m., the dam failed, and the entire structure breached less than six hours later (Figure 16.10A). The resulting flood wave was 16.4–30 ft (5–9 m) high and 6.2 mi (10 km) wide flowing at nearly 31 mph (50 kmph). The outflow of Banqiao Dam was 2.78 million ft³ (78 800 m³) of water per second. In addition, the release from the breached Shimantan Dam was 1.06 million ft³ (30 000 m³) per second which also fed through the Banqiao Dam. As a result, more than 2.48 billion ft³ (701 million m³) of water was released to the flooded area downstream in six hours (Figure 16.10B). The volume of earthfill that was eroded from the Banqiao Dam was greater than 1.3 million cubic yards (1 million m³). From all breached dams, more than 554 billion ft³ (15.7 billion m³) of water was released in total.

The massive release flooded an area 34 mi (55 km) long and 9.3 mi (15 km) wide. The water formed flood lakes up to 4600 mi² (12 000 km²) (Figure 16.10C). Although the government had issued evacuation orders, their delivery was spotty as a result of the heavy rain and flooding as well as poor communications. In order to protect other dams from failure, the army evacuated several flood diversion areas, and the military launched air strikes deliberately destroying several other dams to systematically release water. However, the destruction continued despite these efforts. On 9 August, dams on the Quan River collapsed and the entire downstream region was flooded. The situation continued to worsen with the cresting of the Boshan Dam, holding 14.1 billion ft³ (400 million m³) of water and the floods from the Banqiao and Shimantan dams failures rushing toward it. In response, several more air strikes were launched against a number of other dams in an attempt to protect the Suya Lake Dam, which already held 42.4 billion ft³ (1.2 billion m³) of water. More than nine days later, more than one million people were still trapped by the waters. The trapped survivors were devastated by epidemics and famine.

The results of the flood were devastating. The death toll from the entire event is difficult to estimate because of the number of bodies swept away or buried in the mud. It is estimated that the death toll directly from flood waters was in excess of 26 000. In addition, it is estimated that another 145 000 victims died from epidemics and famine. This total of 171 000 deaths falls within the range of estimates from 85 600 to 240 000 deaths. This would make this event the third deadliest flood in history. It impacted more than 10.15 million people and flooded about 30 cities and counties destroying at least 6.8 million houses. Removal of the hydroelectric generators on the destroyed dams caused the sudden removal of 18 GW of power from the electric grid. This is equivalent to removing nine large modern coal-fired power stations, which is about one-third the peak demand on the United Kingdom National Grid. The cost of the damage was about $6.7 billion in 1975 US dollars.

336 CHAPTER 16 Devastating Floods and Their Aftermath

FIGURE 16.10A Photograph of the remains of the Banqiao Dam after the 1975 river flood. *Source*: Wikimedia Commons/Public domain.

FIGURE 16.10B Photograph of the raging 1975 river floods in China. *Source*: AFP/Getty Images.

FIGURE 16.10C Aerial photo of flooding from the breached dam in the 1975 flood in China.
Source: IntechOpen.

The disaster occurred during the Chinese Cultural Revolution, so it was not publicized. In fact, the Chinese government hid the details of the disaster completely until the 1990s. The only important legacy of the 1975 flood is the redesign of flood-control structures in China to incorporate probable maximum flood potential from maximum precipitation potential.

CHAPTER 17

Droughts and Desertification

CHAPTER OUTLINE

17.1 Deserts 341

17.2 Desertification 344

17.3 Drought and Climate Indices 345

17.4 Droughts and Dust Storms 345

17.5 Megadroughts 357

Earth's Fury: The Science of Natural Disasters, First Edition. Alexander Gates.
© 2022 John Wiley & Sons Ltd. Published 2022 by John Wiley & Sons Ltd.
Companion website: www.wiley.com/go/gates/earthsfury

Words You Should Know:

Deserts – These include extremely arid land having at least 12 consecutive months with no rainfall, and arid land having less than 9.8 in. (25 cm) of annual rainfall.

Drought – In simple terms, it is the absence of water, but there can be (i) meteorological droughts, (ii) hydrological droughts, (iii) agricultural droughts, and (iv) socioeconomic droughts. Meteorological droughts are 75% of normal precipitation or less relative to a 30-year average.

Dust storm – A cloud of dust swept up and transported by a strong wind; includes sandstorms.

Haboob – An intense dust storm driven by a weather front.

Megadrought – A drought lasting 20 years or more.

Wildfire – An unplanned, unwanted, uncontrolled, and destructive fire of vegetation.

Human existence, just as most terrestrial life, depends on the availability of fresh water. Ecosystems develop around having a certain amount of available water. If the source of water decreases, it greatly stresses the ecosystem leading to significant loss of life. This situation also stresses human communities, causing them to respond or experience a natural disaster. It is thought that the migration of human ancestors out of Africa and around the world was in response to a drought. There are several causes to droughts and desertification and effects that characterize it.

17.1 Deserts

Deserts are areas in which the precipitation is less than 9.8 in. (25 cm) per year. It is a misperception that deserts must be hot; only tropical deserts have this requirement. Polar regions are also deserts. They contain frozen water at the surface, but the annual precipitation is low, and the accumulation rate is extremely slow.

Most deserts are around 30° north and south latitudes and at the poles as dictated by the circulation system of the atmosphere (Figure 17.1). Descending air in these locations creates a higher air pressure which produces clear sunny weather with little precipitation. It takes a predominance of high pressure on a long-term basis to form a desert. The sun is more direct in the tropics making the deserts around 30° north and south latitudes tropical or hot deserts. The sun is least direct at the poles and as a result the air is cooler forming polar deserts.

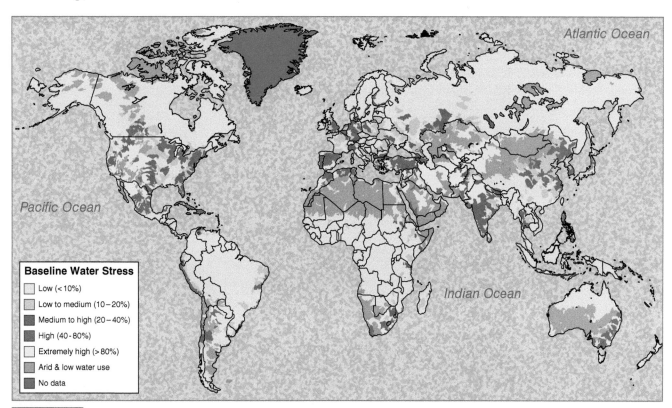

FIGURE 17.1 Map of the world showing the deserts and areas of water shortage. Note the dry areas around 30° north and south latitudes.
Source: Adapted from World Resources Institute.

Other effects may also produce small to large deserts in unexpected areas. If air is forced to ascend a mountain, it rains all of the moisture out on the slopes. As this air continues over the mountain and descends the slopes on the far side, it is dry and cool forming a rain shadow which can also be a desert. If there is too much landmass and too little open water for evaporation, the precipitation can be drastically reduced. These effects can form deserts at any latitude.

In deserts, moving sediment is the main mechanism of erosion. This erosion is primarily accomplished by wind which can sweep unimpeded over the landscape. The wind carries sand and silt in suspension and can carry it aloft at heights up to 1 mi (1.6 km) and spread tens to hundreds of miles downwind of the source. In some cases, the dust can be transported thousands of miles from its source region. For example, dust from Africa is commonly blown across the Atlantic Ocean, all the way to North America (Figure 17.2).

Sand-sized grains can also be moved in saltation where they bounce during transportation. They generally bounce about 20 in. (50 cm), but they can jump up to 6.5 ft (2 m) if they bounce off a rock or other hard surface. The sand is driven by the wind so hard that it can strip the paint from a car or facet a rock. If the wind is strong enough, it can remove the finer sediment in a deflation structure and leave a deposit of pebbles (Figure 17.3). The surface pebbles can set together and be polished by the wind forming desert pavement. The exposed larger grains are polished flat on the wind side forming ventifacts.

Sand can also be transported by water during the infrequent precipitation events. Runoff and surface water flow very quickly and form sheetwash and flash floods in the ephemeral streams in deserts. These floods cut deep, steep-walled channels into the desert that are dry most of the time. They are called dry washes or arroyos.

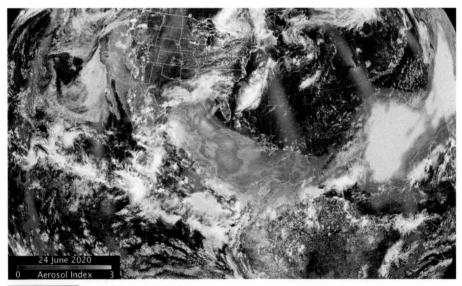

FIGURE 17.2 Map showing the drifting of airborne dust from the Sahara Desert to Central and North America. *Source*: Image courtesy of NASA.

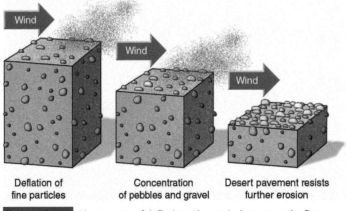

FIGURE 17.3 The process of deflation where wind removes the fine particles leaving a hard pan of the remnant materials.

Wind erosion is the major shape agent of the desert landscape. Sandblasting abrades most rock surfaces quickly removing softer rocks like shale. The contrast in resistance to wind erosion of the different rock types creates the desert landforms. In areas of flat-lying strata, erosion progresses through layers by depth. Erosion proceeds much quicker in less-resistant layers and slower in more-resistant layers. The contrast between the more-resistant layers on top and softer layers beneath results in flat-topped features dropping to cliffs through the softer layer on the sides. Erosion of this feature proceeds by wind abrading away the softer layer until the capping resistant layer is undermined and collapses. Slowly, the capping layer erodes away until the landscape height is reduced. If erosion proceeds around the flat area so that it becomes a raised flatland, it is called a plateau (Figure 17.4). As erosion reduces the size of the flat top, it is a mesa, a small table-shaped feature with dropping cliffs on all sides. As erosion continues, the feature becomes a pinnacle or chimney before it erodes away completely. In some cases, two chimneys may contain a flat slab of caprock across the top forming an arch. A yardang looks like small bush with a larger piece of caprock perched atop a thin pinnacle.

If the resistant caprock is tilted, it erodes into another group of structures. For gently inclined strata, the caprock forms a gentle slope and the edge forms a steep slope. This feature is called a cuesta. If the strata are steep or vertical, the resistant layer forms a spine with the weaker layers and talus forming lower sides. This feature is a hogback.

The most identifiable desert features are sand dunes which are piles of sand shaped by wind movement. Sand dunes are initiated on an obstacle such as a rock. Sand builds on the obstacle until it is buried in a crescent-shaped dune that then can move away in a downwind direction. The dune has a gently inclined slope on the upwind or stoss side and a steeper slope on the downwind or lee side. Sand is carried up the stoss side of the dune which is eroded by the high wind velocity. Once the sand is swept over the top of the dune, it slides down the lee side slope or slip face. This side is protected from the highest-velocity wind by the dune forming a wind shadow. The sand accumulates on this side of the dune.

Dunes migrate by this process of eroding sand from one side and depositing it on the other. The deposited layers of sand on the slip face are asymptotic both on top and on bottom. However, the top of the slip face deposit is eroded as the stoss side migrates over it leaving the slip face deposit sharp on top but asymptotic on the bottom. These deposits are called cross-beds and indicate migrating dunes. The slip face deposits are called the foresets and the flattened off tops are called topsets. The Navajo Formation Sandstones in the Southwestern United States have spectacular desert-dune-generated cross-beds.

Sand dunes can take on a number of shapes depending upon the wind (Figure 17.5). The most common dune is the barchan dune which retains its original formation shape on the obstacle. These dunes are crescent-shaped with the two pointed tips in the downwind direction. They are common in areas of flat topography, a single wind direction, and limited sand supply. If the wind direction is from several directions, several barchan dunes can coalesce into a multiply pointed star dune. Where there is an ample supply of sand and moderate single-direction winds, transverse dunes develop. They have long straight crests like an ocean wave. If the wind speed increases, the straight crests bend and break forming a crescent to horseshoe shapes similar to barchan dunes but called barchanoid dunes. In very-high-velocity winds with ample sand supply, the sand dunes stretch out parallel to the wind direction in elongate ridges called longitudinal dunes or seif dunes, the Arabic word for sword. These dunes can reach heights of 325 ft (100 m) and extend in length for several miles. There is an area in Saudi Arabia called Rub al Khali where seifs reach 975 ft (300 m) high and extend for nearly 120 mi (192 km). Parabolic dunes are similar to barchan dunes, but the tips point upwind instead of downwind. They form in areas where there is vegetation present. Most dunes are about 100 ft (30 m) in height but transverse to parabolic "draas" dunes in Saudi Arabia which can reach heights of 820 ft (250 m) with wavelengths of 2/3 of a mile (1 km) or more.

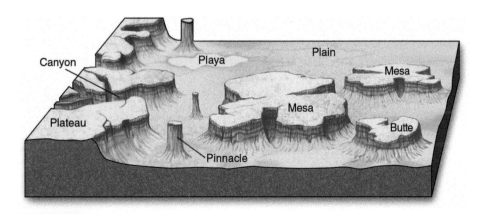

FIGURE 17.4 Geomorphic features produced by wind erosion in an arid environment.

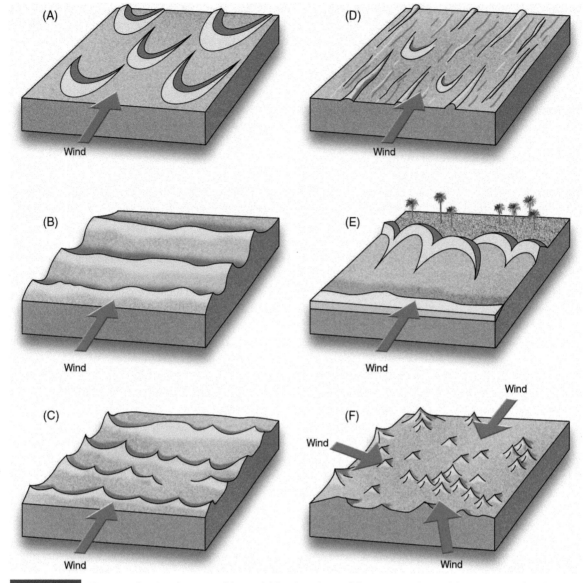

FIGURE 17.5 Diagrams showing the types of dunes: (A) barchan dunes, (B) transverse dunes, (C) transitioning barchan to transverse dunes, (D) linear dunes, (E) parabolic dunes, and (F) star dunes.

17.2 Desertification

The original definition of desertification was the spread of deserts into surrounding productive areas, but it has evolved to mean the conversion of any productive area into desert conditions. Desertification can be a natural process in response to changing climate from wetter to dryer conditions. A major concern, however, is the rapid spread of desertification on a worldwide basis as the result of human activity. The main causes of human-induced desertification are overgrazing by domesticated animals, deforestation in dry areas, soil removal, poor drainage of irrigated lands, poor farming techniques, overuse of water supplies, and human-induced global warming. Desertification advances in patches that expand in an inconsistent ebb and flow. Africa has experienced an alarming loss of productive land as the Sahara Desert spreads southward. This spread is largely the result of overgrazing, which removes the stabilizing properties of the plant roots in addition to climate change. Another popular example is the Aral Sea in Southern Russia. As the result of overuse of water from the Aral Sea for irrigation, it has shrunk to less than one-third its original size and what was formerly water is now desert.

Desertification damages productive soil, water, and air. Infertile and salty sand is blown from exposed desert onto productive soil slowly reducing its productivity. In other cases, sand accumulations migrate over productive areas and bury the productive soil in sand. Uncovered sand and soil with no vegetation cover are prone to erosion during precipitation events, which can run off and choke streams with sediment, compromising the water quality. Winds lift silt and fine sand or dust from deserts compromising air quality with particulate as well. Devastating dust storms are common around deserts which can cause severe health problems. The effects, however, can be much more widespread. Windblown dust from the Sahara Desert crosses the Atlantic Ocean to the Americas on an annual basis and, on occasion, dust crosses the Pacific Ocean from the Gobi Desert in China to North America.

17.3 Drought and Climate Indices

Before an area undergoes desertification, it first increases the frequency and intensity of droughts which are characterized by periods of insufficient precipitation and streamflow. There are several systems to classify the severity of droughts. They share a commonality that they all measure the amount of precipitation in an area relative to normal. Droughts are carefully monitored around the world to different degrees. The United States maintains an online real-time drought monitoring system.

In 1965, the Palmer Drought Index was developed built on decades of earlier attempts to monitor droughts. Palmer's index was innovative in that it measured drought and wet periods in a standardized index that reflected the balance of water supply and water demand. The Palmer Drought Index was later improved to the Standardized Precipitation Index (SPI) which measures water supply and specifically precipitation as well as an imbalance between water supply and water demand. The SPI covers several timescales, ranging from 1 to 24 months, so it can assess both short-term and long-term droughts. It also determines precipitation excess in wet periods.

The United States maintains the US Drought Monitor system which incorporates information from the SPI, the Palmer Drought Index, and US Geological Survey streamflow data and locally reported drought impacts in near real time (Figure 17.6). It partitions the country into areas and percentages based on drought indices D0 to D4. D0 is defined as abnormally dry relative to a history average for that time and includes short-term dryness, slowed planting and growth of crops, and lingering water deficits. D1 is considered moderate drought and is characterized by some crop and pasture damage, developing water shortages and voluntary water restrictions. D2 is severe drought and includes crop and pasture loss and water shortages and restrictions. D3 is extreme drought and includes major crop and pasture losses and widespread water shortages and restrictions. D4 is exceptional drought and includes widespread crop and pasture losses and water emergencies. It also reports whether drought conditions are short-term (less than six months) or long-term (greater than six months).

The most recent drought index is the US Climate Extremes Index (CEI) which was developed in 1995 to quantify changes in climate within the United States. The CEI uses a set of climate indicators including extremes in average monthly maximum and minimum temperatures, one-day heavy precipitation events, drought severity, number of days with and without precipitation, and wind intensity of tropical cyclones upon landfall.

A CEI of 0% indicates that no part of the United States was subject to the extremes of temperature or precipitation considered in the index. A CEI of 100% means that the entire country had extreme conditions during the period for every indicator. In general, CEI values of more than 20% indicate "more extreme" above-average conditions, and CEI values less than 20% indicate "less extreme" below-average conditions.

The CEI is regularly determined for eight specific periods including spring, summer, autumn, winter, annual, cold season, warm season, and hurricane season. CEI indicates that for the annual, summer, warm, and hurricane seasons, the percent of the United States experiencing extreme conditions has been increasing since the early 1970s.

17.4 Droughts and Dust Storms

The lack of rainfall causes a drought. As a result, the vegetation holding the soil together slowly dies off and is removed. This allows the wind to directly abrade the exposed soil and remove the fine material. This is a constant occurrence in deserts, but rare outside of them. The danger in the non-desert areas is that the top layer of the soil contains all of the nutrients and is what makes the soil productive. Its removal severely degrades the ability of an area to produce food. In big windstorms, so much surface material can be removed that it can create a dust storm.

One type of dust storm is called a haboob (Figure 17.7A,B). These form in advance of thunderstorms crossing a desert or very arid areas. Strong downbursts from the thunderstorm produce high-velocity winds in all directions. The winds erode the sand and soil surface, causing the fine particles to become airborne. As the storm quickly advances, it lifts this dust into

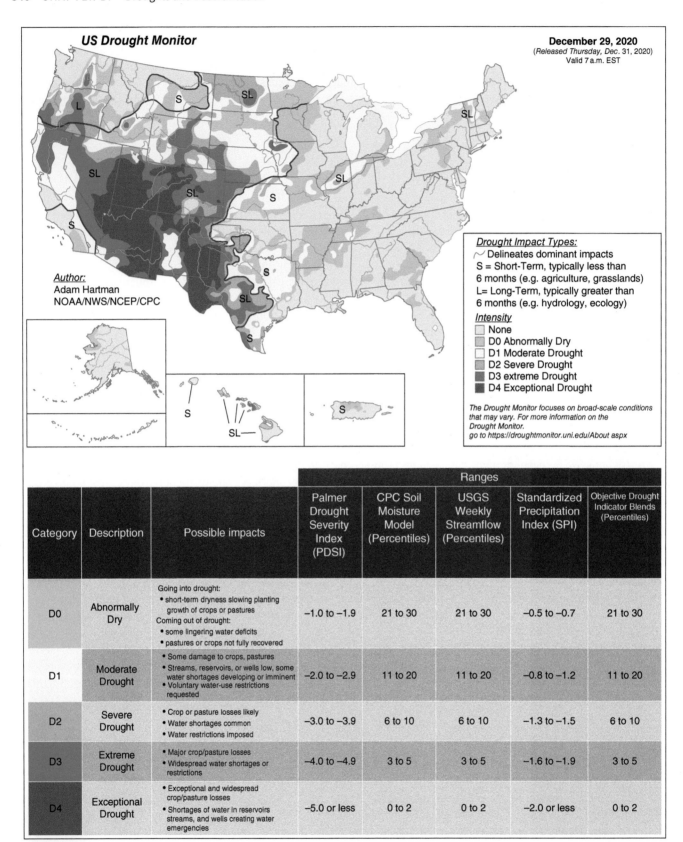

FIGURE 17.6 Drought monitor report for the United States showing areas of relative drought and their duration as reported by the University of Nebraska–Lincoln. Below are the characteristics of each of the drought levels including the standard drought indices.

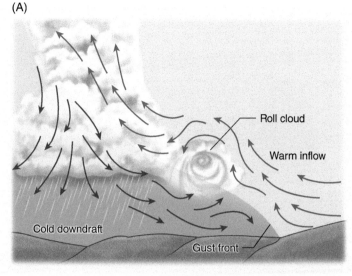

FIGURE 17.7 (A) Diagram showing the formation of a dust storm haboob. *Source*: Adapted from NOAA. (B) Photograph of a haboob in the Southwestern United States. *Source* (B): Photo courtesy of the US National Weather Service/NOAA.

the air in a dark wall of silt and sand that can be a mile (1.6 km) or more high. This dust wall can be more than 62 mi (100 km) wide and travel at speeds of 22–62 mph (35–100 kmph) and strike without warning. Once the rain begins, they can turn into mud storms.

CASE STUDY 17.1 1930s Central United States

Drought and Dust Bowl

When the Central Plains of the United States were opened to settlement, they were covered with deep-rooted, natural grasses. These grasses fixed the soil and had evolved to survive in the dry, cold, and windy conditions more typical for this region than the unusual wet, warm weather of the mid-1800s. Farmers removed these grasses and replaced them with crops such as corn, soybeans, and alfalfa, which made the soil less cohesive and attached to the land. By the early 1930s, all it took to create the Dust Bowl conditions was a drought and wind. (Figure 17.8).

For nearly a century, the Central United States enjoyed a period of above-average rainfall and moderate temperatures. This, combined with the demand for agricultural production that accompanied World War I, increased the demands on farmers to produce more food and to cultivate land that they would have normally left fallow. These less desirable areas were close to stream banks, on steep ridges, or areas covered with trees. The intense level of agriculture included several very poor farming techniques such as no rotation of crops, plowing entire fields after each harvest, and allowing animals to graze on stubble and other crop residues.

CHAPTER 17 Droughts and Desertification

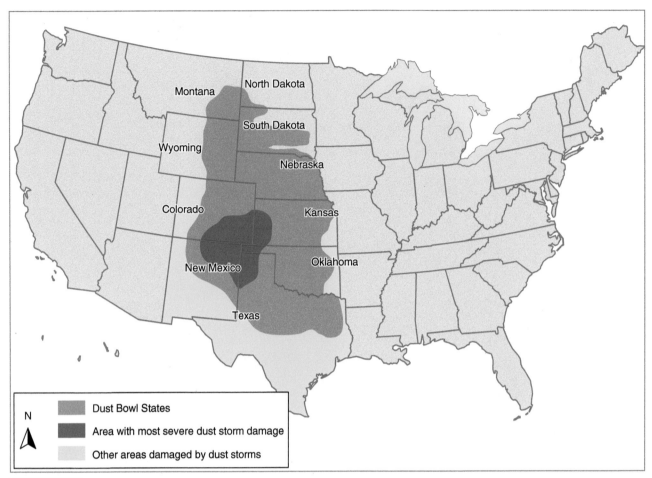

FIGURE 17.8 Map of the United States showing the Dust Bowl states and area damaged by dust storms.

Although periods of low rainfall in the Central United States occurred in 1890 and again in 1910, they were nothing compared to the catastrophic drop in precipitation between 1926 and 1934. Areas that received 30–40 in. (76.2–101.6 cm) of rainfall per year on average were getting as little as 15–20 in. (38–50.8 cm) annually. Research suggests that the reasons for this drought were the changes in the jet stream and ocean currents that may have altered atmospheric circulation and moisture patterns. It was not until the early 1940s that precipitation levels approached pre-drought levels.

As the agricultural fields dried, wind picked up the fertile topsoil and moved it in massive, blowing clouds of dust (Figure 17.8A). With no trees or grass to bind the soil or obstructions to slow the wind, the blowing topsoil drifted like black snow. In some places, it buried farms in layers of 3 ft (1 m) or more and turned previously productive fields into barren wastelands (Figure 17.8B). In 1932, 14 dust storms were recorded in the Central Plains, and by 1933, there were 38 storms. By 1934, the peak of the drought, more than 100 million acres (40.5 million ha) of prime farmland had lost its topsoil to wind erosion (Figure 17.8C). At this point, the Dust Bowl covered much of the United States encompassing an area approximately 399 mi (644 km) long and 298 mi (483 km) wide in parts of Colorado, New Mexico, Kansas, Texas, and Oklahoma (Figure 17.8D). Drought conditions lasted for some seven to eight years in the hardest-hit region of the Central Great Plains.

The worst dust storm took place on 14 April 1935, which was named Black Sunday. With wind speeds of 60 mph (96.6 kmph), dust was lifted so high that it blocked out the sun making the day as dark as night (Figure 17.8E). Traffic came to a standstill on the roads.

Regardless of how well sealed, a layer of dust that could be measured in inches seeped into every home in the area causing severe health effects. The Black Sunday storm blew eastward, depositing a thick layer of black dust over Chicago equal to 4 lb (1.8 kg) for every person living there. It was blown all the way into New England, New York City, and even Washington, D.C., where legislators experienced just a hint of what almost one-third of the United States was suffering through. It was the result of this storm that an Associated Press reporter used the term Dust Bowl for the first time to describe this disaster in the Central Plains.

Human costs associated with the Dust Bowl were enormous. Much of the United States was suffering through the effects of the stock market crash of 1929 and onset of the Great Depression, but the Central Plains was relatively insulated from this economic disaster. Farmers grew their own food and maintained a reasonably unchanged lifestyle. The Dust Bowl literally blew away their livelihood, and they could no longer make mortgage and tax payments on their farms. Many people were forced to abandon their homes and seek a new life somewhere else. The largest mass migration in American history resulted with more than 2.5 million people leaving the Central Plains states and settling elsewhere by 1940. About 15% of the people in Oklahoma left the state, many moving westward and settling in California. Other parts of the country were overwhelmed with the influx of Dust Bowl refugees, collectively called "Okies," even though they were from several states. The Great Depression was in full swing by this time and these transient newcomers stressed local relief services and competed with local residents for jobs, often willing to work for greatly reduced wages.

FIGURE 17.8A Photo showing an approaching dust storm during the Dust Bowl. *Source*: Photo courtesy of NOAA.

FIGURE 17.8B Photo showing farm buildings covered by sand drifts and dunes during the Dust Bowl. *Source*: Photo courtesy of US Department of Agriculture.

Government response to the disaster was slow. Eventually political leaders understood that much of the productive farmland was becoming permanently unproductive and they took action. Congress formed the Soil Conservation Service (later changed to the Natural Resources Conservation Service) in 1935 that investigated and implemented cultivation techniques to reduce topsoil erosion by wind and water. As New Deal legislation started to be enacted, the Works Progress Administration (WPA) and other newly formed government agencies and projects began to provide emergency supplies, funds, livestock feed, and transportation to maintain the livelihoods of farmers and ranchers. Health-care facilities were also established and medical supplies provided to impacted communities to meet emergency medical needs. The government established reliable markets for farm goods, imposed higher tariffs on imported foodstuffs, and established loan funds for farm market maintenance and business rehabilitation. By 1934, congressional appropriations for drought relief exceeded $500 million.

Later, additional soil conservation measures and policies were implemented. Water supply and irrigation systems were improved, and

FIGURE 17.8C Photo showing an agricultural field made unproductive by drifting sand during the Dust Bowl. *Source*: Photo courtesy of US Department of Agriculture.

FIGURE 17.8D Photo showing an agricultural field and farm overtaken by drifting sand during the Dust Bowl. *Source*: Photo courtesy of US Department of Agriculture.

FIGURE 17.8E Photo showing an approaching haboob during the Dust Bowl. *Source*: Photo courtesy of US Department of Agriculture.

a federal crop insurance program was established. The average size of farms increased and low-yield, ecologically sensitive, and buffer lands were removed from production. These and other soil conservation programs helped to mitigate the impact of future droughts and establish a pattern of soil conservation measures. However, the topsoil has not been replenished to much of the Central Plains and the soil is red, acidic, and nowhere as productive as it was prior to the Dust Bowl. Natural processes will take thousands of years to replace what was lost in less than a decade.

CASE STUDY 17.2 1871 Peshtigo Wildfire

Dry Conditions and Wildfires

Droughts and dry conditions dry out vegetation and make it much more flammable. The lack of moisture and surface water allows fires to spread much quicker in drought conditions. For these reasons, wildfires are much deadlier and uncontrollable during droughts and dry conditions. They are among the deadliest threats in droughts. 8–10 October 1871 witnessed the worst outbreak of wildfires in United States history. The Upper Midwest lost millions of acres of woodlands and thousands of lives in the blaze. The deadliest of the fires was the Peshtigo Fire, Wisconsin, but it was overshadowed by the most famous of them, the Great Chicago Fire (Figure 17.9). There have been larger wildfires than these in history, but modern warning systems and firefighting techniques have rendered them generally less deadly.

The Peshtigo and related fires in Michigan and Chicago resulted from several factors. The summer of 1871 was very hot and dry, producing a prolonged drought with no precipitation at all in August or September primarily as the result of large upper-level high-pressure ridge across the region. At this same time, there was extensive logging and clearing of the land for agriculture, leaving flammable dried-up vegetation and logging debris throughout the area. Further, the land-clearing procedure included the use of small, controlled fires that numbered in the hundreds. A strong low-pressure system then developed over the Central Plains, producing powerful southwesterly winds across this region beginning at about 2 p.m. on 8 October. The gale-force winds, warm temperatures, and dry conditions whipped together the small pre-existing fires leading to an uncontrolled spread of a huge blaze.

The Peshtigo Fire began in the dense forest in Wisconsin on Sunday afternoon, 8 October, and quickly spread to the village of Sugar Bush, killing every resident. The gale-force winds sent the now 200-ft (61-m)-high flames rapidly northeast toward Peshtigo. Now a firestorm, temperatures in the blaze reached 2000 °F (1093 °C) causing trees to explode into flames. The fire reached Peshtigo that evening without warning.

It struck Peshtigo with a vengeance. Survivors reported that it moved like a fire tornado, hurling railcars and houses into the air. The speed of the flames consumed all available oxygen suffocating some fleeing people and causing others to burst into flames. Nearly 200 people died in a single tavern and 75 people perished while sheltering in a boarding house. Three people jumped into a water tank and were boiled to death when the fire heated the tank. Many people fled to the Peshtigo River, but the fire crossed the river and burned both sides of the town. Many attempted to escape the flames by immersing themselves in the river, but some of them drowned and others died of hypothermia in the frigid water. The firestorm was through Peshtigo in one hour and on to its next victims.

The death toll in the town of Peshtigo has been estimated at 800. A mass grave of 350 people was dug because the bodies were so charred that they could not be identified. Only two buildings in the entire town survived the fire.

352 CHAPTER 17 Droughts and Desertification

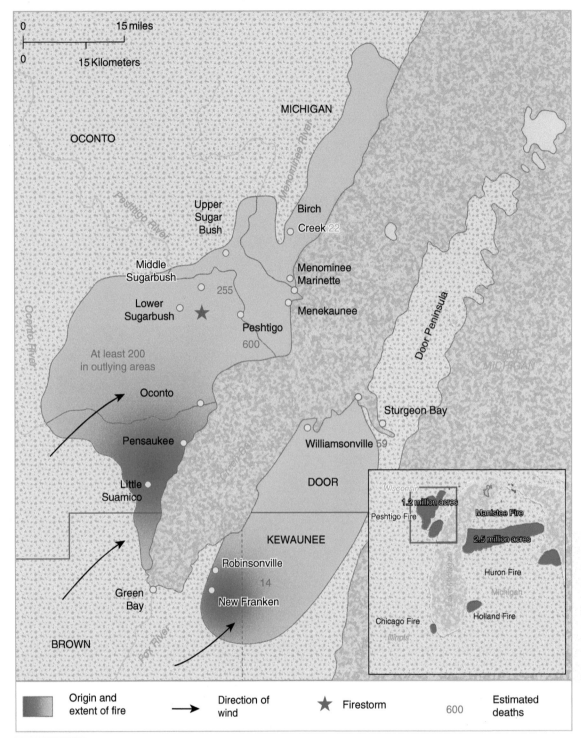

FIGURE 17.9 Map of the 1871 Peshtigo Fire in Wisconsin. Inset: the area of the multiple simultaneous fires in Wisconsin, Michigan, and Illinois and its coverage.

The firestorm continued past Peshtigo, destroying a swath of forest 10 mi (16 km) wide and 40 mi (64 km) long and it obliterated the towns of Peshtigo and Brussels and damaged 16 other towns in just two hours. By the time it was over, the fire burned approximately 1875 mi² (4860 km²) or 1.2 million acres (490 000 ha) of forest. The human death toll was 1152 confirmed and 350 believed dead, though some claim as many as 2400 deaths. More than 1500 were injured and greater than 3000 people were made homeless. The property loss was estimated at $5 million (1871 US dollars) in addition to the loss of 2 000 000 trees and scores of animals, though others claim it was in excess of $160 million.

At the same time as the Peshtigo Fire, the Great Michigan Fire began and included a number of simultaneous forest fires in Michigan. Although some of the sources of fires are unknown, they were fanned by the same gale-force winds that accelerated the Peshtigo Fire. A number of cities, towns, and villages, including Alpena, Holland, Manistee, and Port Huron, were damaged or destroyed. In addition to fires that originated across Michigan, the Peshtigo Fires also crossed the Menominee River and burned Menominee County, Michigan, killing 128 people. Not only was the land left barren, thousands of houses, barns, stores, and mills were destroyed and no lumber was left to rebuild. Hundreds of families were left homeless. The Port Huron Fire burned several cities, towns, and villages including White Rock and Port Huron, and much of the countryside in the "Thumb" region of Michigan for a total of 1.2 million acres (4850 km^2). More than 50 people died because of the Port Huron Fire. More than 2.5 million acres (1 million ha) were burned throughout Michigan in the event. The death

FIGURE 17.9A Map of the area of the Great Chicago Fire of 1871. *Source*: Courtesy of US Library of Congress.

FIGURE 17.9B Photo showing the Great Chicago Fire of 1871. *Source*: Courtesy of US Library of Congress.

toll for the Great Michigan Fire is a matter of some speculation but is roughly estimated at 500.

Perhaps the most famous of the 8–10 October 1871 blazes was the Great Chicago Fire (Figure 17.9A). On 8 October at 2 p.m., the wind turned southwesterly in Chicago and slowly intensified. The fire alarm was rung as fire broke out in the forested areas to the south and southwest sides of the city. Into the evening, the wind increased in velocity, until by midnight it was hurricane force and spread the firestorm quickly into the city (Figure 17.9B). By the time the fire brigades got the blaze under control, it had destroyed more than 4 mi^2 (10.4 km^2) of the city, including the business district, and at least 17 000 buildings (Figure 17.9C). The damage estimate for the fire in Chicago was at $169 million. The death toll for the disaster was 300 and 90 000 of 500 000 people were left homeless (Figure 17.9D).

FIGURE 17.9C Photo showing burned buildings in Chicago in the fire of 1871. *Source*: Courtesy of US Library of Congress.

FIGURE 17.9D Photo showing the burned area in Chicago in the fire of 1871. *Source*: Courtesy of US Library of Congress.

CASE STUDY 17.3 1876–1878 Northern China Drought

Worst Famine in History

Between 1876 and 1878, perhaps the most disastrous large-scale drought in recorded history struck Northern and Central China. It caused unprecedented famine and was accompanied by epidemics and locust plagues. The impacted residents were forced into unimaginable solutions to stay alive, but ultimately more people died in this event than any other famine in history making it likely the worst natural disaster ever.

The drought of 1876–1878 occurred at the end of the cold period of the Little Ice Age and the beginning of rapid warming of the Northern Hemisphere. At this time, parts of Europe and North America were beginning to warm already, but East Asia remained cold. In 1876, severe cold air masses struck North China early with severe snowstorms in Shandong by early November. Frosts in 1877 in North China set records for earliest ever with killing frosts as early as 21 August. The winter of 1877–1878 was marked by intense and frequent cold waves that included more than 60 consecutive days of snow and ice starting from the end of December in Northern China. The drought also occurred during an extremely strong El Niño and a period of few typhoons.

The drought of 1876–1878 impacted parts of 13 provinces in Northern and Central China including Liaoning, Inner Mongolia, Hebei, Shanxi, Shaanxi, Henan, Shandong, Gansu, Ningxia, Sichuan, Hubei, Anhui, and Jiangsu (Figure 17.10A). The event began in January 1876 and ended about mid-August 1878 and was accompanied by famine, locust, and epidemics for each of the three years.

A summer drought began in 1876 in Northern China in the provinces of Hebei, Shanxi, Henan, and Shandong and parts of Liaoning and Inner Mongolia. There was little to no rainfall in these areas from January 1876 to mid-July 1877. After this, the drought was relieved by some rain from south to north. However, in Shandong province, the rain did not occur until late August and in Yingshan of Hubei and Nanchong of Sichuan in the mid-upper basin of the Yangtze River, it did not occur until autumn 1877.

In addition, the drought-stricken area expanded in 1877 and lasted even longer. There was no rainfall in summer or autumn of 1877 in Hebei and Shanxi. In Hengshan, in the Shaanxi province, the drought lasted until the spring of 1878. In the counties of Gansu and Ningxia, no rain fell for an entire year. In the spring of 1878, the drought remained severe throughout Shandong province. By 1878, the drought abated slightly in Hebei, Shanxi, Inner Mongolia, Henan, and Shandong provinces, but the worst impacted area shifted westward toward Shaanxi, Gansu, and Sichuan. There was little or no rainfall in spring and summer in Hebei, Shanxi, Henan, and Shandong, but the drought came to an end in most areas in mid-August 1878.

The documentation of days without rain in the cities is astounding. The town of Huayin had not a drop of rain for 340 consecutive days. Gaoping had no rain for more than 330 consecutive days. Fangcheng and Mianchi had no rain for more than 300 days. Many other towns had similar reports.

The impact of the drought on the local residents was nothing less than apocalyptic. In 1876, the drought drove grain prices sky-high, and shortages were rampant. This initiated a famine that spread quickly through the provinces of Hebei, Shanxi, Shaanxi, Inner Mongolia, Liaoning, and Shandong, and parts of Jiangsu, Anhui, and Henan (Figure 17.10B). In 1877, crop production failed, spreading the famine to an even bigger area. This drove the residents to extremes to find food. There were large areas completely stripped of vegetation with people eating dirt in a poor to unsuccessful attempt to survive. Entire villages died of starvation. People began killing each other for food or trading their children for food (Figure 17.10C). There were numerous macabre stories of widespread cannibalism and even people digging up recently buried corpses for any sustenance they could find (Figure 17.10D). Dead bodies littered the countryside, and it is estimated that 40–50% or more of the residents perished in many of the worst struck areas. More than 50 years after the drought, the populations had still not recovered.

With all of the unburied decaying corpses, disease was widespread by 1877. The main epidemic was in Hebei, Shanxi, and Henan, but also in parts of Shandong and Liaoning. The area impacted by epidemic more than doubled in 1878. To worsen this already horrible situation, the drought-stricken area was also struck by plagues

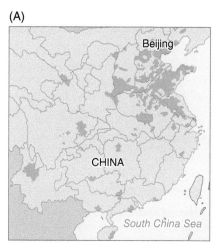 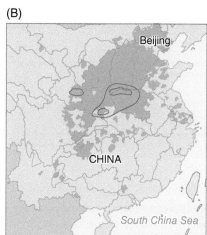 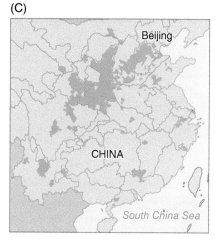

FIGURE 17.10A Drought-stricken areas of Central China (A) in 1876, (B) in 1877, and (C) in 1878. The blue contours show the number of days without rainfall. *Source*: Adapted from Zhang and Liang (2010).

356 CHAPTER 17 Droughts and Desertification

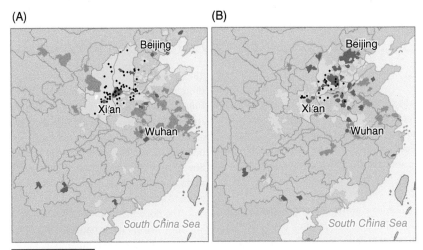

FIGURE 17.10B Map of China showing the distribution of areas of famine (pink), locust plague (brown), and disease (red), and location of documentation of humans surviving on human corpses (cannibalism) (blue dots) (A) in 1877 and (B) 1878. *Source*: Adapted from Zhang and Liang (2010).

FIGURE 17.10C Woodcut showing the heartbreak of the 1876–1878 Great China drought and famine. Left: Children dying of starvation. Right: People committing suicide rather than starve. *Source*: The Graphic (London), July 6, 1878. From the collection of Pierre Fuller.

FIGURE 17.10D Woodcut showing the atrocities of the 1876–1878 Great China drought and famine. Left: Cannibalism among survivors. Right: People starving to death. *Source*: The Graphic (London), July 6, 1878. From the collection of Pierre Fuller.

of locust from 1876 to 1878, but especially in 1877. The locust plague was worst in Gansu, Hebei, Henan, Shandong, Anhui, Jiangsu, Zhejiang, and Hubei. These insects devoured the pitiful crops, exacerbating the famine.

The famine caused by drought in Central and Northern China in 1876–1878 caused the deaths of between 9 and 13 million people, but the actual number of deaths will never be known. This makes the 1876–1878 Northern China drought the worst natural disaster ever. China was not the only area to be devastated by the extreme weather. In India, the monsoon rains failed to materialize from 1876 to 1878 and the resulting drought and famine resulted in the deaths of approximately 5 million people there.

17.5 Megadroughts

Megadroughts are droughts that are unusually severe and last for 20 years or more. There are typically several megadroughts operating around the world at any time. They were also common in the past. By using multiple tree-ring studies, a history of megadroughts has recently been assembled for the North American Midwest and West for the past 1200 years (Figure 17.11). During this time, four major megadroughts appear to have occurred including during the 900 AD, the 1100s–1200s, and the late 1500s. The periods of these megadroughts were 940–985 AD, lasting 45 years; 1100–1247, lasting 147 years; 1340–1400, lasting 61 years; and possibly 1550–1600, lasting 50 years.

The problem is that the American Midwest and far West are currently in a new and emerging megadrought (Figure 17.12). By comparing the drought conditions between 2000 and 2018 with the tree-ring calibration system, this recent period is the second driest of all of the 19-year dry periods over the past 1200 years. This drought abated in 2019, but returned to the same or worse drought conditions in 2020. The 2020 wildfire season was the worst on record for Colorado and California and other nearby states were at near record levels as well. Smoke from these fires dimmed skies all across the country including the East Coast. The advancement in wildfire fighting techniques is the only reason that these fires did not develop into large-scale natural disasters.

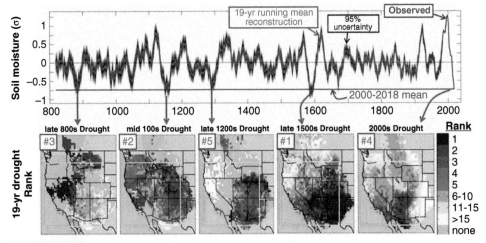

FIGURE 17.11 Top: Graph of the results of the tree-ring and interpreted soil-moisture study for the past 1200 years in the Western United States. Bottom: Maps showing the distribution of low soil moisture during the interpreted five worst droughts over the past 1200 years. *Source*: From Williams et al. (2020), DOI: 10.1126/science.aaz9600.

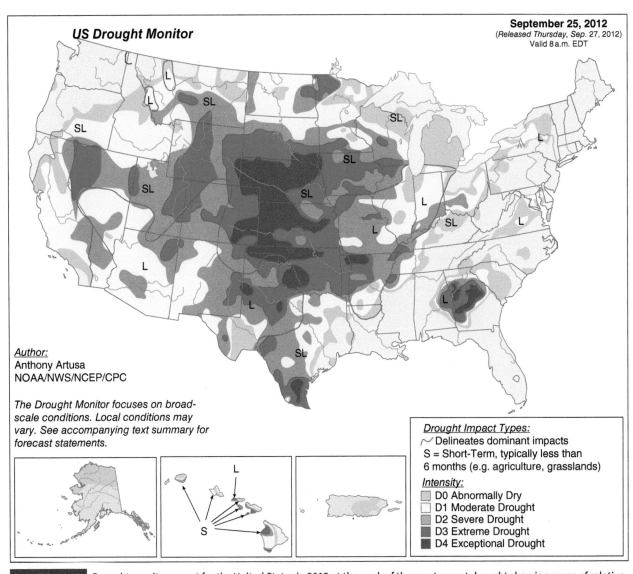

FIGURE 17.12 Drought monitor report for the United States in 2012 at the peak of the most recent drought showing areas of relative drought and their duration (from University of Nebraska–Lincoln). Compare to Figure 17.6 to see the shift of dry areas to the west.

Another disturbing finding of the tree-ring calibration study is that the twentieth century was determined to be the wettest century in the entire 1200-year record. This means that the conditions that dictated the rapid development of the Western and Southwestern United States were actually a historical fluke in terms of precipitation. It is likely that we will not be able to count on precipitation necessary to maintain the communities that were developed over the past century. If this future is combined with other droughts such as the Australian drought ending in 2019, which was the worst in 800 years, and Brazil's current drought, which is the worst in history, the shortage of freshwater may become one of the most urgent issues facing the human race.

References

Williams A.P., Edward R. Cook, Jason E. Smerdon, Benjamin I. Cook, John T. Abatzoglou, Kasey Bolles, Seung H. Baek, Andrew M. Badger and Ben Livneh (2020). Large contribution from anthropogenic warming to an emerging North American megadrought; Science, 368 (6488), 314–318, DOI: 10.1126/science.aaz9600.

Zhang, D., and Y. Liang, 2010: A long lasting and extensive drought event over China in 1876–1878. Adv. Clim. Change Res., 1, DOI: 10.3724/SP.J.1248.2010.00091.

CHAPTER 18

Impacts: Collisions from Space

CHAPTER OUTLINE

18.1 The Solar Nebula 363

18.2 Objects Colliding with Earth 364

18.3 Potential Impacts 364

18.4 Impacts with Earth 366

Earth's Fury: The Science of Natural Disasters, First Edition. Alexander Gates.
© 2022 John Wiley & Sons Ltd. Published 2022 by John Wiley & Sons Ltd.
Companion website: www.wiley.com/go/gates/earthsfury

Words You Should Know:

Asteroids – An extraterrestrial body of rock that is larger than a meteoroid but smaller than a planetoid.
Bolide – A very bright meteor or fireball.
Comets – An extraterrestrial body of asteroid size composed of ice and rock debris and that orbits the sun.
Impact – The collision of extraterrestrial bodies or an extraterrestrial body with the earth.
Impact crater – A depression left on the surface of the earth or other body as a result of an extraterrestrial impact.
Meteorites – The extraterrestrial rocks that fall to the earth as meteors.
Meteoroids – Free-floating celestial fragments that are larger than dust but smaller than asteroids.
Meteors – A meteoroid that is plunging through the atmosphere and glowing from the heat.
Solar system – The sun and all planets and other bodies held in its gravitational field.

18.1 The Solar Nebula

The origin of the vast majority of the meteorites and asteroids that collide with the earth was in the formation of the solar system. The nebular theory explains how the solar system was formed. About 4.57 billion years ago, there was a cloud of debris that was spinning as the probable result of the explosion of a nearby supernova. The spinning cloud flattened into a disk and mass was forced toward the center in a gravitational collapse. When the center of the cloud achieved critical mass, a fusion reaction began, forming a star that would be the sun. The spinning debris disk began segregating into belts similar to the rings around Saturn we see today. One of these rings still exists as the asteroid belt between Mars and Jupiter (Figure 18.1).

As these belts of debris rotated around the sun, eddies formed and the pieces of debris collided, building protoplanets. These early impacts involved large bodies at great speed destroying and rebuilding the protoplanets for a long period of time. Planets and other bodies that did not have their surfaces reworked by later processes still contain evidence of this period. Mercury and the moon still contain the huge craters from this period of development. Based on the ages of lunar samples, the period of heaviest collisions lasted until about four billion years ago, but they slowly decreased in intensity after that rather than abruptly ending. The evidence for the disassembling of planets by impact includes meteorites found on the earth that originated from Mars. There is also good evidence that the moon originated as part of the earth and was removed by large impact about 4.4 billion years ago.

We know that most of the meteorites that strike the earth are part of the debris that built the solar system because isotopic dating of them yields the same age of about 4.5 billion years. There are several other sources of meteorites such as comets and

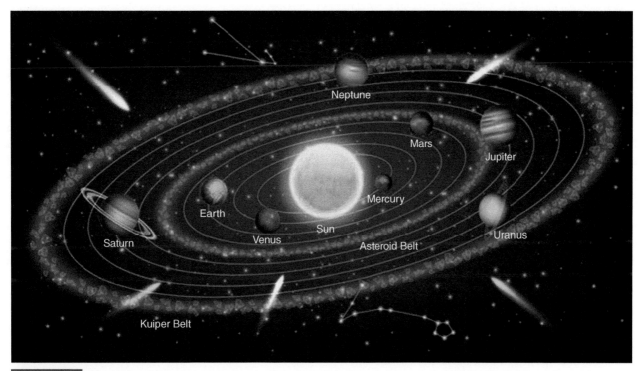

FIGURE 18.1 Illustration of the solar system showing the relative locations of planets, asteroid belt, Kuiper Belt, and possible comets. *Source*: D1min/Adobe Stock.

asteroids that can still impact the earth. However, the frequency of impacts continually decreases, and big impacts become less likely. Nonetheless, an impact of a large extraterrestrial body with the earth would generate the largest natural disaster in history with the capacity to cause the extinction of the human race. For this reason, even though they are unlikely, extraterrestrial impacts must be considered as a potential natural disaster.

18.2 Objects Colliding with Earth

Although seeing a meteor or shooting star is a relatively rare experience for most people, extraterrestrial material is raining down on the earth all of the time. It is estimated that an average of between 48.5 and 60 tons (44 000–54 400 kg) of extraterrestrial material falls on the earth every day. This is a very rough estimate, and it certainly varies greatly, especially during meteor showers and other events. Some estimate that it could be as much as 300 tons (276 000 kg) in a single day. Fortunately, the vast majority of the material is dust.

There are many similarly named terms for extraterrestrial objects, but all have different meanings. Meteoroids are any natural object floating in space and they range in size from dust to small asteroids. Upon entry into the earth's atmosphere, these meteoroids descend to the earth at phenomenal speeds estimated at 6.2–25 mps (10–40 kmps) or an average of about 12.4 mps (20 kmps) or 44 600 mph (71 777 kmph). The friction with the atmosphere causes the larger objects to be heated to exceptional temperatures as high as 2900–5430 °F (1600–3000 °C), causing them to burn a glowing trail across the sky called a meteor. Exceptionally bright meteors are also called bolides, or fireballs.

A few meteors can typically be seen per hour on a given night. If there are a lot of meteors, it is most likely a meteor shower. Some meteor showers are annual or regular and predictable occurrences as the earth intersects the trail of debris left by a comet or even an asteroid. Meteor showers are commonly named for the body that emits the debris. Even though a lot more material strikes the earth during these events, they are no more dangerous than typical days.

The vast majority of the material that strikes the earth is dust and smaller than 0.004 in. (0.1 mm) in diameter. This material falls slower and is too small to glow. About pebble size is the minimum for glowing. This dust is a combination of cosmic dust, debris from comet and asteroid collisions, or from the Kuiper Belt (Figure 18.1). Even most of the larger particles burn up upon entry to the atmosphere or land in the oceans.

18.3 Potential Impacts

There are several types of extraterrestrial bodies or meteoroids that could collide with the earth. They are classified by size and composition. The largest of the dense bodies are asteroids. Asteroids are almost exclusively from the asteroid belt around the sun between Mars and Jupiter. Asteroids range from 33 ft (10 m) to 621 mi (1000 km) in diameter and are classified as stony and carbonaceous (C-type), metallic (M-type), or some mix of the two (S-type), though there is evidence that some are ice-bearing. There are about 700 000 to 1.7 million asteroids with diameters of 0.6 mi (1 km) or more in the asteroid belt. The largest asteroid by far is Ceres, which should probably be classified as a planetoid (Figure 18.2). It is 578 mi (933 km) across and contains nearly 25% of the mass of all the asteroids combined. It is almost twice as large as the second-largest asteroid. The next largest asteroids are Pallas, Vesta, and Hygiea, all of which range from 248 to 325 mi (400–525 km) in diameter. Ceres contains nearly 50% of the total mass of the entire belt. All of the other asteroids in the asteroid belt are less than 210 mi (340 km) across.

Several asteroids have orbits in the area relatively close to the earth. Collisions among the asteroids are relatively common. These collisions have the potential to knock the asteroids out of their orbit and propel them into the path of other planets, including the earth. Asteroids are strongly affected by the gravity of the massive planet Jupiter. Any perturbations in Jupiter's gravity field, even including regular variations, can also dislodge an asteroid from the belt and send it into another orbit. Asteroid impacts with the earth are, by far, the most dangerous of possible extraterrestrial impacts and they can happen at any time.

The next largest of the extraterrestrial bodies are comets (Figure 18.3). Comets range from less than 10 ft (3 m) to more than 124 mi (198 km) in diameter. They are essentially dirty snowballs, composed primarily of ice with debris of rock and metal randomly mixed in. However, some comets are thought to have stony cores. All comets are thought to have originated at the edge of the solar system in the Kuiper Belt or far out of the solar system in the Oort cloud. Some comets are ejected from these areas and form huge orbits around the sun. Halley's Comet is the most famous of these comets, appearing every 77 years, which is its orbit interval.

Comets consist of a head, or coma, and a gaseous tail as much as thousands of miles long. The head consists of a solid nucleus surrounded by a nebulous coma as large as 1.5 million miles in diameter. The tail is an elongate curved vapor trail arising from the coma when sufficiently close to the sun, and it is proposed to consist largely of ammonia, methane, carbon dioxide, and water. Comets are observed only where their orbits are relatively close to the sun. As comets approach the sun, they melt and form the tail of gases and debris. The comet is the nucleus of this streak.

Extraterrestrial bodies less than 32 ft (10 m) that are not composed of ice are regular meteoroids as described. The meteoroid is the core of the object falling to the earth. Once a meteoroid enters the earth's atmosphere, it is a meteor and most commonly

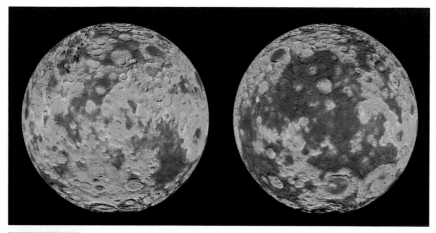

FIGURE 18.2 Color-enhanced photo of the largest asteroid or planetoid Ceres in the asteroid belt. *Source*: Courtesy of NASA.

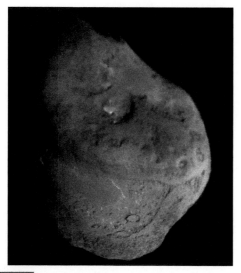

FIGURE 18.3 Photograph of the Comet Temple. *Source*: Courtesy of NASA.

FIGURE 18.4 Photographs of meteorites: (A) Achondrite meteorite from Mars (1.6 in. [2 cm] across). (B) Iron meteorite on the surface of Mars (3.2 in. [4 cm] across). *Source*: Courtesy of NASA.

burns up. On average, less than 5% of the original meteoroid reaches the ground. If a meteoroid survives the trip through the atmosphere and impacts the ground, it is called a meteorite. Meteorites are no more radioactive than ordinary terrestrial rocks, and no meteorite has been found to contain any life or element that does not occur naturally on the earth. Larger meteorites have left craters around the world both on land and on the ocean floor.

There are three main types of meteorites: iron meteorites making up about 6%, stony meteorites making up nearly 94%, and stony–iron meteorites at less than 1%. Of these, it is estimated that 99.8% are from asteroids and the 0.2% are split between meteorites from Mars and the moon. It is predicted that there may also be meteorites from Venus and Mercury, but none have been yet identified. The over 60 Martian meteorites are considered to have been removed from Mars by exceptional meteoroid impacts early in its history (Figure 18.4A).

The composition of meteorites most closely reflects the primitive parts of planets including the mantle and core of the earth. Iron meteorites are actually composed of iron–nickel alloys with minor sulfides all in crystalline form (Figure 18.4B). Although numerically minor, they are the most common older meteorites. The iron–stony meteorites occur in two types. Pallasite is composed of olivine crystals encased in iron–nickel alloy. It is speculated that they represent rocks from the core–mantle boundary. Mesosiderite is breccia of mixed iron and stony meteorite fragments formed in a collision. Stony meteorites are 86% chondrites and 8% achondrites. Chondrites are composed of chondrules, droplets of silicate rock or glass mixed with iron–nickel sulfides. There are also carbonaceous chondrites that contain water, sulfides, and organic material. They are considered to record the beginning of the solar system. Achondrites appear like regular igneous rocks from the earth, but they are from the moon, Mars, or other bodies.

18.4 Impacts with Earth

There are about 190 impact craters thus far identified on the earth. The most famous and best-preserved exposed crater is the Barringer Meteor Crater near Winslow, Arizona (Figure 18.5). It is almost 1 mi (1.6 km) in diameter and has been dated as approximately 50 000 years old. It was created by the impact of an iron–nickel meteorite about 150 ft (45.7 m) across, and weighing roughly 300 000 tons (272 155 mt). It is estimated that the meteoroid was traveling at a speed of 40 000 mph (64 375 kmph). The impact caused an explosion equivalent to 20 million tons (18.1 million mt) of trinitrotoluene (TNT) and has left behind a chasm 570 ft (174 m) deep.

Much of what we learned about the features and processes of cratering on the earth comes from the Barringer Crater. In the 1960s and 1970s, the Apollo astronauts were trained at the Barringer Crater before they went to the moon. It is marked by a rim of crushed debris that sits about 165 ft (50 m) higher than the surrounding plains (Figure 18.6). Around the rim, the plains are covered with meteoritic iron debris over an area about 7.5–9.3 mi (12–15 km) in diameter. So far, more than 30 tons (27 mt) of iron–nickel spherules have been found. They were formed during the vaporization of the asteroid when it struck. The pre-existing bedrock in the area of the crater was ejected and flipped over and now sits upside down for a distance of 0.62–1.2 mi (1–2 km). Within the crater, there are vast zones of breccia or crushed rock from the impact. There are half-melted slugs of meteoritic iron mixed with fragments of local rock. There are minerals that can only form under extremely high-pressure conditions such as the minerals coesite and stishovite, rare forms of quartz. It even contains hints that it may have had somewhat of a central peak.

These and other features helped scientists identify the Chicxulub Crater just off the north shore of the Yucatán Peninsula in the Gulf of Mexico. This feature is interpreted to be a 65-million-year-old asteroid or comet impact scar which is

FIGURE 18.5 Photos of Barringer Crater, Arizona: (A) Satellite image of the crater. *Source:* Courtesy of NASA. (B) Photo of the crater. (C) Side view of the crater. *Source:* Courtesy of the US Geological Survey.

proposed to have brought about the extinction of the dinosaurs, and about 70% of life on the earth. The impact would have raised a global cloud of dust and sulfur gases that blocked sunlight from penetrating the atmosphere and subjected the earth to a decade or more of near-freezing temperatures. Small glassy sediments dating 65 million years called tektites, which

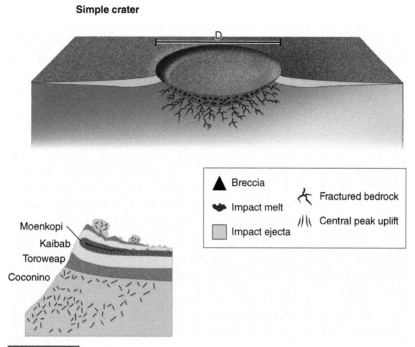

FIGURE 18.6 Illustrations of the parts and features of the Barringer Crater including overturned beds along the edges.

are of impact origin, have been found all around the Gulf of Mexico and into the Southern United States. The crater is estimated to be 120 mi (193 km) across and has since filled with ocean sediment.

When the Chicxulub impact became a realization in the late 1970s and 1980s, the US Congress held hearings on the potential environmental damage of extraterrestrial impact disasters. The new phrase introduced in these debates was "nuclear winter," which describes the effects of both nuclear war and an asteroid impact. A large impact would pulverize the asteroid and an enormous amount of rock, sediment, and soil at the impact site and raise a huge dust cloud high into the atmosphere. The particulate would encircle the planet, drastically reducing sunlight for periods up to several years. The particulate would be toxic to life on the planet and the lack of sunlight would eliminate photosynthesis causing a worldwide famine.

The hearings on impacts sparked a worldwide interest and search for ancient craters. Many were identified including a very large one discovered in 1983 near to Washington, D.C. in the Chesapeake Bay (Figure 18.7). A crater 53 mi (85 km) wide and 0.8 mi (1.3 km) deep is buried 985–1640 ft (300–500 m) beneath the southern Chesapeake Bay area in Southeastern Virginia. One of the first hints of this structure was the identification of an 8-in. (20-cm)-thick layer of ejecta found in a drilling from Atlantic City, New Jersey, some 170 mi (275 km) north. The layer contains tektites (fused glass beads) and shocked quartz which can only be made by an impact. Drill cores into the crater found shocked quartz, melted rocks, rubble, and a 0.81-mi (1.3-km)-thick layer of breccia. The age of the impact is about 35 million years and the estimated size of the asteroid or comet is about 1.9–3.1 mi (3–5 km). It is estimated that this bolide had an impact speed of about 37.3 mps (60 kmps). The body was vaporized on contact and the basement rocks were fractured to 5 mi (8 km) depth. There is a deep crater with a central peak in the center that is 24 mi (38 km) across and surrounded by collapsed blocks forming ring faults.

During this search for impacts, a very fortuitous event took place. The Hubble Telescope picked up a comet about 0.9–1.2 mi (1.5–2 km) across headed directly toward Jupiter. As the comet entered Jupiter's atmosphere, it broke into at least 20 pieces of various sizes. Soon after, in July 1994, humans had their first visual evidence that large bodies could impact planets as the fragments of the Shoemaker–Levy 9 Comet collided with Jupiter in a long chain (Figure 18.8). The impacts were catastrophic to Jupiter with the equivalent explosive power of 300 million atomic bombs. Each impact created plumes in the atmosphere that were 1200–1900 mi (2000–3000 km) high. It is estimated that they heated the atmosphere to temperatures as hot as 53 000–71 000 °F (30 000–40 000 °C). These impacts left scars in Jupiter's atmosphere that were visible for 14 months afterward. Although Jupiter is a huge planet compared to the earth with a much stronger gravitational field to capture space bodies, the impact caused humans to take notice and devote more effort to scan the skies for extraterrestrial objects.

The earth has experienced a few small impacts and near misses in recent history. The 1908 Tunguska Event in Siberia occurred when a 200-ft (61-m) asteroid or comet weighing about 100 000 tons (90 718 mt) exploded 5 mi (8 km) above the earth, releasing approximately 2040 mt of energy. An area the size of Rhode Island was demolished in the explosion, killing all wildlife for 20 mi (32 km), and starting fires that burned for weeks. On 23 March 1989, an asteroid named 1989 FC with a diameter about 0.3 mi (0.48 km) and a kinetic energy of over 1000 one-megaton hydrogen bombs passed within 430 000 mi (692 020 km) of the earth.

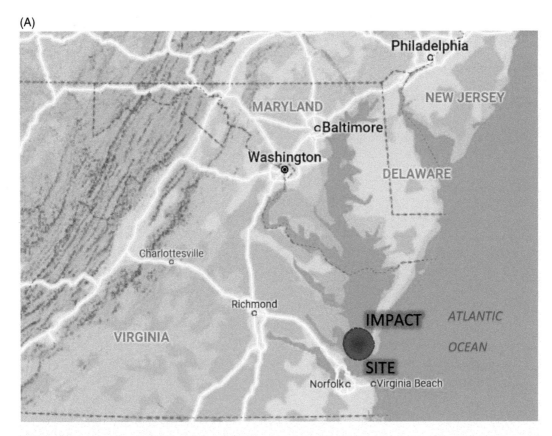
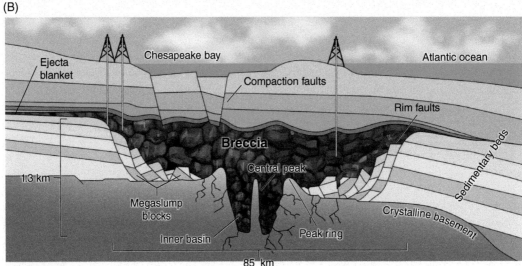

FIGURE 18.7 Illustrations of the Chesapeake impact crater: (A) Map of location of the impact site. *Source*: Base map is Google Map. (B) Cross section of the interpreted impact crater. *Source*: Adapted from US Geological Survey Fact Sheet 049-98.

This asteroid was not discovered until it had passed its point of closest approach, and only after calculating its orbital path backward. Since then, several bodies of similar sizes have been measured as coming close to the earth.

The history of bolide impacts with the earth will continue to grow as new discoveries of hidden craters are made. Nearly all known craters have been recognized only since 1950, and several new structures are found each year. Evidence suggests that there have been many thousands of other impacts over the course of the earth's history, most of which have yet to be discovered.

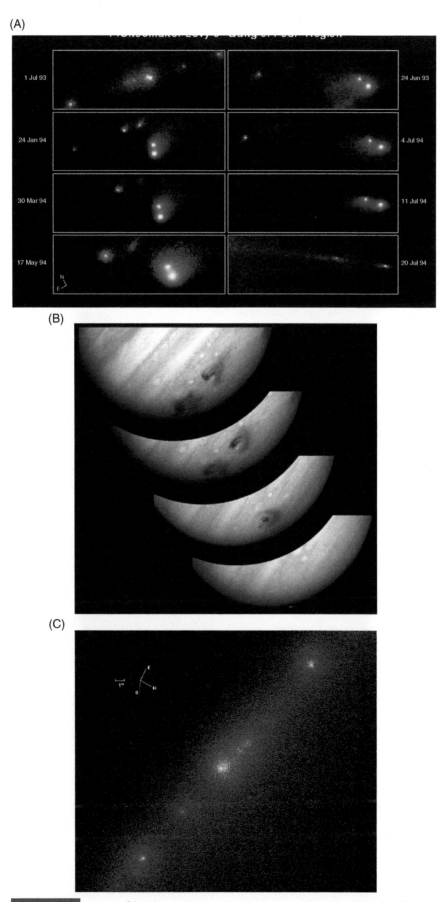

FIGURE 18.8 Images of the Shoemaker–Levy 9 comet impact with Jupiter: (A) Series of images showing the fragmentation of the comet. (B) Successive images of Jupiter showing the scars of the impact (maroon spots in the Southern Hemisphere) fading away over a 14-month period. (C) Image showing the chain of comet fragments at impact. *Source*: Courtesy of NASA.

CASE STUDY 18.1 The Chicxulub Impact

Earth-Changing Event

Walter Alvarez was a geologist who set out to determine why the dinosaurs went extinct. He traveled around the world sampling the sedimentary rock sequences that crossed the Cretaceous–Tertiary time transition, which is when the dinosaurs went extinct. He found anomalous amounts of the element iridium in these rocks, and on the basis of further investigation, he proposed that a large asteroid collided with the earth at that time. This led to a concerted worldwide effort that documented one of the largest extraterrestrial impacts during the time when there was abundant life on the earth. Whether the impact caused the extinction of the dinosaurs, is still debatable.

The main reason that this catastrophic event was not identified prior to this is that the impact crater is on the seafloor. The crater was identified on the Yucatán Peninsula in the southern Gulf of Mexico using geophysical techniques that imaged the underwater and underground structure. These methods located the buried semicircular structure as large as 93–112 mi (150–180 km) in diameter and 12 mi (20 km) deep (Figure 18.9). The crater had actually been drilled by oil companies who identified melt rocks and breccias, but interpreted them as originating from volcanic activity. Borehole samples from depths of 0.75 and 0.8 mi (1.2 and 1.3 km) into the crater were re-examined and found to contain compelling evidence of an impact origin. Another well penetrated 820 ft (250 m) of breccia (crushed rock) followed by 1247 ft (380 m) of melt rock and the well bottomed in anhydrite at 5350 ft (1631 m) depth.

The crater was named Chicxulub which is Mayan for "tail of the devil." The crater is buried under 0.6 mi (1 km) of sediments, which is why it took time and effort to explore. The Chicxulub Crater is a peak-ring-type impact structure. If a crater exceeds 1.2–2.4 mi (2–4 km) in diameter, there is a central peak on the crater floor and the crater walls collapse into a series of benches. The damage zone beneath the crater penetrates into the continental crust about 6.2–18.6 mi (10–30 km). The age of the impact was dated at 66 million years old, the Cretaceous–Tertiary (K-T) boundary.

The impact generated deposits all around the Gulf of Mexico and beyond. For example, there is an odd deposit of that age in Haiti that is 17 in. (0.5 m) thick. It is composed of melt spherules and a large amount of shocked quartz, both of which indicate it was ejected from an impact crater. Indeed, shocked quartz of this age is found as far away as the New Jersey coast.

Arroyo el Mimbral in Mexico also contains complex sediments that are several yards (meters) thick of that age. The lower part is composed of altered impact melt spherules overlain by sand with abundant plant debris. The seafloor at that location was under 1625 ft (500 m) of water, so plant debris should not have been there unless something swept it there. The top of the deposit is enriched in iridium, another indicator of impact. There are other thinner deposits in Texas and Alabama that also reflect a major disruption.

The Chicxulub impact was modeled based on the size and shape of the Chicxulub Crater, the deposits around the Gulf of Mexico, and observation of other impacts especially on Jupiter. It is thought that the Chicxulub impactor was a fragment of a larger asteroid of about 106 mi (170 km). The fragment that caused the Chicxulub Crater was 7–50 mi (11–81 km) across and had a mass of 1.1×10^{12} to 5.1×10^{14} tons (1.0×10^{15} to 4.6×10^{17} kg). There is evidence that other fragments struck across the Atlantic Ocean in a row similar to the observed impact of the Shoemaker–Levy impact on Jupiter in 1994. The velocity of impact is estimated at 12.4 mps (20 kmps) or 44 640 mph (72 000 kmph). The explosion of impact delivered an estimated energy between 1.3×10^{24} and 5.8×10^{25} joules, which is the equivalent of 21–921 billion Hiroshima atomic bombs or 100 million megatons of TNT.

The explosion generated shock waves about equivalent to a magnitude-10 earthquake. The temperature at the core of this explosion was in excess of 18 000 °F (10 000 °C). It would have produced a shock wave and air blast that radiated in all directions with winds of more than 621 mph (1000 kmph) at the impact site. The pressure pulse and winds scoured soils, shredded vegetation, and killed all animals in a blast zone 1118–2237 mi (900–1800 km) across. Debris from the impact rained down across the Gulf of Mexico and entire Caribbean region. Depending upon distance, it ranged from rocky rubble to

(A)

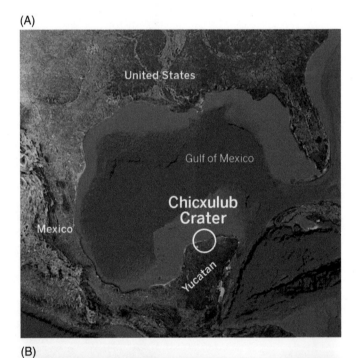

(B)

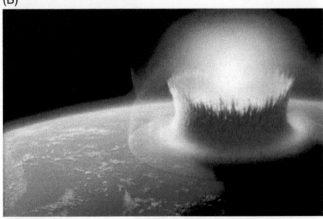

FIGURE 18.9 (A) Map showing the location of the Chicxulub impact crater in the Yucatán Peninsula, Mexico. (B) Illustration of the impact of the interpreted Chicxulub impact. *Source*: National Science Foundation.

impact-melted spherules. This impact ejecta was more than 1000 ft (305 m) thick near the impact site and decreased with distance away.

Impact-generated tsunamis shot across the Gulf of Mexico, devastating nearby coastlines, but they also continued across the developing Caribbean and Atlantic basins. The wave height estimates range from 164 to 984 ft (50–300 m) depending upon source and distance from the impact. They crashed onto the Gulf of Mexico shorelines and penetrated as much as 62 mi (100 km) inland before sweeping continental debris back into the sea. The waves and backwash deeply eroded the seafloor and the shock waves caused widespread landslides on the seafloor.

The impact also produced extensive environmental effects including acid rain, wildfires, dust and aerosols, ozone destruction, and greenhouse gases. Acid rain was from nitric acid produced by heating the atmosphere and sulfuric acid directly from the asteroid and from vaporizing sulfur-bearing minerals beneath the impact. The effect of acids on rocks and life is observed in the sedimentary deposits of that time. Excessive soot with geochemistry of burning biological material is also abundant in these deposits, but the location and extent of these fires is still a matter of speculation.

The dust and aerosols from the impact and soot from wildfires choked the atmosphere. The fossil record shows that photosynthetic plankton which comprise the base of the marine food chain suffered a worldwide extinction event at this time. It is suggested that sunlight was unable to reach the earth's surface through this particulate, shutting down photosynthesis. Shutdown of photosynthesis was also evident in plants on land, and, in turn, the herbivores that fed on the plants and the carnivores that preyed on them all underwent a huge extinction event at this time. This is equivalent to the nuclear winter hypothesized in an atomic war.

The temperature of the atmosphere initially cooled several degrees because of the particulate and aerosols contributing to the extinctions. However, once the dust settled back to the earth, the temperature increased significantly. The cause was the increase in greenhouse gases from the vaporization of the materials at the impact site, and also from the wildfires. The estimates from this global warming range from 2.7 °F (1.5 °C) to as much as 13.5 °F (7.5 °C), which would cause a huge disruption of the remaining ecosystems.

The controversy is on the absolute effects of this disaster, not the disaster itself. There is no question that it produced a worldwide extinction of non-avian dinosaurs as well as numerous marine and terrestrial animal and plant groups. The question is whether this event was responsible for the extinction of all dinosaurs. The dinosaurs did go extinct around this time and there is no question that Chicxulub contributed to their demise. However, there is evidence that some of the major dinosaur groups existed for more than 10 000 years after the disaster. In addition, many animals that should have gone extinct from this calamity by virtue of their fragility, such as frogs, did not. This controversy may never be resolved.

CASE STUDY 18.2 1908 Tunguska Asteroid Event, Siberia

Modern Atmospheric Explosion

There was an even larger impact event than the recent 2013 Chelyabinsk event, but it was in 1908 in a remote area of Siberia (Figure 18.10). It was long enough ago that the level of scientific detail that could be obtained paled in comparison to the level that can be achieved today. Further, the area was so sparsely populated that there were fewer personal observations than desirable. In addition, this was an apparent

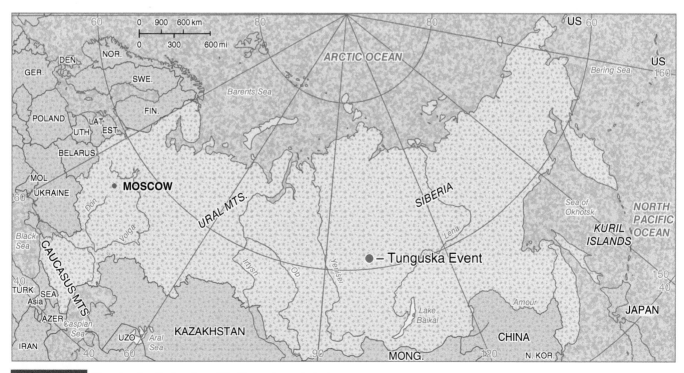

FIGURE 18.10 Map showing the location of the Tunguska explosion.

FIGURE 18.10A Photo showing the trees blown down by the Tunguska explosion. *Source*: Bettmann/Getty Images.

atmospheric explosion rather than a surface impact. For that reason, much of the information on size, speed, altitude, and other parameters had to be derived from modeling rather than direct observation or sensing. However, this is the best even approximate information available on such an event. Finally, because this area is so sparsely populated, this is not really a natural disaster for humans, though a large amount of wildlife perished.

On 30 June 1908, at 7:17 a.m., residents northwest of Lake Baikal in the Eastern Siberian taiga of Russia witnessed a column of blue light crossing the sky, as brilliant as the sun. At about 7:27 a.m., there was a blinding flash and a loud explosion. Nearby eyewitnesses reported that the source of the sound was moving from east to north across the area. There was also a shock wave that emerged from the explosion that knocked people down and broke windows for hundreds of miles in all directions.

The blast leveled an area of forest that was 830 mi^2 (2150 km^2), in a butterfly shape with a wingspan of 43 mi (70 km) and a body length of 34 mi (55 km). This area included an estimated 80 million trees (Figure 18.10A). The air blast from the explosion registered on sensitive barometers as far away as England. Dense clouds of dust developed over the area at high altitudes which reflected sunlight from beyond the visible range. As a result, the night skies glowed in Europe and Asia. There were even reports that the glow was so bright that people in Asia could read newspapers outdoors until midnight.

There were numerous expeditions to the area to find the crater and asteroid. Several dozen pothole-shaped bogs between 33 and 164 ft (10–50 m) in diameter were found and thought to be meteorite craters. They were not. A detailed aerial photographic survey of 97 mi^2 (250 km^2) of the area of leveled forest was conducted, but still, no evidence of the impact crater or asteroid/meteorite was ever found.

Modeling exercises were conducted based on the observations and condition of the blast zone to determine the mechanics of the event. The most accepted scenario is that the explosion was caused by the airburst of a stony meteoroid about 160–620 ft (50–190 m) in diameter and 220 million lb (100 million kg) in mass. The asteroid entered the earth's atmosphere at a speed of about 34 000 mph (55 000 kmph). The heat generated by air compression in front of the asteroid was immense causing it to heat to more than 44 500 °F (24 710 °C) and the asteroid to explode in midair. It is estimated that the asteroid exploded at an altitude from 4 to 6 mi (6–10 km) to as much as 15 mi (24 km).

It is estimated that the power of the explosion was equivalent to 10–15 megatons of TNT (42–63 pJ) to as much as 30 megatons of TNT (130 pJ) depending upon the model. The 30-megaton (130-pJ) estimate is approximately equal to the energy of about 2000 Hiroshima atomic bombs. The shock wave from this airburst would have registered a magnitude 5.0 on seismographs. An explosion of this magnitude would be capable of destroying a large metropolitan area. Fortunately, this has never happened. Recent estimates suggest that Tunguska-sized events occur once every 1,000 years. By contrast, an average of 5-kt airbursts are annual events.

Another model that closely matches the observed event and results is for an iron-rich asteroid as large as 650 ft (200 m) in diameter to have traveled at 7.1 mps (11.2 kmps) and glancing off the earth's atmosphere before returning into a solar orbit. The explosion was the glancing blow of the asteroid rather than its destruction. This would explain why no debris was found.

When the Tunguska asteroid exploded over Siberia, it likely generated as much as 33 million tons (30 million metric tons) of nitric oxide (NO) into the stratosphere and mesosphere. Modeling of the photochemical effects of the event indicates that as much as 45% of the stratospheric ozone in the Northern Hemisphere may have been lost by the nitric oxide cloud in 1909 and significant reduction of ozone may have persisted until 1912 or longer.

CASE STUDY 18.3 2013 Chelyabinsk Oblast Event

Minor Impact on a City

The earth is continuously bombarded by meteorites, the vast majority of which are very small and nearly 75% land in oceans. Although extraterrestrial impacts have the potential to be the most destructive natural disasters, they are so infrequent that they are rarely even considered. Asteroid impacts on the earth with potential for major disasters strike once every 100 000 years and comets strike approximately every 500 000 years. There are very few in recent history that had any impact at all.

A very recent exception to that occurred at about 9:20 a.m. on 15 February 2013 in Chelyabinsk Oblast, Russia (Figure 18.11). A small asteroid, about 66 ft (20 m) across and having a mass of about 13 000–14 000 tons (12 000–13 000 metric tons), exploded in the sky above Korkino about 25 mi (40 km) south of Chelyabinsk and at a height of about 14.5 mi (76 000 ft, 23.3 km). The asteroid was traveling about 41 750 mph or 11.6 mps (about 67 000 kmph or 18.6 kmps), which is about 60 times the speed of sound. The friction with the atmosphere at this exceptional speed caused the body to superheat which, in turn, generated the explosion.

The explosion was accompanied by a bright flash that lasted about five seconds (Figure 18.11A). This turned the body into a brilliant superbolide meteor that was brighter than the sun and visible as far as 62 mi (100 km) away over the Southern Ural region. The explosion had an equivalent blast yield of 400–500 kt of TNT (about 1.4–1.8 pJ), which is 26–33 times the energy released from the Hiroshima atomic bomb. It generated a cloud of hot dust and gas that penetrated to 16.3 mi (26.2 km) altitude (Figure 18.11B). It also generated an airburst causing a powerful blast wave that traveled 18.5 mps (30 kmps) and impacted the earth's surface. When this blast wave hit the ground, it produced a seismic wave which measured a magnitude 2.7 on seismographs. These sonic blast waves circled the globe for distances as far as 53 000 mi (85 000 km).

The blast wave reached the Chelyabinsk area in about 2.5–3 minutes. The explosion broke the asteroid into two large fragments and numerous small ones. The large fragments flared at about 15 mi (24 km) altitude, one breaking up at 11.5 mi (18.5 km) and the other continuing to glow to a height of 8.5 mi (13.6 km). This large meteoroid continued its trajectory into frozen Lake Chebarkul, where it punched a hole in the ice. This part of the path was recorded on video by local residents.

The Chelyabinsk Oblast area was pelted with asteroid fragments measuring 0.39–1.97 in. (1–5 cm) spread around 0.62 mi (1 km) from Chebarkul. However, these and a large fragment in the lake did not cause appreciable damage. The shock wave, on the other hand, caused considerable damage. Approximately 7200 buildings in six cities across the region sustained damage from the shock wave (Figure 18.11C). The buildings included about 6040 apartment blocks, 293 medical facilities, 718 schools and universities, 100 cultural organizations, and 43 sport facilities. This damage was valued at more than 1 billion rubles or approximately $33 million in 2013 US dollars.

The blast also caused injuries to 1491 people in Chelyabinsk Oblast. Most of the injuries were from flying glass as the shock waves burst most widows across the blast zone. However, there were other injuries from collapsing buildings. In total, there were 112 people hospitalized including two people in serious condition. No one perished making the designation as a disaster questionable, but it is the best documented and studied asteroid impact.

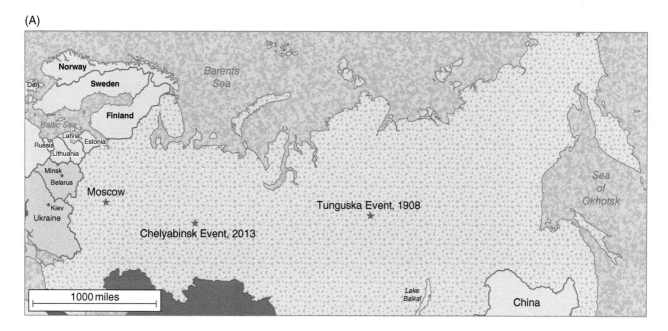

FIGURE 18.11 (A) Map showing the location of the Chelyabinsk impact. (B) Map showing damaged areas around the Chelyabinsk impact site (B) Map showing damaged areas around the Chelyabinsk impact site. *Source*: Adapted from Popova et al. (2013).

(B)

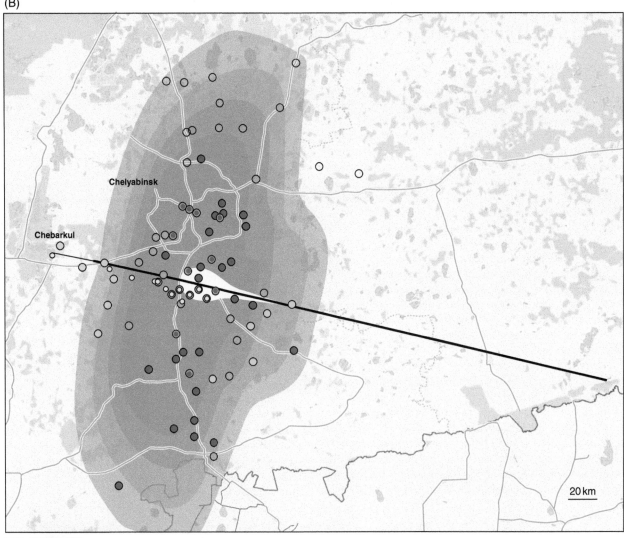

FIGURE 18.11 (Continued)

FIGURE 18.11A Photo of bright glare of the Chelyabinsk bolide in the sky over Siberia. *Source*: Image courtesy of Phys.org.

FIGURE 18.11B Photo of the trail left by the Chelyabinsk bolide in the sky over Siberia. *Source*: Image courtesy of NASA.

FIGURE 18.11C Photo of a building destroyed by the explosion of the Chelyabinsk bolide in Siberia. *Source*: Image courtesy of AFP/Getty Images.

Reference

Popova, O.P. and 50 others, (2013) Chelyabinsk Airburst, Damage Assessment, Meteorite Recovery, and Characterization; Science, 342 (6162), 1069–1073, DOI: 10.1126/science.1242642.

CHAPTER 19

Climate Change Dynamics

CHAPTER OUTLINE

19.1 What Is Climate Change? 379

19.2 Can Volcanoes Change Climate? 379

19.3 Global Cooling 379

19.4 Global Warming 386

19.5 Volcanoes During Modern Human Development 388

Earth's Fury: The Science of Natural Disasters, First Edition. Alexander Gates.
© 2022 John Wiley & Sons Ltd. Published 2022 by John Wiley & Sons Ltd.
Companion website: www.wiley.com/go/gates/earthsfury

Words You Should Know:

Aerosols – A suspension of very fine solid particles or liquid droplets in the air.

Climate change – A long-term change in global or regional weather or climate patterns.

Dead zones – Areas of the ocean where overfertilization causes algal and microbial proliferation that reduces the oxygen content enough to kill or drive away most marine life.

Greenhouse gas – A gas in the atmosphere that permits the input of sunlight, but which traps heat in the troposphere.

Hypoxia – Low levels of oxygen in surface waters.

Industrial effect – The replacement of radiogenic carbon by dead carbon from the burning of fossil fuels.

Stratosphere – A layer in the atmosphere above the troposphere where aerosols can dissipate incoming sunlight.

19.1 What Is Climate Change?

Climate change has become our greatest current environmental concern, and one of the greatest controversies. Without a doubt, the earth's climate naturally changes with time. There were several ice ages in recent geologic history separated by warmer periods. The polar ice caps completely melted at various times, and at other times the whole earth was significantly cooler, called "snowball earth." The current controversy centers on whether humans are causing the current pronounced global warming, and it generally breaks along political and ideational lines.

Humans are increasing the amount of CO_2 in the atmosphere. CO_2 from biological sources has a radiogenic component used in C^{14} dating, but it is absent from the burning of fossil fuels. This was discovered in the early 1950s and termed the "industrial effect." The quantity of anthropogenic CO_2 can be determined air from anywhere on the earth that is not receiving direct input from the burning of fossil fuels. This is done by analyzing it for radiogenic versus non-radiogenic carbon. This proves that the radical increase in CO_2 in the atmosphere is from burning of fossil fuels. Compounding this is the reduction of carbon sinks that would remove CO_2 in the atmosphere and store it. Deforestation and destruction of marine environments among other human activities are exacerbating the situation.

It is important to understand the short-term and medium-term natural sources of climate change to put the anthropogenic changes into perspective. This chapter focuses primarily on volcanic input to the atmosphere, but also the lessons we can learn from natural climate change. There have also been many other sources of natural climate change in this book such as impacts, but they will not be reiterated.

19.2 Can Volcanoes Change Climate?

Global climate change concern focuses on human influence on the chemistry of the atmosphere, but volcanoes are capable of an even greater effect. In addition to lava and solid ejecta, volcanoes also emit large amounts of gas composed of several chemical compounds depending on the volcano. Some of these compounds, such as methane and carbon dioxide, are strong greenhouse gases that cause global warming. Greenhouse gases increase the temperature of the atmosphere by allowing sunlight to reach the surface, but not allowing the heat produced to escape, similar to a glass greenhouse.

Other volcanic gases, such as sulfur compounds, form aerosols and cause global cooling by diffusing and reflecting incident sunlight. The injection of large quantities of sulfur-, chlorine-, and nitrogen-bearing compounds to the stratosphere through a summit volcanic eruption can deplete ozone which further impacts global temperature over a longer period. Volcanic ash is added to the atmosphere by the same mechanism and blocks sunlight from reaching the surface, thereby cooling the atmosphere as well. The complex chemical and physical impacts have short-term effects lasting up to four or five years for recent volcanoes, but there are potentially longer-term effects for extended periods of volcanism. There are well-documented examples of both cooling and warming as a result of volcanic eruptions.

19.3 Global Cooling

The best example of global cooling resulting from a volcanic eruption was from Tambora in 1815. Tambora is a large, active stratovolcano in Indonesia. It produced a massive summit eruption in April of 1814 that produced 38 mi³ (160 km³) of ejecta. A portion of the ash was shot 27 mi (43 km) into the stratosphere where it spread around the entire earth. It also injected sulfur aerosols in the stratosphere which, in addition to the ash, diffused enough incident sunlight to lower the average temperature of the earth significantly. The results of this cooling produced natural disasters in the Northern Hemisphere for years.

CASE STUDY 19.1 1815 Tambora, Eruption

The Year Without a Summer

The 13 000-ft (3960-m) Mount Tambora in Indonesia was believed to be an extinct volcano, but it became active sometime in 1814 when it began to emit minor ash eruptions (Figure 19.1). In April 1815, the eruptions grew much more vigorous. During the night of 5 April, a strong earthquake struck the area and was followed by several explosions and an eruption of ash. The explosions were heard as far as 870 mi (1400 km) away. Ash darkened the sky for 300 mi (483 km) away and fell from the sky as far as 800 mi (1287 km) away. By the following morning, the eruption had become small detonations and ash emissions that drifted around the area. The eruption, however, intensified again on 10 April when explosions were heard 1600 mi (2600 km) away and mistaken for cannon fire by foreign governments. Fragments of pumice as large as 7.9 in. (20 cm) were expelled around 8 p.m. and rained down on the area followed by ash emissions and fallout at around 9–10 p.m. Three separate eruption columns merged into a single huge column that extended 26.7 mi (43 km) high (Figure 19.1A). Ash fallout occurred as far as 810 mi (1300 km) from the volcano. Glowing pyroclastic flows sped down the slopes of the volcano as well. These explosions continued on 11 April and the ash eruptions continued until 17 April before ending.

The 1815 eruption of Tambora was the largest in recent history. The Volcanic Explosivity Index (VEI) of the eruption was 7 and it produced an estimated 38 mi³ (160 km³) of pyroclastic ejecta. A tsunami was produced from the explosion with heights up to 13 ft (4 m) and the eruption itself caused the deaths of about 11 000 people (Figure 19.1B). The ash was highly sulfurous, and between the blanketing effect of the ash and sulfur damage, it caused famine and disease that led to the deaths of 50 000–80 000 additional people over the ensuing months.

The greatest impact of the Tambora eruption was the climate change it caused during the following years. The sulfur-rich emissions were injected high into the atmosphere where sulfurous aerosols were spread around the earth. Estimates of the amount of sulfur compounds released during the eruption vary from 10 to 120 million tons (9–108 million MT) based on ice core studies. The ash and gas were injected to 33 000–98 000 ft (10–30 km) altitudes in the atmosphere. These stratospheric aerosols diffused incoming sunlight, so it could not reach the earth. This caused average global cooling by 0.7–1.3 °F (0.4–0.7 °C) over the next year, but it locally changed up to 5.4 °F (3 °C) and the effects persisted for a decade. In addition, there was a persistent "dry fog" in the lower atmosphere in the Northeastern United States during the spring and summer of 1815.

Tambora caused the strongest documented global cooling from a single eruption. 1816 has been called the "year without a summer" because of the extreme extended cold temperatures. Europe and North America experienced freezing temperatures well into June. There was snow in New England into June and freezing temperatures as far south as Virginia until early July. There was even a snowstorm up to 13 in. (32 cm) in August. The late snows shortened the growing season in New England to less than 80 days and killed crops and livestock crippling the food supply. There were food shortages and grain prices in some areas increased as much as sevenfold as a result. Food shortages in Europe

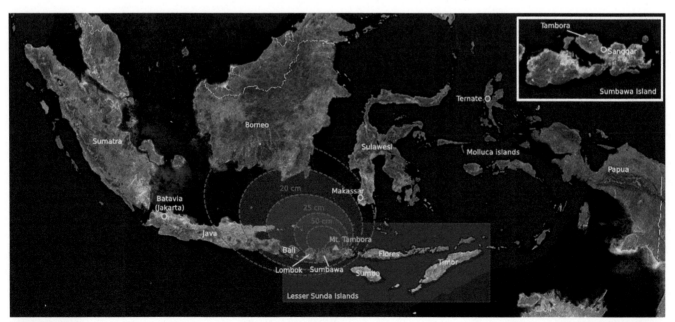

FIGURE 19.1 Map of Indonesia and the surrounding areas showing Mount Tambora and the eruption clouds. Inset is an enlargement of the Island of Tambora. *Source*: NASA/Wikimedia Commons/Public Domain.

FIGURE 19.1A Satellite image of the caldera at the summit of Mount Tambora. *Source*: Courtesy of NASA.

FIGURE 19.1B Photo of the remains of natives killed in the pyroclastic flow from the Mount Tambora eruption. *Source*: Courtesy of Rik Stoetman.

resulted in famine, civil unrest, and death. Approximately 200 000 people died in Europe as a result of the cold weather. In the Asian subcontinent that year, monsoon rains were later and heavier than usual resulting in a severe cholera epidemic. The global death toll resulting from the climate effects of Tambora is estimated in the millions.

During the winter of 1817, temperatures were 30 °F (34 °C) below normal in central and northern New York where lakes and rivers normally used to transport supplies froze over. Cold temperatures and heavy precipitation caused harvests to fail in the British Isles. Famine in Wales and Ireland resulted from failure of wheat, oat, and potato harvests. Crop failures in Germany resulted in high food prices and demonstrations, riots, arson, and looting at markets, and they spread to other European cities. This caused the worst famine of the nineteenth century.

Tambora was the most extreme documented case of global cooling from a volcanic eruption in recent history, but there are other examples. The first recognition of volcanoes impacting climate was in 1783 with the large eruptions of Laki in Iceland and Asama in Japan. The Laki eruption was the largest outpouring of basalt in historical times. It emitted 3.4 mi³ (14 km³) of lava. It also emitted a large amount of fluorine that killed approximately half of the livestock in Iceland and caused the deaths of one-quarter of the human population. Benjamin Franklin described a persistent haze that developed across Europe that year. He was the first to suggest that the source of the weather problems was the Laki eruption. In 1783–1784, the crops failed in Europe because of the acidic deposition from the 120 000 000 tons (1.2 × 10^{11} kg) of sulfur that had been emitted and produced the haze. Adding to the situation was an extraordinarily hot summer that was followed by a harsh winter. It is unclear which of the two eruptions had the greatest impact on climate, or if it was a combination of the two. These extreme weather conditions continued until 1788. The famines that resulted from the extreme weather are estimated to have caused six million deaths and resulted in civil unrest in much of Europe. The unrest led to the French Revolution in 1789.

More recent volcanic eruptions also resulted in global cooling. The eruptions of El Chichón in Mexico in 1982 and Mount Pinatubo, Philippines in 1991 both caused global cooling. El Chichón erupted from 29 March to 4 April 1982 with a VEI of 5. It injected 7.7 million tons (7 million MT) of sulfur dioxide (SO_2) and 22 million tons (20 million MT) of particulate as high as 12–16 mi (20–25 km) into the stratosphere where it spread around the planet. It should have caused an approximate global cooling of 0.5 °F (0.3 °C) over the next year by causing a 16% reduction in incident solar radiation. However, the effects were masked because it coincided with one of the strongest El Niño events of the twentieth century. Three weeks after the El Chichón eruption, New York City was struck by the largest April blizzard on record and the following summer was one of the coldest in the Northeast United States in the twentieth century.

Mount Pinatubo volcano erupted on 15 June 1991 with a VEI of 6 making it among the most powerful in the twentieth century. The eruption released huge amounts of sulfur compounds and ash into the atmosphere which resulted in a 12% reduction in incoming solar radiation. This reduction caused global cooling of about 0.9–1.3 °F (0.5–0.7 °C) over the next year or two.

CASE STUDY 19.2 1991 Mount Pinatubo

Recent Global Cooling

The 15 June 1991 eruption of Mount Pinatubo in the Philippines was the second largest volcanic eruption of the twentieth century (Figure 19.2). It was also the largest eruption to affect a densely populated area with nearly three million people in Central Luzon. It also severely damaged two American military bases which were ultimately abandoned. However, the most significant effect of this eruption was the global cooling it caused over the next year.

The Philippines is the location of numerous natural disasters. Not only is it in the main track of typhoons, it is also the site of a convergent margin that produces major earthquakes and volcanoes. The sequence of events that led up to the eruption began on 16 July 1990 when a magnitude-7.8 earthquake struck with the epicenter about 60 mi (100 km) northeast of the previously inactive Mount Pinatubo on the island of Luzon in the Philippines. The earthquake caused a landslide on Mount Pinatubo and an increase in steam emissions from a nearby geothermal area. However, in March and April 1991, magma began rising toward the surface from 20 mi (32 km) below the volcano, triggering small earthquakes and causing steam explosions on the north flank of the volcano. The small earthquakes continued beneath Mount Pinatubo through May and early June, and sulfur dioxide gas was emitted by the volcano.

Eruptions of Mount Pinatubo began in 7–12 June 1991. At first, they were slow, forming a lava dome. This changed on 12 June, when gas-charged magma caused an explosive eruption. This was just the beginning. On 15 June, the volcano underwent an even more explosive Plinian to ultra-Plinian eruption with a VEI 6 in size (Figure 19.2A). It produced greater than 2.4 mi³ (10 km³) of ejecta. It produced an ash cloud 28 mi (40 km) high. The ash was dispersed through the area by an incoming typhoon, resulting in the spread of volcanic ash and larger pumice lapilli over the countryside. Ash was deposited as far away as the Indian Ocean, and satellites tracked the ash cloud several times around the earth. Then extensive pyroclastic flows raced down the slopes of Mount Pinatubo at 60 mph (100 kmph), filling the surrounding valleys with volcanic deposits as thick as 660 ft (200 m) (Figure 19.2B). At the end of the eruption, the summit collapsed forming a large

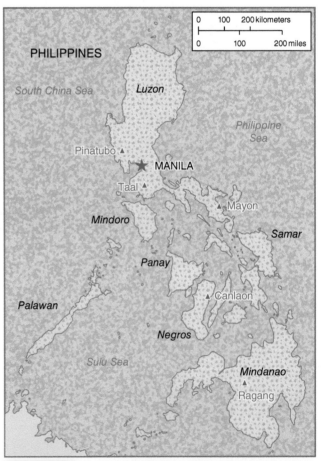

FIGURE 19.2 Map of the Philippines showing Mount Pinatubo.

FIGURE 19.2A Photo of the 1993 eruption of Mount Pinatubo. *Source*: Courtesy of NOAA.

caldera 1.6 mi (2.5 km) wide. From July to October 1992, a second lava dome slowly built in the caldera.

Considering the high population density of the area, the eruption of Mount Pinatubo could have resulted in a catastrophic human death toll. However, the large Clark U.S. Air Force Base sat just on the edge of the slopes of Mount Pinatubo. Further, the Subic U.S. Naval Base was just on the edge of the eruption impact area. For this reason, the eruption was very carefully monitored by the military. As soon as the eruption accelerated, not only did the military remove their expensive weapons, they also recommended evacuation of 200 000 residents from the entire area and 60 000 people complied. This saved a minimum of 5000 human lives. However, even though they were warned about the hazard of the Pinatubo ash cloud, 16 commercial aircraft well west of the Philippines were impacted by flying through an ash cloud and sustained $100 million in damage (Figure 19.2C). The eruption cost $700 million in total damage.

However, the threat was not over, and it was deadly. The nearby Typhoon Yunya struck the area the next day dumping torrential rain on the freshly deposited and very hot ash creating a boiling lahar (Figure 19.2D). The mudflows caused 847 deaths. But this threat was not over. After more than five years, the thick, pyroclastic-flow deposits had kept most of their heat. They still had temperatures up to 900 °F (500 °C) in 1996. Stream water and groundwater still exploded upon contact with the ash and spread fine ash downwind. Ash deposits were also remobilized by heavy rains to form lahars for years.

About 19 000 000 tons (17 000 000 mt) of sulfur dioxide were injected into the stratosphere in the eruptions. This is the largest volume of SO_2 recorded from an eruption by modern instruments. The injection of these aerosols into the stratosphere was the largest since the 1883 eruption of Krakatoa in Indonesia and no eruption has yet exceeded it. The spreading of this stratospheric aerosol cloud around the world caused global temperatures to decrease from 1991 through 1993. The decrease of temperatures in the Northern Hemisphere averaged 0.9–1.1 °F (0.5–0.6 °C) and on a global level was 0.7 °F (0.4 °C).

FIGURE 19.2B Photo of a house covered by ash from the eruption of Mount Pinatubo. *Source*: Courtesy of NOAA.

FIGURE 19.2C Photo of an airplane damaged by ash from the eruption of Mount Pinatubo. *Source*: Courtesy of NOAA.

FIGURE 19.2D Aerial photo of the lahar produced by the 1993 eruption of Mount Pinatubo. *Source*: Courtesy of NOAA.

CASE STUDY 19.3 534–536 AD Unknown Eruption

Invasion of the Huns

Among the greatest impacts of volcanically induced climate change on society may have resulted from an eruption that was not observed. In 534–536 AD, significant changes took place in the global climate, possibly leading to the Dark Ages. This time was marked by signs of a potential major volcanic eruption such as long-term, low-hanging dry haze, reported "days of darkness," and strange weather patterns as reported in historical journals. Tree rings from many places around the world were exceptionally thin indicating slow growth which is typically associated with low temperatures (Figure 19.3A). There are reports of snowfall in Europe and China throughout the summer, massive droughts, and resulting crop failures for many years in some areas and great floods in other areas.

The ensuing chaos led to the downfall of major cities and the demise of cultures in South America, Persia, and Arabia. The Huns fled Central Asia and a major drought to overthrow the Gupta Empire in India, and the Byzantium Empire defeated the Vandal Kingdom in

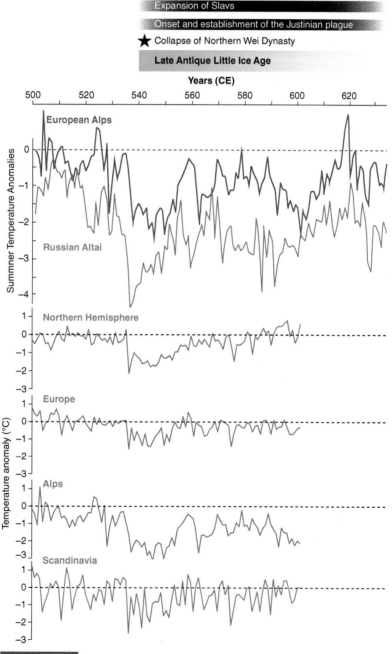

FIGURE 19.3A Graphs of temperature variations at various locations showing a sharp drop in temperatures in the late 530s AD with the timing of several shifts in history also plotted. *Source*: Adapted from Newfield (2018).

Africa. The Roman Empire was divided and Islam expanded, whereas Christianity contracted. A major plague swept across the earth for the ensuing 10 years which caused the deaths of millions of people.

The evidence that the event was volcanic is a huge spike in sulfates in ice cores from Greenland ice sheets in 534–536 AD (Figure 19.3B). High concentrations of sulfates result from the sulfur released during a volcanic eruption, although one study suggests the spike was from an extraterrestrial impact. If the source is a massive volcanic eruption, the source is yet unknown. The suspected volcano has changed with time. The early suspect was a proto-Krakatoa volcano, but it has also been proposed for Mount Rabaul in New Britain, Ilopango in El Salvador, somewhere in North America, and the current main suspect in Iceland. Modeling exercises on the possible eruption predict a VEI of 6 or 7 ultimately including a caldera collapse that resulted from flooding of seawater into the base of the volcano. The phreatomagmatic explosion would have generated a plume in the range of 16 to >32 mi (25 to >50 km) high carrying 12–24 mi³ (50–100 km³) of sulfate aerosols into the stratosphere

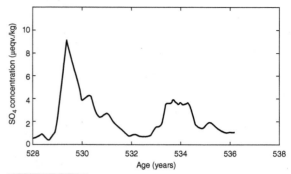

FIGURE 19.3B Graph of spikes in SO_2 in ice cores from Greenland in 530 and 534 AD. *Source*: Adapted from Larsen et al. (2008).

in addition to 18 mi³ (75 km³) of fine ash. This is postulated to have possibly been the largest volcanic eruption in the past 2000 years. A volcanic eruption of this size could have cooled the earth by as much as 5.4–9 °F (3–5 °C) over the next five or more years.

19.4 Global Warming

In addition to global cooling, volcanic eruptions release significant quantities of greenhouse gases such as carbon dioxide, methane, water, and some aerosols that cause global warming. Greenhouse gases allow the passage of incoming short-wave solar radiation, but absorb infrared radiation that is released back to the atmosphere from the earth. This heats the atmosphere causing global warming. The Mount Pinatubo eruption caused warming at 10–15-mi (16–24-km) altitude in the lower stratosphere of up to 3.6–5.4 °F (2–3 °C) within the first four to five months along a band between the equator and 20° north latitude. This complex system of aerosols cooling the atmosphere and gases heating it causes the stratosphere to warm and the troposphere to cool. The troposphere only warms slightly in the winter because circulation in winter is stronger than the aerosol cooling in summer. The overall effect at the surface is that cooling dominates over warming.

Overall cooling is especially pronounced as a result of explosive intermediate and felsic volcanoes that inject ash and gas high into the stratosphere. Eruptions of basalt volcanoes, however, are not explosive and release the majority of their emissions to the troposphere. In the short term, they can cause global cooling as was the case with the 1783 Laki eruptions. The ash and aerosols are washed from the atmosphere by natural processes leaving only the greenhouse gases behind. Individual eruptions inject just a small fraction of the annual greenhouse gases produced by humans (less than 1/1000), but long periods of eruptions, like from flood basalt provinces, can add enough gas to produce significant global warming.

The most devastating global warming produced by volcanic activity was at the end of the Permian Period, 245 million years ago. There was one ocean at the time, and it had poor circulation, making it unable to absorb excess greenhouse gases as is usually the case. A huge flood basalt province in Siberia shortly followed a smaller one in China. The massive release of gas caused a catastrophic collapse of the natural systems that keep the oceans and atmosphere in chemical balance. Ultimately, the collapse caused the extinction of 95% of marine organisms and 75% of terrestrial organisms making it the most catastrophic extinction event ever in the history of the earth.

CASE STUDY 19.4 Natural Global Warming

Permian Extinction Event

The great concern about the current climate change may be well-founded. The Permian extinction, the greatest mass extinction event in earth history, was probably the result of climate change that cascaded into a global catastrophe (Figure 19.4A). This collapse of earth's biogeochemical systems caused the extinction of 96% of marine organisms and nearly 75% of terrestrial organisms. Although a specific set of global conditions were required, the ultimate cause of the catastrophe was erupting volcanoes.

The Permian Period was at the end of the Paleozoic Era, which was characterized by distinct forms of life. When the Permian Period ended about 245 million years ago, there was one continent called Pangea and one ocean called Panthalassa (Figure 19.4B). Circulation in the large ocean was much weaker than it is today. At this time, a major volcanic province developed. First, the Emeishan flood basalt province of South China was extruded (Figure 19.4C). This province covers about 96 525 mi² (250 000 km²). Quickly following this activity was the Siberian flood basalt province or "Siberian Traps" which filled the West Siberian Basin. The volcanic deposits cover about 1.5 million mi²

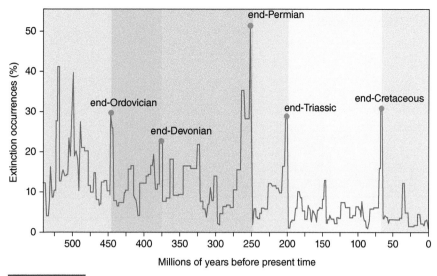

FIGURE 19.4A Graph of mass extinctions over the past 600 million years with the Permian extinction being the largest.

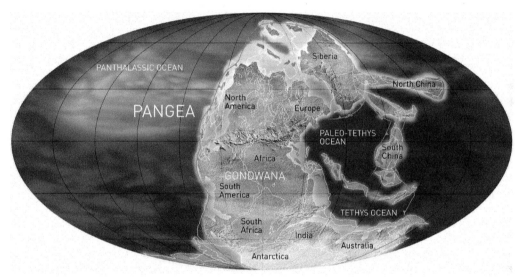

FIGURE 19.4B Map of the world during the Permian Period showing the continent Pangea and Panthalassa Ocean. *Source*: C.R. Scotese and the Paleomap project, USGS.

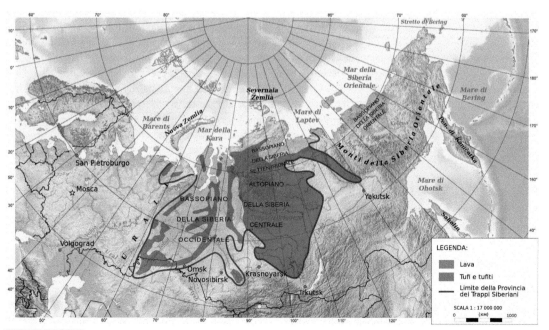

FIGURE 19.4C Map of the location of the Siberian Traps during the Permian Period. *Source*: Creative Commons Attribution-Share Alike 3.0 Unported.

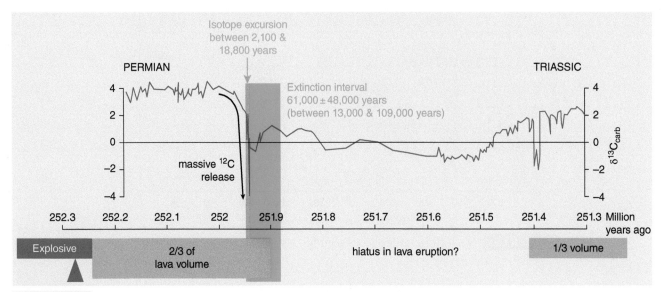

FIGURE 19.4D Graphs of carbon isotope variations through time showing a sharp flooding of the atmosphere in volcanic CO_2 at the end of the major eruption of the Siberian Traps. *Source*: Adapted from Burgess and Bowring (2015).

(3.9 million km²), which is about 15 times the area of the United Kingdom. The province comprises a 2.2-mi (3.5-km)-thick pile of basalts and pyroclastics intruded by gabbro. The volume of magma to produce this sequence was about 0.3–0.6 million mi³ (1.2–2.5 million km³).

The extrusion of the Siberian Traps alone would have released enough CO_2 to cause major global warming. One 96-mi³ (400-km³) flow of basalt degases about 7 gigatons (6.4 GMt) of CO_2 over about a decade (Figure 19.4D). Over the entire formation of the Siberian Traps, about 11 000 gigatons (10 000 gigaMT) of CO_2 was released. This is equivalent to about 5000 ppm of CO_2 added to the atmosphere as compared to the modern-day atmospheric concentration of 370 ppm. This volcanism and degassing took place over 165 000 years, which is geologically quick. The Siberian volcanism alone appears to have led to a doubling of atmospheric CO_2 which would have led to a global temperature increase of 2.7–8.1 °F (1.5–4.5 °C) in addition to that released by the volcanism in China. Increased ocean temperatures may have caused the release of methane that was sequestered on the ocean floor. Methane is a very strong greenhouse gas that further increased temperature of the atmosphere.

The volcanism also released significant amounts of toxic gases including chlorine, sulfur, and fluorine, all of which produce acid precipitation. This would have added stress to the earth's biotic systems. The reduction of incoming solar radiation as a result of aerosols and volcanic haze further stressed the systems. This chemistry promoted the reduction of oxygen solubility in seawater and the release of methane hydrates from the seafloor, further reducing the relative efficiency of global carbon sinks. The lack of ocean circulation allowed these unhealthy areas to build up without being dissipated as usual. Early algal blooms followed by proliferation of oxygen-consuming bacteria resulted in large areas of ocean hypoxia to anoxia otherwise known as "dead zones." The result was the demise of sessile and benthic marine organisms and later, as the dead zones expanded, all marine life was strongly impacted. Only organisms able to tolerate low oxygen conditions survived this event. These catastrophic marine extinctions occurred in just 10 000–30 000 years signifying the collapse of the earth's biologic systems and failure of the regulatory systems.

The reason for the demise of terrestrial biologic systems is less certain but demonstrates a strong interdependence of the marine and terrestrial systems. The reduced sunlight and acid precipitation from volcanic emissions stressed and reduced terrestrial plant life. Prior to this event, vegetation had been in its greatest boom in earth history. The lack of availability of food from the oceans caused the only other mass extinction of insects. Vegetation and insects comprised the primary diet of larger animals of the time. Without food, many of those species went extinct, especially the larger herbivores which had become dependent on the overabundance of vegetation. There was also a major transgression of the ocean which may have contributed to terrestrial system damage. The primary cause of this catastrophe was the massive volcanism and damage it did to the ocean-atmosphere system.

19.5 Volcanoes During Modern Human Development

Modern society began about 10 000 years ago after the retreat of the last ice age. It has been characterized by recording historical events and advancement of technology, religion, and political systems. There have been a few major volcanic eruptions during this time that altered the climate for a short period such as the 535 AD eruption and the 1815 Tambora eruption. The climate responses from these events wreaked havoc on human civilization causing significant death and destruction and shifts in society. These eruptions, however, were small compared to those in Earth's history. Tambora released 38 mi³ (160 km³) of ejecta into the atmosphere and caused significant short-term climate impacts. In comparison, the last Yellowstone supervolcano eruption some 600 000 years ago released 244 mi³ (1000 km³) of ejecta into the atmosphere. The eruption is more than six times as large as Tambora and the climate impact must have been devastating. More recently, the larger eruption of Toba, Indonesia released 700 mi³ (2800 km³) of ejecta into the atmosphere about 74 000 years ago likely causing an even stronger climate impact. It is possible that such a catastrophic eruption could collapse society.

There are no currently active flood basalt provinces on the earth. Emission of quantities of greenhouse gases that could cause significant global warming through volcanic eruptions is therefore unlikely for the foreseeable future. The Siberian volcanism took 165 000 years to accumulate the greenhouse gases to catastrophically alter the climate necessary to cause the Permian extinction. The current climate concerns should focus on the two main catastrophic changes that drove the Permian extinctions: (i) global warming from the release of greenhouse gases and (ii) large scale marine hypoxia. Humans release greenhouse gases to the atmosphere through combustion of fossil fuel at a far greater rate than even the largest volcanic eruptions.

The overuse and runoff of agricultural fertilizer coupled with the fallout of nitrogen-compound air pollution created extensive "dead zones" in coastal waters worldwide. These hypoxic areas have become epidemic over the past 25–30 years. Human impact on the environment is creating the conditions that caused the greatest extinction event of all time but at a far faster rate.

Eutrophication of freshwater water bodies is widespread. Oceans were previously considered to be too large and have too good circulation to be impacted by eutrophication. However, in the 1960s, a "dead zone" developed in the Black Sea, USSR. It was termed a "dead zone" because the hypoxic conditions killed the sessile and limited-mobility fauna, and the mobile fauna fled leaving the area lifeless. The cause of the Black Sea problem was runoff of the excessive agricultural fertilization. After precipitation, nutrients flooded the streams leading to the Black Sea. These nutrients triggered blooms of phytoplankton increasing the increased food supply. When the phytoplankton died, they sank, and bacteria proliferated from the increase in food supply and used up the oxygen in the water. In these dead zones, oxygen concentrations are <0.5 ppm, compared to seawater which contains 10 ppm.

References

Burgess, S.D. and Samuel A. Bowring (2015) High-precision geochronology confirms voluminous magmatism before, during, and after Earth's most severe extinction; Science Advances, Vol 1, Issue 7, Doi: 10.1126/sciadv.1500470.

L. B. Larsen, B. M. Vinther, K. R. Briffa, T. M. Melvin, H. B. Clausen, P. D. Jones, M.-L. Siggaard-Andersen, C. U. Hammer, M. Eronen, H. Grudd, B. E. Gunnarson, R. M. Hantemirov, M. M. Naurzbaev, K. Nicolussi (2008) New ice core evidence for a volcanic cause of the A.D. 536 dust veil; Geophysical Res. Letters, Vol. 35, No. 4; https://doi.org/10.1029/2007GL032450.

Newfield, T.P. (2018). The Climate Downturn of 536–50. In: White, S., Pfister, C., Mauelshagen, F. (eds) The Palgrave Handbook of Climate History. Palgrave Macmillan, London. https://doi.org/10.1057/978-1-137-43020-5_32.

Index

NB Page locators in *italic* type indicate illustrations, those in **bold** indicate tables

acid precipitation, 45, 91, 371
Africa, 191, 235, 344
    desertication, 344
    source of hurricanes, 268–269, 273, 276, 285
    tsunamis, 166
    volcanoes, 98
aftershocks, 104, 116
Agadir, Morocco, 1960 earthquake, 138–140, *142*
    death toll, 139
    effects, 138, *139*, *140*
    epicenter, 138
    shaking intensity, 138
    magnitude, 138
air pressure, 223, 341
Aitape earthquake and tsnuami *see* Sissano, Papua New Guinea, 1998 tsunami
Alaska, 83, 86
    ash fields, 83–85, *84*
    tsunamis, 175
    see also Good Friday 1964 earthquake, Alaska; Nova Erupta, Alaska, 1912 eruption
Aleutian Subduction Zone, 142–143
alluvial fans, 215, 217, 218, **218**
Altai Mountains, China, 91
Alvarez, Walter, 370
American Association for the Advancement of Science (AAAS), 194
Anak Krakatau, *78*
Ancash, Peru, 1970 earthquake, 149–153, *150*
    death toll, 151
    effects, 149–151, *151*, *152*, *153*
    epicenter, 149
    magnitude, 149
    shaking intensity, 149
Ancient Greece, 52–53
andalusite, 44
Andean margins, 44
Andean orogeny, 23
andesite, 33, 34, 38, 42, 44
    volcanoes, 51
Andes Mountains, 44
Apennine range, Europe, 191
Appalachian Mountains, 30, 245–246
Armero, Colombia, 79, 81
asteroids, 363–364, 367–368, 370–372
asthenosphere, 17, 18, *18*, 104
Atlantic Ocean, 22, 23, 175

atmosphere, 17, 222
    acid rain, 91
    aerosols, 45, 371, 379, 385
    carbon dioxide in, 45, 379
    humidity, 222, 227, 268
    temperature, 222, *222*
    volcanic effects on, 45, 57, 81, 91
atmospheric pressure *see* air pressure
Australia, 23, 261
avalanches, 10, 203, 209, 210
    debris, 209, 210
    ice-rock, 150, *151*, *152*, *153*
    snow, 209, 211–214
    velocities, 210

Bangladesh, 11, 293
Ban Ki-moon, 13
Banqiao Dam, China, 1975, 334, *335*
    disaster, 335–337, *336*, *337*
barrier islands, 251, *253*, 258
Barringer Crater, Arizona, 366, *366*, *367*
basaltic volcanism, 35, 54, 94, 386
    rocks, 33, 35–36, 37, *91*, 389
bathymetry, 251, *252*
beaches, 251, 253, 258, 260
Black Sea, 389
Boxing Day tsunami *see* Indian Ocean, 2004 tsunami
breakwaters, 260, *260*
brittle deformation, 103–104
Bruce, Victoria, *No Apparent Danger*, 70

Calcutta, India, 1737 disaster, 14
calderas, 51, 68, 75
California, 9, 11
    plate tectonics, 23
Cameroon volcanic line, 98, *98*
Canary Islands, 175, *175*
carbon dioxide
    as greenhouse gas, 379, *387*, 388
    and fossil fuels, 379
    volcanic, 45, 98, 100, *387*
Caribbean plate, 12
Cascade Mountains, North America, 23
Cascadia, British Columbia, 1700 earthquake, 109–110
    magnitude, 110
    and orphan tsunami, 109, *109–110*

Earth's Fury: The Science of Natural Disasters, First Edition. Alexander Gates.
© 2022 John Wiley & Sons Ltd. Published 2022 by John Wiley & Sons Ltd.
Companion website: www.wiley.com/go/gates/earthsfury

Cascadia Subduction Zone, 23, 105, *109*, 110
Central America, damage from Hurricane Mitch, 276–278, *277*, *278*
Central American Subduction Zone, *129*, 130
Central United States, 347–348, *348*
    agriculture, 347–348, 351
    see also Dust Bowl disaster, United States, 1930s
Ceres asteroid, 364, *365*
Chelyabinsk Oblast, Russia, 2013 bolide event, 371, 373–375, *373–374*
    casualties, 373
    effects, 373, *374*, *375*
Chesapeake Bay crater, 367, *368*
Chicxlub Impact, Mexico, 370–372
    crater, 366–367, 370
    modeling, 370–371
China, 27, 331, 355
    Cultural Revolution, 337
    dam construction, 334, *335*
    drought events, 355–357, *355*
    earthquake events, 8, 9, 29–30, 136–149, 190–194
    flooding events, 331–337
cinder cones, 49, *50*, 51
Clark, William, 108
climate, effect of altitude on, 222, 223
climate indicators, 345
climate change, 379–389
    ancient, 366–367, 386–389
    anthropogenic, 379
    and atmospheric circulation, 260, 329
    evidence for, 383, 385, *386*, 387
    glaciation, 19, 225
    global cooling, 45, 91, 371, 379–386, *385*
    global warming, 45, 91, 355, 371, 386–389
    natural, 379–389
    and ocean circulation, 260, 261
    precipitation, 329
    and volcanic eruptions, 91, 92, 379–389
climate systems, global, 19, 223–226, *225*
    convection, 223–225, *224*
    ocean currents see ocean currents
Clinton, Bill, 13
clouds, *224*, *226*, 227, 228, 232, 297–299
    and derechos, 241
    and fronts, 227, 228, 233, 297
    and hurricanes, 269
    lightning, 232, 235
    pyroclastic see pyroclastic surges
    and thunder, 228, 229
coastlines, 252, *253*, 258, *259*, 260, *260*
    concave, 252, 256
    continental shelf, 251, *252*
    convex, 252, 256
    jagged, 252
    tidal flats, 253
    and wave energy, 252–253, *255*
    see also beaches; breakwaters; deltas; littoral drift
columnar joints, 35–36
comets, 363, 364, *365*, 367
Concert for Bangladesh, 1970, 293, *294*
conduit pipes, 50, 51
continental crust, 18, *18*, 43
continental rift, 17, 20, 35, 36

convection cells, *22*, 225–226, *226*, 297
    at convergent margins, 22
    at divergent margins, 20
    at transform margins, 23
    see also Coriolis Effect, the
convergent margins, 17, 22–23, 36–37, 103
    types, *21*, 23
Coriolis Effect, the, 222, 223, *223*, 225, 269
crustal contamination, 41, 43, *44*
Cuba, 12, 245, 270, *270*
Cumbre Vieja Volcano, Canary Islands, 175–180, *175*
    hypothetical tsunami, 175, *177*
    volcanic eruptions, 175, 176
Cyclone Bhola, 1970, 268, 291–294
    death toll, 293
    disaster response, 291, 293
    effects, 291, 293, *293*, *294*
    political effects on Pakistan, 293
    storm surge, 291
    track, 291, *291*, *292*
Cyclone Nargis, 2008, 288–291, *289*
    death toll, 289
    development, 288
    disaster response, 289, 291
    effects, 289, *290*, 291
    storm surge, 289
    track, 288–289, *288*
    wind speeds, 288, 289
cyclones, 14, *266–267*, 268
    types, 268
    see also hurricanes

dacite, 33, 34, 44
Dashiqaio, China, 191
dead zones, marine, 379, 388
    and agricultural practices, 389
death tolls, accuracy of, 8, 9, 10–11
Decade Volcanoes, 49, 52–53, *53*, 68
decompression melting, 33, 37, 38
Deep-Ocean Assessment and Reporting of Tsunamis (DART) program, 164
deformation, of rocks, 103–104
    plotting, 120, *122*
    types, 103–104
Delta Flight 191 air crash, 1985, 232–234, *234*
deltas, 251, 253, 288
Denver, 2017 hailstorm, 237–241, *239*, *240*
    effects, 240–241, *240*, *241*
    hailstone size, 240, 240
derechos, 222, 241–243
    formation, 241–242, *242*
    radar echoes, 241
    types, 242
desertification, 342–343, 344–345
deserts, 341–344, *341*
    erosion, 342–343, *342*
    location, 341
    polar, 341
    sand dunes, 343, 344
    weather patterns, 341–342
Devil's Postpile, California, 35, *36*

devolatilization, 33, 38
differential stress, 103–104
diorite, 34, *34*, 38
dip-slip faults, 29, 103, 104, 120
disaster reduction building techniques, 9, *9*, 11, 130
    base isolation, 187, 198–199, *199*
    pilings, 199
    shear walls, 196, 197
    structural reinforcement, 187, 196
    tuned mass dampers, 198, *198*
disaster reduction engineering, 3, 187, 195
    building techniques *see* disaster reduction building techniques
    failures *see* poor building practices, and natural disasters
    structural design, 196, *197*, 198
disaster responses, 13, 96–97, 108
    evacuation, 42, 53–54, 94, 96, 189, 285, 383
    international aid *see* international aid, for disaster victims
    preventative action, 146
divergent margins, 17, 20–22, 37
    development of, 20–22, *21*
Dolomite Mountains, Italy, 211–212, *211*
Dominican Republic, 11, 12, *12*, 270
doublet earthquakes, 128, 136
downbursts, 232, 241
    and air traffic, 232, *233*
    forecasting, 232
drainage systems *see* rivers
drawback, tsunami, 158, 159
    Indian Ocean 2004, 166, *168*
Dronka, Egypt, 1994 lightning strike, 235, *235*, 236, *237*
drought, 341–359
    and famine, 355–357
    and migrations, 341, 348
    monitoring, 345, *346*, 358
    and wildfires, 351, 357
ductile deformation, 103, 104
Dust Bowl disaster, United States, 1930s, 347–351, *348*
    causes, 347–348
    disaster responses, 349–350
    effects, 348, *349*, *350*
    human cost, 348–349
dust storms, 341, 345–346, *347*, 348, *349*
    haboob, 341, 345–346, *347*, *351*

early warning systems, 8–9, *8*, 130, 195
earth architecture, 17–19, *19*, 33
    asthenosphere, 17, 18, *20*, 104
    atmosphere *see* atmosphere
    hydrosphere, 17
    lithosphere *see* lithosphere
    magnetic field, 18
    mantle *see* mantle, the
    stratosphere, 64, 372, 379, 386
    tectonic plates *see* tectonic plates
    troposphere, 17, *222*, 386
    see also continental crust; liquid core; oceanic crust
earthquake causes, 103–110, *126–127*
    igneous activity, 110
    intraplate faults, 105, 108, 187
    megathrusts, 108, 110
earthquake hazard reduction, 194–200
    liquefaction, 199–200
    mass wasting, 194–195
    shaking, 196
    tsnuamis, 195
    see also disaster reduction engineering; earthquake precursors, study of; earthquake prediction
earthquake hazards, *126–127*
    ground failure, 142–143
    infrequent activity, 138, 139
    liquefaction *see* liquefaction
    mass movements *see* mass movement hazards
    reduction techniques *see* earthquake hazard reduction
    shaking *see* shaking, as hazard
    slope failure *see* slope failure
earthquake light, 136, 190
earthquake magnitude, 5, 116–117, **117**, 119
    historical estimates, 187
    and recurrence intervals, 188
earthquake precursors, study of, 187, 188–191, *189*
    radon measurements, 193
    satellites, 200
earthquake prediction, 187–193, 200
    difficulty, 189
    satellite technology, 200, *200*
    timing, 187–188
    use of records, 187
    see also early warning systems
earthquakes, 3, *4*, 104, 113–123
    aftershocks, 104, 116
    causes *see* earthquake causes
    clusters, 116
    epicenters, 113, 119, *126–127*
    and fire hazards, 25–26, 128, 133
    foreshocks, 10, 24, 104, 116, 190
    magnitude *see* earthquake magnitude
    main shocks, 104, 106, 116
    mapping, *188*
    precursors *see* earthquake precursors, study of
    prediction *see* earthquake prediction
    recurrence intervals, 187, 188
    scales, 117–118
    shaking intensity, 113, 116–117, 118, **118**
    shallow-focus, 11, *126–127*
    swarms, 110
    tornillo, 70, 110
    tsunami generation, 158–159, 181
    wave types *see* seismic waves
Eastern Mediterranean region, 9, 27
    see also Greece
East Pakistan *see* Bangladesh
ejecta, 38, 49, 70
    in craters, 367, *367, 368*
    igneous rock types, 34
    supervolcanoes, 57, 58, 83, 388
    tephra emissions, 94, 96
elastic deformation, 103, 104
El Chichón, Mexico, 1982 eruption, 382
Eldfell, Iceland, 1973 eruption, 94–98
    effects, 94, 96, 97
    explosivity, 94
    disaster response, 94–96

El Niño, 260, 261–263
    effects, 261, *262*, 263, 331
    event characteristics, 261
    and global cooling, 382
    and global warming, 261
    and Southern Oscillation, 331
Enhanced Fujita Scale, 295, 299
Enriquillo-Plantain Garden fault system, 11–12, *12*
Environmental Seismic Intensity Scale (ESI 2007), 161
erosion, 342–343
    and desertification, 342–343
    dust bowl, 351
    and resistant rock types, 343
    wind *see* wind erosion
eruption columns, 42–43, *42*, 45, 64
    examples, *41*, 75, 79, 87–91, 94, 380
    lateral, 65, 67
escape tectonics, 27, *28*, 29–30, 136
estuaries, 251, 253, *254*
Etna *see* Mount Etna
Europe
    climate, 225, *226*, 355, 380–381, 385, *385*
    and Eyjafjallajökull eruption, 43
    and Laki fissure eruptions, 90, 91, 382
    and Pangea, 29, 36
    tornadoes, 318
    see also individual cities and countries
eutrophication, 389
    see also dead zones, marine
extraterrestial impacts, 363, 364–367, 373
    effects, 371
    potential for disaster, 367, 373
extrusion *see* escape tectonics
Eyjafjallajökull eruption, Iceland, 41–43, *42*, *43*
    disaster response, 42
    effects, 42–43

famine
    and drought, 261, 350, 355–357, *356*, *357*
    and flooding, 334, 335
    and global cooling, 91, 380, 381
    see also Northern China, 1876–1878 drought disaster
faulting, 103–105, 108, *121*, 187
    dip-slip, 103, 104, 120
    strike-slip, 29, 103, 105, 120, 136
    see also deformation, of rocks; fault planes; plate margins
fault mapping, 120, 122, *122*, 187
fault planes, 103–104, 109, *121*, 213
Federal Emergency Management Agency (FEMA), 313
feldspar, 34, 37, 43
felsic rocks, 33, 34, 36
fire hazards, 13, 128, 133, 134, 352
    and earthquakes, 25–26, 128, 133
    and lightning strikes, 236
    wildfires, 341, 357
firestorms, 128, 134, 352
fissure eruptions, 49, 50–51, *50*, 88–89
flash floods, 215, 217, 232, 323
    and lahars, 79
    in urban areas, 325–326

flood basalts, 35, 389
flooding, 323, 326, 327, *327*
    see also flash floods; flooding events; inundation
flooding events, 216, *278*, 323, 327–337
    death toll accuracy, 8, 329
    and hurricanes, 275, 277–278, 285
    and volcanic eruptions, 42
    waterborne epidemics, 334, 335
Florida, 7, *7*, 220, 236
    Delta air disaster, 232–234
    Hurricane Katrina, 279
    Storm of the Century, 245–247
fluorine, 45, 54, 91, 382
focal mechanism analysis *see* fault-plane solutions
foreshocks, 10, 24, 104, 116, 190
fossil fuels, industrial effect of, 379
fractional crystallization (fractionation), 33, 39
Franklin, Benjamin, 382
Fujita, Ted, 309–310
Fujita Scale, 297, 298–299
Fukushima nuclear plants, Japan, 172, *174*
fumaroles, 64, *84*, 85, *85*
funnel clouds, 228, 297, 299

Galeras, Colombia, 68, *69*, 70
Galveston, TX, 1900 hurricane, 268, 270–273
    death toll, 271
    disaster responses, 271
    effects, 270–271, *271*, *272*
    remedial action, 271
    storm surge, 270
    track, *270*
Gansu, China, 1920 earthquake, 140–143
    death toll, 142
    effects, 140, 141, 142, *142*
    magnitude, 140
    shaking intensity, 140
    source, 140–141, *141*
gas emissions, volcanic, 64, 70, 88, *88*, 382
    clouds *see* nuée ardente
    fluorine *see* fluorine
    and human disasters *see* toxic gas emissions
    sulfur *see* sulfur dioxide emissions
    see also tephra emissions; volatiles
gauging stations, 323, 325, *326*
geophones *see* seismographs
Gharm, Tajikistan, 10
Giants' Causeway, Ireland, 35–36, *36*
Giuliani, Guampaolo, 193
glacial lakes, 92
glaciation, 19, 227
glaciers, 194, 210, 329
    alpine, 225
    ice age, 19
    and mass movements, 150, *178*, 179
    melting, 79, 209
    and volcanic eruptions, 43, 50, 92, *93*
Global Earthquake Satellite System (GESS), 200
global positioning systems (GPS), 200
global warming, 45, 91, 355, 371, 386–389

Good Friday 1964 earthquake, Alaska, 142–146
    death toll, 143
    effects, 143, *144*, *145*
    magnitude, 142
    source, 142–143, *143*
    tsunami, 161
Gorham, IL, 316
granite, 34, *34*
    see also felsic rocks
gravity, science of, 45, 64, 71, 203–204, 251
    building construction, 196
    earth's atmosphere, 17, 223
    and faulting, 104
    glaciers, 225
    mantle, 20
    planetary, 255, 364
    tidal, 251
    water, 18, 323
Great Chicago Fire, 1871, 351, 352, *353*, 354, *354*
Great Chilean tsunami *see* Valdivia, Chile, 1960 earthquake and tsunami
Great East Japan earthquake, 2011 *see* Tōhoku, 2011 earthquake, 171–174, 195
Great Kantō, Japan, 1923 earthquake, 14, 133–135, *133*
    death toll, 135
    effects, 133–134, *134*
    epicenter, 133
    magnitude, 133
    shaking intensity, 133
    source, 133
Great Michigan Fire, 1871, 352–353, 354
Great New England Hurricane, 1938, 273–276
    cleanup operation, 277
    death toll, 273–274
    development, 273
    effects, 274, *275*, *276*
    eye, 274
    storm surge, 274
    track, *273*
    wind speeds, 273, 274, *274*
Great San Francisco Earthquake, 1906, 24–27, *25*
    effects, 14, 25–26, *26*, 27
    shaking intensity, 25
    magnitude, 24
    rebuilding, 26
Greece
    Ancient, 52–53
    geology of, 52
greenhouse gases, 379, 386, 388
    and climate change, 379, 386, 389
    effect of extraterrestial impacts, 371
    and volcanic eruptions, 389
Griggs, Robert, 85
Grímsvötn, Iceland, 1996 eruption, 92–94, *93*
    effects, 92, *92*, *93*, *94*
groundwater, 195, 219, 324–325
    and base flow, 325
    and earthquake prediction, 189, 190, *190*
    interaction with streams, 324–325
    and lava, 98

    and liquefaction, 128, 131, 199
Guatemala, 1972 earthquake, 12
Gulf of Mexico, 271
    Chicxlub crater, 366–367, 370
    storms, 245, 270, *270*, 276, 279
    tsunamis, 371
    weather fronts, 226, 227, 238, 315
Gulf Stream, 225, *226*, 260

Hadley cells, 224
Haicheng, China, 1975 earthquake, 190–194
    death toll, 191
    disaster response, 191
    effects, 190–191, *191*
    epicenter, 190
    foreshocks, 190–191
    magnitude, 190
    precursors, 190–191
    predictions, 189–190
    shaking intensity, 190
Hail Alley, 238, *239*
hailstorms, 237–241, 297
    development, 238–239
Haiti, 13, 14
    disaster preparedness, 12–13
    2010 earthquake *see* Haiti earthquake, 2010
Haiti earthquake, 2010, 5, 11–12, 13–14
    death toll, 11, 13, 194
    effects, 12, *12*, *13*, *14*
    epicenter, 11, *12*
    reconstruction plan, 14
    shaking intensity, 11, 12
    source, 11
Hawaii, 39, 41
    aa lava, 40, *40*
    acid rain, 45
    pahoehoe lava, 40, *40*
    volcanoes in, 39–40, 51
Hebgen Lake, 1959 earthquake, 58
Heimaey, Iceland, 94, *95*, *96*, 97
Henan, China, flooding event, 1975, 334–337
    causes, 334–335
    death toll, 335
    effects, 335, *336*, *337*
Herculanium, 5, *6*
Himalayan orogeny, 23, 27, 29, 136
Hispaniola, 11, *12*
hot spots *see* mantle plumes
Houston, TX, 271
Huascarán, Peru, 1962 avalanche, 210, *210*
humidity, 222, 227, 269
Hurricane Hugo, 1989, 54
Hurricane Irma, 2017, 285
Hurricane Katrina, 2005, 5, 268, 279–282, *280*
    death toll, 280
    development, 279–280
    disaster response, 280–281
    effects, 280–281, *280*, *281*
    storm surges, 279–280
    track, *279*

Hurricane Maria, 2017, 285–287, *286*
    death toll, 287
    development, 285
    disaster responses, 285–287
    effects, 285–286, *287*
    storm surge, 285
    track, *286*
    wind speeds, 285
Hurricane Mitch, 1998, 276–278
    death toll, 278
    development, 276
    effects, 277–278, *277*, *278*
    precipitation, 277
    track, *276*
hurricanes, 3, 5, 54, *266–267*, 268–294
    classification, 269, *282*
    Coriolis Effect, 269
    development, 268
    eye, *266*, *267*, 268, 269
    eyewall, *266*, 268, 269, *269*, 285
    monitoring, 8
    predicting, 269
    rainbands, 268, 269
    storm surges *see* storm surges
    wind speed, *266*
    see also cyclones; typhoons
Hurricane Sandy, 268

ice ages, 19, 225, 379
Iceland, 22, *87*, 94, *95*
    volcanic eruptions, 41–43, 87–96
igneous activity, 22, 38, *39*, 57
    as cause of earthquakes, 110
igneous rocks, 34–35
    components, 34
    types, 33, 34, *34*, 35, 36
    see also felsic rocks; intermediate rocks; mafic rocks
Illinois
    Mid-West derecho, 242, 243
    tornadoes, 307, 308, 315, 316
impact craters, 363, 366–367, *366*, *368*
India, 23, 27, 225–328, *328*
    drought, 357
    regional weather changes, 329
    see also Calcutta, India, 1737 disaster; Uttarakhand floods, India, 2013
Indiana
    Mid-West derecho, 242, 243
    tornadoes, 307, 308, 315, 316
Indian Ocean, tsunami warning sensors, 9, 143, 195
Indian Ocean, 2004 tsunami, 166–170
    death toll, 166, 170
    dispersion, *168*
    effects, 166, *168*, *169*
    media coverage, 170
    and tourists, 166, *168*
Indochina, 27, *28*
Indonesia, 23, 76, 261, 263
    tsunamis, 76, *76*, *166*, 169–170
    volcanic eruptions, 57, 379, 383, 388
        Krakatoa *see* Krakatoa, Indonesia, 1883 eruption
        Tambora *see* Tambora, Indonesia, 1815 eruption

Integrated Tsunami Intensity Scale (IT IS 2012), 161
Inter-American Development Bank, 13
Interferometric Synthetic Aperture Radar (InSAR), 200
intermediate rocks, 33, 34–35, *35*
international aid, for disaster victims, 8, 11, 14, 291
    benefit concerts, 54, 293
    Haiti, 2010, 13–14
    Montserrat, 1995–2013, 55
    Myanmar, 2008, 291
International Association of Volcanology and Chemistry of the Earth's Interior (IAVCEI), 52, 68
International Decade for Natural Disaster Reduction, 52, 68
International Monetary Fund, 14
intraplate earthquakes, 105, 108, 187
intraplate faults, 103, 105
inundation, 158, 159, 209, 331
    height, 158, *159*, 285
    and storm surges, 256, 271, 285
isostasy, 18, *18*
isostatic rebound, 19
Italian Alps, 206, 211–212

Jamaica, 12
Japan, 14, *166*, 171–174, 196
    natural disaster engineering, 9, 195
    1700 tsunami, 109–110
    1964 earthquake *see* Niigata, Japan, 1964 earthquake
Johnston, David, 65
Joplin Tornado, 2011, 300–303, *301*
    death toll, 300, 301
    development, 300
    effects, 300, *300*, 301, *302*, 303
    track, 300–301, *300*
Juan de Fuca plate, 110

karst topography, 203, 219–220, *219*
Khait, USSR, 1948 earthquake, 10–11, *10*
    destruction, 10, 11
    effects, 10, *11*
    magnitude, 10
    monument to, 11
    shaking intensity, 10
Kīlauea, Hawaii, 51
kinetic energy, 104, 367
Kobe, Japan, 1995 earthquake, 196
Krakatoa, Indonesia, 1883 eruption, 75–79, *76*, *77*
    caldera collapse, 75–77
    death toll, 77
    effects, 75–77, *78*, *79*
    explosivity, 76

lahars, 55, 64, *67*, *81*, *82*, 209
    examples, 79, *80*, 81, 278, 383, *384*
    and tsunamis, 73
Lake Champlain, 19, *19*
Lake Monoun, Cameroon, 100
Lake Nyos, Cameroon, 1986 disaster, 98–100, *99*, *100*
    causes, 98
    death toll, 99
    effects, 98–100, *99*, *100*
Lake Pontchartrain, Louisiana, 279

Lake Sarez, Tajikistan, 10
Lake Texcoco, Mexico, 128
Laki, Iceland, 1783 fissure eruptions, 87–91, *90*, *91*
    death toll, 91
    gas emissions, 91
    international effects, 91, 382
landslides, 10, *11*, 142, 146, 175, 203
    aftermath of earthquakes, 10
    debris flows, 203, 213
    earthflows, 203, 213, *214*
    rock, 203
    slumping, 143, *145*, *146*, 213–214
La Niña, 260, 261
L'Aquila, Italy, 2009 earthquake, 191–194
    aftershocks, 192
    death toll, 191–192
    disaster response, 191, 194
    effects, 191, *192*, 193, *193*
    epicenter, 191, *192*
    foreshocks, 192
    magnitude, 191
    prediction, 193–194
    seismic history, 191, 192
lateral blasts, 45, 64, *66*
    see also fissure eruptions
lava, 33, 39–40, 45, 49
    magma production, 33, 38, 45
    tubes, 40
    see also lava domes; lava flows
lava domes, *50*, 51, 52
    resurgent, 51
lava flows, 55, 64, 94, *96*, *97*
    combatting, 96, *97*, *98*
    see also pyroclastic flows
lightning, 222, 229, 232
lightning strikes, 235–236, *236*
liquefaction, *126–127*, 128, 130–132, 199
    and groundwater, 128, 131
    hazard reduction, 199
liquid core, 18, *18*, 33
lithosphere, 17, *18*
    effect of glaciation, 19, *19*
    see also tectonic plates
Little Ice Age, 355
littoral cones, 50
littoral drift, 258, *259*, 260
Lituya Bay, 1958 megatsunami, 175, 177–180, *179*, *180*
    death toll, 180
    earthquake 175, *178*
    rockslide, 175, 179
Lituya Glacier, 179
loess, 10, 141–142, *141*, 194
Loma Prieta earthquake, California, 11, 194, 200
Long Island, New York, 19
    see also Great New England hurricane
longshore currents, 251, 258

maars, 49–50, 98
Machu Picchu, Peru, 9
mafic rocks, 33, 34, 35, 37, 39
magma, 22, 33–45, 49, 110
    assimilation, 33, 41, 44
    change to lava, 45
    chemistry of, 45
    contamination, 45
        see also crustal contamination
    production, 22, 23, 33
    types *see* magma types
    and volcanic eruptions, 22, 33, 45, 51
magma chambers, 50, 51, 58, 110
magma types, 33–34, 51
    felsic, 36, 39, 43
    intermediate, 43
    mafic, 36, 39, 41, 43
main shocks, 104, 106, 116
mantle, the, *17*, 18, *18*, 20, 36
    convection cells *see* convection cells
    and plate tectonics, 20–21
    plumes *see* mantle plumes
    upwelling, 20–21, 38
mantle plumes, 38, 41
    rock melting, 38–39, *39*
    see also eruption columns
Mao Tse-tung, 29
Martinique, *71*
    see also Mount Pelée, 1903 eruption
mass extinctions, 370, 371, *387*
mass movement hazards, 206–209
    avalanches *see* avalanches
    landslides *see* landslides
    low-speed, 213–214
    mudflows *see* mudflows
    rockslides *see* rockslides
    slumping, 143, *145*, *146*, 213–214
megadroughts, 341, 357–359
megathrusts, 103, 108, 110, 166
megatsunamis, 158
melting, of rocks, 33
    decompression, 33, 37, 38
    partial, 33, 38, 41
    wet, 33, 37–38, *38*
meteorites, 363, 364–366, *365*
meteors, 363, 364
methane, 364, 379, 388
Mexico City, 1985 earthquake, 81, 128–130, *129*, *130*
    death toll, 128
    effects, 128, *129*, *130*, 196, *196*
    epicenter, 128, *129*
    magnitude, 128
Miami, Florida, 7, *7*
microbursts, *229*, 231, 232, 234
microclimates, 225
microtsunamis, 159
mid-ocean ridges, 17, *18*, 252, 260
    and divergent margins, 20, 22, 23
Midwest derecho, 2020, 241, 243–244, *242*, *243*
    costs, 243, 244
    effects, 243–244, *243*, *244*
    windspeed, 243
migrations
    and climate change, 385–386
    and famine, 341, 348

Minoan civilization, Greece, 52
Mississippi Delta, 253, *254*
Missouri
    earthquakes, 106, 108
    tornadoes, 300, *300,* 303, 315, 316
Modified Mercalli Scale, 118–119, **118**
    historic measurements, 5, 10, 11, 12, 25, 106–107
Modified Sieberg Scale, 160–161, **160**
    relationship to Modified Mercalli Scale, 160
Mohorovičić discontinuity (Moho), 18
moment magnitude, of earthquakes, 117–118, 136
Montserrat, 55, *56*
Moore, Oklahoma, Tornado, 1999, 311–314, *312, 313*
    death toll, 312
    development, 311
    effects, *314*
    multiple nature, 311–312
    path, *312*
    wind speed, 312
mountain-building *see* orogeny
Mount Etna, 9, 52, *53*, 96
Mount Katmai, Alaska, 83, *86*
Mount Marmolada, Italy, 211
Mount Pelée, Martinique, 1902 eruption, 71–75
    death toll, 73
    disaster response, 73
    effects, 71, *72*, 73, *73, 74*
    explosivity, 73
    tsunami, 73
Mount Pinatubo, 1991eruption, 53, 382–384, *383*
    and climate change, 386
    effects, 382–383, *383, 384*
    disaster response, 383
    military monitoring, 383
Mount St. Helen's, 1980 eruption, 57, 64–67
    death toll, 65
    effects, *65, 66, 67*
    explosivity, 65
Mount Toba eruption, 57, 388–389
mudflows, *80*, 150, 203, 209
    volcanic *see* lahars
Myanmar, 288
    see also Cyclone Nargis
Mycenae civilization, Greece, 9, 52

Naples, Italy, 5, *6*
    proximity to Mount Vesuvius, 5, *6*
National Weather Service, 247, 308, 312
Native Americans, 109
natural frequency, of tall buildings, 128, 198
Nazca ocean plate, 44, 149
Netherlands flood defences, 9
Nevado del Ruiz, Colombia, 1985 eruption, 79–82
    death toll, 81
    effects, 79, *80*
    explosivity, 79
    lahars, 79, *80*, 81, *82*
Nevados Huascarán, Peru, 149–150, 151, *153*, 194
    1962 avalanche, 210
New Madrid, Missouri, 1811–1812 earthquakes, 105–108
    death toll, 108
    disaster response, 108
    effects, 107–108, *106, 107*
    magnitude, 105
    shaking intensity, 105
New Orleans, Louisiana, 5, 279
New York City, 36, 382
New Zealand, 23, 163
Niigata, Japan, 1964 earthquake, 131–133, 199
    death toll, 133
    effects, 132, *132*
    epicenter, 131
    magnitude, 131
    shaking intensity, 131, *131*
Nixon, Richard, 308
North American ice sheets, 19
North American plate, 12, 23, 57, *143*
Northern China, 1876–1878 drought disaster, 355–357, *356*
    causes, 355
    death toll, 355
    effects, 355, *356*
    plagues, 357
Nova Erupta, Alaska, 1912 eruption, 83–86
    effects, 83–84, *84, 86*
    explosivity, 83
    monument to, 85
    source, 83

Obama, Barack, 14
obsidian, 34, 45
ocean basins, 251, *252*, 256, 260
ocean currents, 225, 251, 260–261, *261*
    see also Gulf Stream, the
ocean floors, *39*, 40, 251–252, 256, 388
    see also mid-ocean ridges
oceanic crust, 18, *18*, 22, 37, 251
    wet, 37–38
ocean processes, 255–263, 388
oceans, 251–263
    currents *see* ocean currents
    dead zones *see* dead zones, marine
    fetch, 251, 256
    floor *see* ocean floor
    tides *see* tides
    waves *see* waves, ocean, 251
    see also coastlines; mid-ocean ridges; ocean processes; weather systems
Okeechobee lake, Florida, 7, *7*
Oklahoma City, 312, 313
Oman ophiolite, 39
Oregon, 51, *110*
orogeny, process of, 10, 17, 23, 27, 29
orphan tsunamis, 109

Pacific Ocean, 27, 162, 163, 252, 260
    trade winds, 261
    see also El Niño; Pacific Tsunami Warning System (PTWS)
Pacific Plate, 24, 142, *143*
Pacific Tsunami Warning System (PTWS), *8*, 9, 143, 163–164
    US centers, 162, 172

Pakistan, 291, 293
palisades, 37
Palmer Drought Index, 345
Pamir Mountains, Asia, 10
Pangea, *29*, 30, 36, 386, *387*
Papua New Guinea, 1998 tsunami *see* Sissano, Papua New Guinea, 1998 tsunami
partial melting, 33, 38, 41
passive margins, 251–252, *252*
Pasto, Colombia, 68
Peloponnesian War, 5
Permian Extinction Event, 386–388
Peshtigo, WI, 1871 wildfire, 351–354, *352*
    causes, 351
    death toll, 351, 352
    related fires, 352–354
Philippines, 54, 382
    Super Typhoon Yolanda *see* Typhoon Haiyan
    see also Mount Pinatubo eruption, 1991
phreatic explosions, 40, 55, 64
phreatomagnetic explosions, 53, 64, 386
pillow basalts, 39–40, *91*
plate margins, faulting on, 105, 187
    see also convergent margins; divergent margins; transform margins
plate tectonics, theory of, 17, 20–21, 30
    see also escape tectonics; tectonic activity
Plinian eruptions, 51, 58, 64, 75
plutons, 33, 34
Plymouth, Montserrat, 55, *56*
poison gas emissions *see* toxic gas emissions
Pompeii, 5, *6*
poor building practices, and natural disasters, 138, 139, 149, 329
population growth, 3, 5
  and urban development, 5, *7*
Port-au-Prince, Haiti, 5
    1946 tsunami, 5
    see also Haiti earthquake, 2010
precipitation, hazards of, 45, 222, 277, 341–342, *341*
    flooding *see* flooding
Prince William Sound, Alaska, 142
Puerto Rico, 12
    Hurricane Maria, 285–287
pumice, 34, 45, 76
P-waves, 113, 114, *114*, 115, 116, 119
pyroclastic density currents (PDCs) *see* pyroclastic surges
pyroclastic flows, 45, 55, *57*, 64, 71, 75
pyroclastics, 49, 70
    see also pyroclastic flows; pyroclastic surges
pyroclastic surges, 71, 73, 75
    see also nuée ardente

quartz, 34, 37, 43
Quijiang Fault, 29

radar weather forecasting *see* weather radar
radar hook, 297
radiocarbon dating, 109
radon emissions, 188, 193

Red River Fault, 29
Rhode Island, NY, 274
rhyolites, 33, 34, *34*, 37
rhyolite volcanoes, 37, 52
Richter, Charles, 116
Richter Scale, *117*
rivers, 323–325
    dam construction, 334, 335
    drainage basins, 323, *324*
    drainage patterns, 324, *325*
    flooding *see* flooding
    gauging stations, 323, 325, *326*
    tributary streams *see* tributaries
    urbanization effects, 326
rockfalls, 177, 203, 204–205, *205*, 206
rockslides, 175, 179, 205–206, *205*, *206*
run-up height, tsunami, 159

Saffir-Simpson Scale, 160, 268, 270
St. Lawrence Seaway, 19, *19*
St. Pierre, Martinique, 71, *71*, 73, *73*, 74, 75
San Andreas fault, 23, 24, *24*, 106, *121*
sand dunes, 343, *344*
San Francisco, California, 11, 26
    earthquake-resistant buildings, 9, *9*
    see also Great San Francisco Earthquake, 1906
Santorini, 52–53, *53*
seiche waves, 105, 158, 180
seismic waves, 114, 196
    body, 113, 114, 117
    surface, 113, 114, *114*, 115, *126—127*
seismograms, 70, 113, 114–115
    determining epicenters, 119–120, *120*
    measuring magnitude, *115*, 117
seismographs, 81, 110, 113, 115, *115*
    early, 115
    records *see* seismograms
Seismologial Society of America, 194
seismologists, 10, 193–194
    techniques, 119–123, *121*, *122*
Shaanxi, China, 1556 earthquake, 194
ShakeMap, 123
shaking, as hazard, *126–127*, 128, 142
    and building design *see* disaster reduction engineering
    hammering, *126–127*, 128
shaking intensity, 5, 10, 11, 25, 105, 131, 143
    measurement of, 123
    see also Modified Mercalli Scale
shield volcanoes, 49, *50*, 51, *62–63*, *87*
    eruption, *62–63*, 87–99
Shoemaker-Levy 9 Comet, 367, *369*
Siberian Traps, 386, 388
Sichuan, China, 2008 earthquake, 146–149
    aftershocks, 146
    death toll, 146–147
    effects, 146, *147*, 148, *149*
    epicenter, 146
    magnitude, 146
    shaking intensity, 146
    source, 146, *147*

Sicily, 9
Sieberg scale, 160–161
    see also Modified Sieberg Scale
Sierra del Ávila mountains, Venezuela, 214
sinkholes, 219–220, *220*
Sissano, Papua New Guinea, 1998 tsunami, 181–183
    death toll, 181
    effects, 181, *182*, *183*
    source, 181, *181*, 182
skyscrapers, 9, 198
slope failure, *126–127*, *128*, 140–143, 146–149
    angle of repose, 203, *203*
    protection against, 194–195, *204*
    repair, 204, *204*
    and vegetation, 194–195, 204
solar system, the, 363–364, *363*
Soufrière Hills, 1995–2013 eruption, 55–57, *55*
    death toll, 55
    effects, 55, *56*, *57*
    explosivity, 55
South Alpine Fault, 23
South American plate, 44
South American Subduction Zone, 162
South China Sea, 27
Southern Oscillation, 331
space objects, 363–364
    bolides, 363, 364, 369
    collisions with earth, 364–375
    meteoroids, 364
    see also asteroids; comets
Sparta, 464 BCE earthquake, 5, 9
    death toll, 5
    and Helot rebellion, 5, 9
    magnitude, 5
    shaking intensity, 5
Standardized Precipitation Index (SPI), 345
Storm of the Century, 1993, 245–247, *245*
    areas affected, 246–247
    blizzards, 245–247, *246*, *247*
    forecast, 247
    storm surge, 245
storms
    and air traffic, 232–233
    generation, 230–231
    hail see hailstorms
    hemispheric differences, 269, 298
    predicting, 269
    supercell see supercell storms
    tropical, 268, 270, 271, 277
    see also cyclones; derechos; hurricanes; thunderstorms; typhoons
storm surges, *267*, 268, *269*, 270, *284*
    effects of, 256, 270–271, 283, *285*, 289, 291
    examples, 245, 256, 270, 279–280
strain see deformation, of rocks
stratosphere, 65, 372, 379, 386
stratovolcanoes, 49, *50*, 51, 52, 54
    eruptions, 53, 55, 64–86, 68, 379
stress, 103–104
    and earthquakes, 104, 116
strike-slip faults, 29, 103, 104, 120

sturtzstorms, 209
subduction processes, 17, *21*, 23, 37, 191, 252
subduction zones, 17, 23, *24*, *38*, 108
    East Asian, 27, *28*
    fault planes, 108
    igneous activity, 38
    intraplate earthquakes, 108
    plate contacts see megathrusts
    see also individual subduction zones
submarine mass failures (SMFs), 182
submarine volcanoes, 22, 50
    black smokers, 50
sulfur, in magma, 45
sulfur dioxide emissions, 54, 88, 382, 383, 385
    and acid rain, 88, 371
    earthquakes, 190
    effects, 54, 379
    volcanic, 379, 380, 382, 383, 386
summit eruptions, 64, 65, *66*, *69*
supercells, 297
supercell storms, 232, *232*
    tornadoes in, 232, *232*, 298
supercontinents, 29, *30*, 35, 386, *387*
super typhoons, 263
    see also Typhoon Haiyan
supervolcanoes, 41, 49, 57–59, 388
surface waves, 113, 114, 115, 117
    types, 114, *126–127*
S-waves, 113, 114, 115, 116
    travel-time curves, 119, *119*

table volcanoes, 50
Tajikistan, 10
Tambora, Indonesia, 1815 eruption, 380–382
    death toll, 381
    effects, 380–381, *381*, 388
    explosivity 380
    map, *380*
Tangshan, China, 1976 earthquake, 8, 136–137, 190
    death toll, 137, 190
    effects, 136–137, *137*
    epicenter, 136, *136*
    magnitude, 136
    shaking intensity, 136
    source, 136
Tangshan Fault, 136, *136*
tectonic activity, 9, 11, 12, 17
    see also continental rift; earthquakes
tectonic plates, 12, 17, 18
    boundary movement see tectonic activity
    density, 18, 23
    and mantle, 20
    margins see plate margins
    movement, 20, 22–23
    see also plate tectonics, theory of
teletsunamis, 158, 175
tephra emissions, 93, 94
Thailand, 166, *168*, *169*, 288
Thera, island of, 52
    eruption, 1600 BCE, 52–53

thunderstorms, 229, 230–231, *231*
Tian Shan, China, 9, 10
tidal range, 255–256, *256*
tides, 251, 255–256
    influence of the moon, 255, *255*, *256*
    and storm surges, 256
Tōhoku, Japan, 2011 earthquake, 171–174
    death toll, 172
    effects, 171–172
    epicenter, 171, *171*
    magnitude, 171
    nuclear disaster, 171, 172–173, *174*
    source, 171
    tsunami, 171, *172*, *173*, *174*
Tokyo, Japan, *134*, 135, *135*, *171*
Tonghai, China, 1970 earthquake, 29
    death toll, 29
    effects, 29
    magnitude, 29
Tornado Alley, *6*, 313
tornadoes, *4*, 223–224, 245, 295–299
    classification, 298–299
    debri clouds, 299
    and derechos, 241
    disaster preparation for, 309–310
    disaster reduction engineering, 9
    examples, 300–319
    formation, 297, *298*, 299
    forward speed, 299–319
    measurement of, 295, 299
    safe rooms, 312–313
    speed of strike, 9
    in supercells, 232, 297, *297*
    super outbreaks, 301–311
    and wall clouds, 232
    waterspouts, 245, 318
    and weather fronts, 229, 297
tornado myths, 311
tornado shelters, 9, 312–313
Tornado Super Outbreak, 1974, 307–311, *309*
    death toll, 308
    development, 307
    effects, 308–309, *309*, *310*, *311*
    number of tornadoes, 307, 308
    paths, 307, *308*
Tornado Super Outbreak, 2011, 303–307, *304*
    death toll, 306–307
    effects, *305*, 306, *307*
    forecast, 303
    number of tornadoes, 303, 306
    satellite image, *305*
    tornado paths, 303–305, *304*, *306*
tornado waterspouts, 245, 318
tornillo earthquakes, 70, 110
toxic gas emissions
    carbon dioxide, 97, 99
    earthquakes, 190
    effects on livestock, 71, 83, 97, *98*, *99*
    Eldfell volcano, 94
    Lake Nyos disaster, 97, *98*

Laki volcano, 87–88
Permian Extinction Event, 388
pyroclastic *see nuée ardente*
sulfur dioxide *see* sulfur dioxide emissions
Transamerica building, San Francisco, 9, *9*, 198, *199*
transform margins, 17, 23, *24*, 27
tree-ring studies, 108–109, 357, *358*, 359, 385
triangulation, 119, *121*
tributaries, 324–325
    effluent, 324, *324*
    influent, 324, *324*
Tri-State Tornado, 1925, 315–317
    death toll, 316
    development, 315
    effects, 316, *316*, *317*
    path, 315–316, *315*
    wind speed, 316
tropical depressions, 268
    see also thunderstorms
tropical storms, 268, 270, 271, 277
tropical waves, 268
troposphere, 17, *224*, 386
Tsunami Environmental Effects Scale (TEE-16), 161
tsunami gates, 187, 195
tsunami generation, 159–160, 181
    earthquakes, 159
    extraterrestrial impacts, 159, 371
    sea floor uplift, *156*, 158, 159, 181
    underwater landslides, 181, 182
tsunamigenic events, 158, 162–183
tsunami hazards, 182, 195
    see also inundation
tsunami intensity, quantifying, 160–161
    see also Integrated Tsunami Intensity Scale (ITIS 2012); Modified Sieberg Scale
tsunami reach, 158
tsunamis, 3, *4*, 154–161, *156–157*, *160*, 162–183
    amplitude, 158
    causes *see* tsunami generation
    coastal protection, 9
    deep water, 159
    drawback, *see* drawback, tsunami
    early warning systems, 8–9, *8*, 130, 195
    flow depth, 158, 159, *160*
    forerunners, 158
    historical, 52–53, 75, 108, 133
    intensity *see* tsunami intensity, quantifying
    local, 160
    meteorological, 283
    open water waves, 158
    origin of term, 158
    orphan, 108
    reach, 158
    remote, 160
    run-up height, 158, 159
    shallow water waves, 158, 159
    small, 22, 160
    velocities, 159
    wavelength, 158
    wave train, 158

tuff cones *see* maars
Tunguska Event, Siberia, 1908, 367, 371–372, *371*
    effects, 371
    map of area, *371*
Turkey, 27
tuya, 50
Typhoon Haiyan, 2013, 282–285, *283*
    development, 282
    disaster response, 283
    effects, *284, 285*
    storm surge, 283, *283*, 284
    track, *282*
    upgrade to super typhoon, 282
Typhoon Nina, 334–335
Typhoon Yolanda *see* Typhoon Haiyan
typhoons, *266–267*, 268
    destructive effects of, 11, 263
    see also Typhoon Haiyan; Typhoon Nina

ultramafic rocks, 35, 36
United Nations, disaster response, 13–14
US Climatic Extremes Index (CEI), 345
US Geological Survey, 70
    earthquake database, 187
    observation, 65
    predictions, 108
USSR, 10–11
US Weather Bureau, 270, 271, 273
    see also National Weather Service
Uttarakhand floods, India, 2013, 327–331, *328*
    causes, 327–329
    death toll, 329
    effects, 329, *329*, 330

Vaiont Dam, Italy, 1963 disaster, 206–208, *207*
    death toll, 208
    disaster response, 207
    earthquakes, 206
    effects, 206–208, 208, *208*
Valdivia, Chile, 1960 earthquake and tsunami, 162–166, *164*
    death toll, 162
    earthquake, 162, *163*
    effects, 162, *164*, *165*, *166*
    tsunami, 162, *164*
Valley of 10 000 Smokes, Alaska, 83–85
Vargas, Venezuela, 1996 mud and debris flow, 215–219, *216*, *217*, *218*
    death toll, 218–219
Venezuela, 215, *215*, *216*
    flooding, 215–217
    population density, 215
    and 2010 Haiti earthquake, 12, 14
    Vargas disaster *see* Vargas, Venezuela, 1996 mud and debris flow
Venice, Italy, *318*
    see also Venice Waterspout, 1970
Venice Waterspout, 1970, 318–319
    death toll, 318
    development, 318
    effects, 318, *319*
Vestmannaeyjar, Iceland, 93, *94*, *95*
Vesuvius, 5, 51, 52
    79 AD eruption, 5, 53

volatiles, 33, 38, 45
volcanic arcs, 23, 54
volcanic craters, 50
    maars, 50
    see also calderas
volcanic emissions, 70
    lava *see* lava flows
    solid *see* ejecta
    tephra, 93, 94
    toxic *see* toxic gas emissions
    see also volcanic hazards; pyroclastic flows; sulfur dioxide
volcanic eruptions, 5, 41–43, 45, 52–57, 64–96
    and air traffic, 43
    emissions *see* volcanic emissions
    explosivity, 38, 42–43, 45, 55, 68, 382
    hazards from *see* volcanic hazards
    lateral *see* lateral blasts
    types, 51–52
    Vesuvius, 5, 53
    and volatiles, 38, 45
volcanic explosivity index (VEI), 49
    criteria, *52*
    and types of volcano, 51–52
volcanic hazards, 45, 49, 53, *65–67*, 67, 70
    bombs, 70
    climate effects, 88, 90, 379–389
    ejecta *see* ejecta
    lahars *see* lahars
    nuée ardente see nuée ardente clouds
    pyroclastic flows *see* pyroclastic flows
    toxic gases *see* toxic gas emissions
    see also lateral blasts
volcanic islands, 23, 52, 75, 93
    see also volcanic arcs
volcanic rock *see* igneous rocks
volcanoes, 3, *4*, 49
    classification by shape, 49–50
    classification by threat, 51–52
    features, 50–51, *51*
    monitoring, 8
    submarine, 22, 50
    types, 49–59
    see also shield volcanoes; stratovolcanoes; supervolcanoes
volcanologists, 65, 67, 68, 70
    field risks, 67, 70
    safety precautions, 70

wall clouds, 232, 297, 298
Watchung basalts, New Jersey, 36
water, fresh
    drainage systems *see* rivers
    and ecosystems, 341, 359
    ground *see* groundwater
    in magma, 45
    shortages *see* drought
water table, the, 199, 324, 331
wave amplitude, 116, *117*, 158, 256, *257*
    tsunami, 159
wave base, 256–257
waves, ocean, 251, 256–258
    and currents, 258

mechanics, 257–258, *257*
parts, *257*
size, 256, 257
storm, 257
tidal *see* tsunamis
weather fronts, 224, 228–230
cold, 229, *229*, 230, 297
occluded, 229–230, *230*
and storms, 230–231
warm, 229, *230*
weather hazards, 230
air masses, 224, 229
downbursts *see* downbursts
hail, 224, 232, 237–238, *238*, 240
lightning strikes, 235–236, *236*
microbursts *see* microbursts
rain, 232
thunder, 224, 229
tornadoes *see* tornadoes
wind shear, 233, 234
weather patterns, 224–225
at altitude, 224
desert, 341–342
El Niño/Southern Oscillation, 261, 331
humidity, 224, 229, 269
precipitation, 224, 232, 359
temperature, 224–225, *224*
see also climate change; climate systems; Little Ice Age
weather radar, 233, 234, 240, 297
derecho echoes, 241–242, *242*, *243*

dust movement, *342*
images, *239*, *243*, *245*, *313*, *328*
wet melting, 33, 37–38, *38*
White Friday Avalanches, 1916, 211–213
death toll, 212
and war, 211, 212, *212*, *213*
Williams, Stanley, 68, 70
Surviving Galeras, 70
wind erosion, 342–344, *342*
effects, 343, *343*
wind speed measurements, 295, 299
Wizard Island, Oregon, 51
World War I, 211–212, *211*, *212*
Wuhan, China, 331, 334

Xenia, OH, 308, *310*
xenoliths, 41

Yangtse-Huai river flood, China, 1931, 331–334, *332*
causes, 331
death toll, 334
effects, 331–332, *332*, *333*, 334
Yellowstone, Wyoming, 41, 57–58, *58*
ancient eruptions, 57–58, *59*, 388
potential as supervolcano, 57
Yucatán Peninsula, 366–367

Zhou Enlai, 29
Zimbabwe, 235